Regional Satellite Oceanography

Regional Satellite Oceanography

S. V. VICTOROV

St. Petersburg, Russia

Taylor & Francis
Publishers since 1798

UK   Taylor & Francis Ltd, 4 John Street, London WC1N 2ET
USA Taylor & Francis Inc., 1900 Frost Road, Suite 101, Bristol, PA 19007

Copyright © Serge Victorov 1996

All rights reserved. No part of this publication may be reproduced, stored in a retrieval system, or transmitted, in any form or by any means, electronic, electrostatic, magnetic tape, mechanical, photocopying, recording or otherwise, without the prior permission of the copyright owner.

British Library Cataloging in Publication Data

A catalogue record for this book is available from the British Library.

ISBN 0-7484-0273-X (cased)
ISBN-0-7484-0274-8 (paperback)

Library of Congress Cataloging in Publication data are available

Cover design by Hybert Design and Type, Waltham St Lawrence, Berkshire.

Typeset in Times 10/12pt by Santype International Ltd, Salisbury, Wilts.

Printed in Great Britain by T.J. Press (Padstow) Ltd.

To the dear memory of my parents

Contents

Preface	ix
Acknowledgments	xiv

1 Oceanography and remote sensing   1

    1.1 Satellites and oceanographers   1
    1.2 Regional and global satellite oceanography   6
    1.3 The concept of regional satellite oceanography   11

2 Information used in regional satellite oceanography   21

    2.1 Introduction   21
    2.2 Matching the requirements (?)   21
    2.3 Satellite data for regional oceanographic research   29
    2.4 Autonomous satellite data acquisition systems   51

3 Methodological aspects of regional satellite oceanography   57

    3.1 Introduction   57
    3.2 Definitions relevant to collection of subsatellite data   59
    3.3 Some field activities at the test areas in the selected seas   61
    3.4 Complex Oceanographic Subsatellite Experiments (COSE)   66
    3.5 The study of 'allowed interval of non-synchronicity' (AINS) between oceanographic observations at different levels   80
    3.6 Calibration of satellite sensors using natural 'standard' test sites   87
    3.7 The use of satellite imagery of IR band in cloudy situations   91
    3.8 The data processing, database and GIS approach in regional satellite oceanography   99

4 Regional satellite oceanography in action   107

    4.1 Regional systems   107
    4.2 Use of regional integrated GIS and satellite databases for marine applications and coastal zone management   121
    4.3 Topic-oriented use of satellite data in regional marine and coastal zone environmental and development activities   131

5 Regional satellite oceanography: case study of the Baltic Sea   153

    5.1 Introduction   153
    5.2 The 'Baltic Europe' geographical information system   155
    5.3 Dynamical processes in the Baltic Sea as revealed from satellite imagery   172
    5.4 Satellite monitoring of biological phenomena and pollution in the Baltic Sea   214
    5.5 Monitoring of the marine and coastal environments in the Neva Bay   225
    5.6 Concluding remarks   264

Conclusion   267

References   277

Index   301

Preface

The principal objective of this book is to introduce for the first time a new interdisciplinary science comprising remote sensing and oceanography – regional satellite oceanography. The author's intention is to describe the development of up-to-date satellite oceanography using the systems analysis approach and to review, in this context, the plans of the former USSR to create the Earth Observation System with its oceanographic segment.

Satellite imagery in visible, infrared and microwave bands and side-looking radar images will be shown to be a reliable source of scientific and managerial information for the oceanography of regional seas, environmental monitoring of coastal zones and marine-oriented elements of the national economy, provided that the satellite data are properly processed together with relevant *in situ* data, and the results are presented in formats suitable for end-users.

A brief review of world activities covered by the term 'regional satellite oceanography' will be given, and some results of a more detailed case study of the Baltic Sea will be presented, with a forecast of the future development of the activities in 'Baltic Europe' in the European context.

The book is addressed to academic members of the remote sensing and oceanographic communities, regional and local authorities dealing with the marine and coastal environments and industrial development on coastal zones, to those who, in various ways, are involved in the monitoring and control of the activities in regional seas, exclusive economic zones, offshore and nearshore zones, to experts working for various national equivalents of 'coastguard', 'seawatch', 'coastwatch', and the relevant segments of national navies. The book may also be of some interest to historians of science and technology.

It is assumed that the reader is already familiar with the basics of remote sensing and at least knows that among many oceanographic parameters quite a few can be determined using various spaceborne instruments operating in different bands of the electromagnetic spectrum. (The problems of integrating remote sensing in electromagnetic radiation with remote sensing based on marine acoustics are beyond the scope of this book.) Those tools can be 'passive' – just looking at the sea and collecting solar radiation transformed within the upper layer of the sea and/or reflected from the sea surface – or they can be operated in 'active' mode – in this case they themselves transmit electromagnetic waves down to the sea surface and analyse the response signal.

A matrix is presented showing what oceanographic patterns, phenomena and features can be recorded, analysed, determined and studied using different types of satellite instruments (Figure 0.1). This matrix was compiled by the author specifically in the context of regional satellite oceanography using the design approach often used by NASA and NOAA decision-makers (see, e.g., Sherman 1993).

Many books have been published in which the basics of remote sensing have been presented, including:

- the physics of the interaction of electromagnetic radiation with various substances and surfaces, including sea water and the sea surface;
- corrections for the atmosphere (effects occurring when the radiation from the sea surface presses through the atmosphere to a satellite);

together with descriptions of satellite-borne sensors design, explanations of the

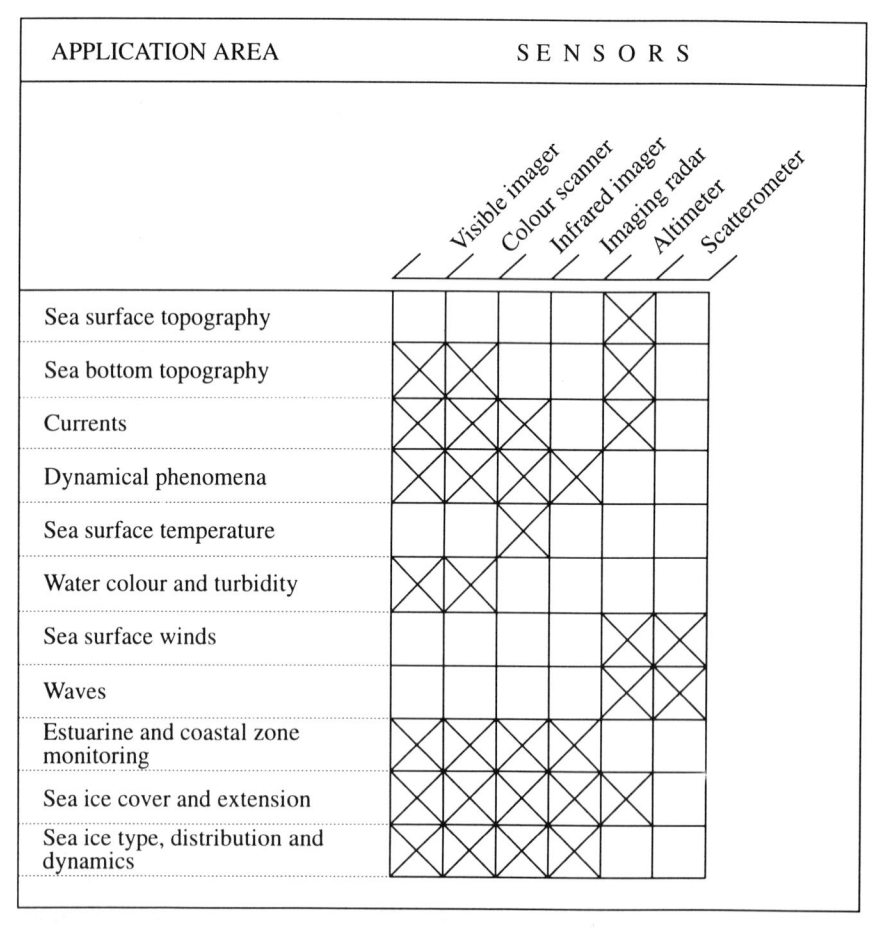

APPLICATION AREA	Visible imager	Colour scanner	Infrared imager	Imaging radar	Altimeter	Scatterometer
Sea surface topography					×	
Sea bottom topography	×	×		×		
Currents	×	×	×	×		
Dynamical phenomena	×	×	×	×		
Sea surface temperature			×			
Water colour and turbidity	×	×				
Sea surface winds				×	×	×
Waves				×	×	×
Estuarine and coastal zone monitoring	×	×	×	×		
Sea ice cover and extension	×	×	×	×		
Sea ice type, distribution and dynamics	×	×	×	×		

Figure 0.1 Matrix of satellite sensors and marine application area for regional satellite oceanography. Those sensors capable of providing useful data for a certain application area are indicated by a cross (×).

basics of digital telecommunications, presentation of algorithms for satellite data processing and relevant procedures for image analyses and pattern recognition. When these topics are covered in detail, there is, in some books, no room left for comprehensive applications of the analyses to particular areas. In some books the oceanographic applications of satellite imagery are just mentioned, along with other areas such as land use, forestry, geology, hydrology, etc.

So it was not the author's intention to include in this book either the basics of remote sensing or the instrumentation of satellites with explanations of sensor performance. I have tried to avoid discussion of general processing procedures of satellite data, as nowadays they have been discussed thoroughly elsewhere and have become common knowledge. In advanced regional satellite oceanography many algorithms and software packages are used as modules and no longer contain any peculiar features. In this book the emphasis is set on the discussion of how to tackle these prefabricated modules of satellite data processing (which are becoming mere standard products), and on how the regional and local peculiarities of the sea regime or meteorological conditions should be accounted for.

Hence here to a large extent a geographical and marine science- and services-orientated presentation will be offered. Bearing in mind that the sea is a dangerous place, and that the temporal variability of the state of the sea is very high, the operational side of satellite data processing, analysis and use is considered, where appropriate.

While writing this book, several times I found myself thinking that I was inevitably comparing certain aspects of the activities in regional satellite oceanography elsewhere with those in my country. Thinking it over, I came to the conclusion that, after all, it was quite natural that I should try to share my knowledge wherever it seemed plausible, because I know first-hand what the situation in this area in the former USSR was, and is today in Russia; and it seems reasonable to inform the world remote sensing and oceanographic communities about certain features of satellite oceanography in the former USSR. It is probably my mission to broaden the horizons of members of the relevant communities in this particular topic, bearing in mind that information on the past may be useful in the course of development of regional satellite oceanography in other countries. Lessons have to be learnt! Both positive and negative results are still results, and it is possible that after this pioneering book some other may appear presenting an insight from the inside on various aspects of the development of science and technology in the former USSR, where a tremendous amount of manpower and resources has been invested in various programmes, and the time now seems appropriate to assess whether the objectives of those projects have actually been achieved.

I called this book **Regional** *Satellite Oceanography* for a good reason. In the mid-1990s, the term 'regional' acquired yet another meaning, different from the original sense of 'of a region'. The wave of *regionalisation* rolled over Europe in the 1990s. The former Yugoslavia fell to pieces, the 15 former 'union republics' of the former USSR anxious to break their ties with the Centre became independent. Even some 'autonomous republics' and 'oblasts' (first levels of territorial and administrative division) of Russia attempted to get a kind of independence from the Centre. Moscow had become by that time the symbol of the Centre, the omnipotent, omniscient monster giving orders ... It seemed that striving for decentralisation and *regionalisation* in the world became the mark of the end of the century, perhaps the end of the millennium.

When, a decade ago, I was thinking of the title for the new interdisciplinary science (and now for the book), *regional* satellite oceanography seemed to me to reflect the real situation, not only because it deals with regional marine applications of satellite data, but also because it highlights the domination of regional needs over the demands of the Centre, though the new preference is accompanied by regional responsibilities and the need to base one's activities on regional resources and local facilities.

What kind of process occurred in the former USSR in relation to remote sensing applications to marine science and services? The overcentralised system (the Agency/Committee for Hydrometeorology and Monitoring of the Environment located in Moscow, the Moscow Main Satellite Data Acquisition Centre, the Moscow Centre for Remote Sensing) for a number of years proved to be unable to provide regional users with the required satellite information and even with raw (unprocessed) data. Regional users searched for a way out. When it became possible for them to receive AVHRR NOAA data on the spot, using simplified home-made autonomous data acquisition stations, when the scientists were able to schedule themselves what satellite data to receive, when to record, what to store – a miracle happened.

Regional users felt that they did not need Moscow services any more or, to be more precise, the centralised services appeared no longer to be the only source of satellite data, and Moscow stopped being the only monopoly feeder. It turned out that it was possible, if there was a will, to get a lot more satellite images than the haughty and pompous Moscow could ever give, and the air of freedom gave the oceanographers self-confidence. There appeared, within the constraints imposed by the available technological facilities, a series of regional investigations, a wave of publications in the top magazines of the former USSR on various topics in the oceanography of the Baltic Sea. That is what was actually the AVHRR scanner data directly transmitted in APT mode from the NOAA series of satellites! And not only for us in Leningrad, but also for our colleagues in Sevastopol, Odessa and Kerch on the coast of the Black Sea and for operational fishery companies in the seas of the Pacific near the Russian coast who have been using the NOAA data for years.

Another aspect of the problem was the inability of the Centre to render the regions any methodological expertise and assistance. What recommendations of the Centre could be effective in solving *regional* problems, in studying *regional* oceanographic phenomena and processes? The leading role here was the *region's*, as the sea-truth data were collected in the *region* by *regional* staff on the *regional* network maintained by *regional* offices. Besides, the central institutions had much better computers and other facilities which at a certain stage led to a sharp contrast with the technological level of *regional* organisations where the real experts and users worked. (It is worthwhile to note that Moscow is situated a thousand kilometres from the nearest sea.) So even the advanced digital procedures worked out in the Centre could not have been implemented in the regions because of the lack of hardware. The Centre has definitely made a lot of mistakes, and those mentioned above were just a few of them, but they definitely caused tension between the Centre and the *regions*; and probably these notes can be related to other work, not only to satellite oceanography.

Because of the deterioration of the national economy there remained for Russian oceanographers fewer means to study the world ocean as a whole, fewer opportunities to take part in the international scientific co-operation on, e.g., global climate change issues. It become impossible to run cruises to oceans on 'big white steamers' with very poor equipment. The State Oceanographic Institute (Moscow) lost its research fleet: when the Ukraine declared its independence of Russia, the research vessels located in the harbour of Odessa on the Black Sea appeared to be abroad.

So through force of circumstances and particularly because of economic constraints, the oceanographers of the former USSR tend to concentrate on the *regional* problems of their *regional* seas. Thus scientists in the former USSR found themselves in the situation typical of the rest of the world, namely, when people are much more concerned with the state of their region, their sea, their bay, than with the probably intriguing, but remote, global problems. It is the responsibility and the privilege of a few prosperous states to deal with global issues.

Regional satellite oceanography originated a decade or two ago as a scientific and technological response to the requirements of marine sciences and the growing activities in the regional seas and coastal zones of oceans worldwide. This interdisciplinary science has passed a difficult period of childhood and it becomes more and more mature with time. Satellite oceanography of the seas or regional satellite oceanography has indeed become an effective instrument of marine research and marine-based branches of the national economy.

That is why I am sure this book will attract its readers.

S. V. Victorov
Sheffield – Dunblane

Acknowledgments

I gratefully appreciate creative contributions to this research for more than a decade made by Dr. Irene Bychkova and Mrs. Leontina Sukhacheva. Gratitude is due to the colleagues who, in different periods, worked in the Laboratory for Satellite Oceanography (St. Petersburg) – the late Dr. Nikolai Lazarenko, the late Dr. Vasiliy Vinogradov, Dr. Vladimir Smirnov, Mrs. Marina Demina, Mr. Victor Lobanov, Mr. Vladimir Losinskiy, Mr. Vasiliy Smolyanitskiy, Dr. Vadim Drabkin and Dr. Valery Gashko. The joint efforts of those named and others made it possible to set the cornerstone and develop regional satellite oceanography in the former USSR. Unfortunately not all of them could continue as researchers in present-day Russia.

It is my pleasure to thank Prof. Arthur Cracknell (Dundee) for discussions of the idea of this book and Dr. Robin Vaughan (Dundee) and Mr. Hans Dahlin and Dr. Bertil Hakansson (Norrkoping) for their suggestions and remarks on the content of Chapter 4.

Special thanks are due to Dr. Igor Berestovskiy (Moscow), Dr. Hans-Jurgen Brosin (Rostock-Warnemunde) and Dr. Josas Dubra (Klaipeda) and their colleagues for fruitful co-operation in the organisation and conducting of international complex oceanographic subsatellite experiments which became an important part of satellite monitoring of the Baltic Sea in the 1980s, and also to colleagues in Moscow and Lesnoye for providing satellite data.

I am grateful to Dr. Gaele Rodenhuis, Dr. Hans van Pagee and Dr. Peter Glas (Delft) for discussions of problems related to the satellite monitoring of the Neva Bay and providing illustrations for Section 5.5.

I would like to thank Prof. Robert Russell (Sheffield), Mr. and Mrs. Martin Davies (Dunblane) and Dr. Peter Jones (Sheffield) who in various ways provided support while I was writing this book.

This book could not have been written without encouragement from my wife Natalia whose contribution is deeply appreciated.

I am also grateful to the staff of Taylor & Francis Ltd, especially Mr. Richard Steele, for help in tackling the manuscript and illustrations.

S. V. Victorov

CHAPTER ONE

Oceanography and remote sensing

1.1 Satellites and oceanographers

1.1.1 Introduction

Thinking now – in the mid-1990s – of this rather philosophical issue, I believe it is worthwhile to recollect one article which appeared in Russian in 1982 in some not very widely known proceedings and therefore, most probably, it was not noticed anywhere except in the former USSR (Victorov 1982). In this article entitled *Satellite oceanography: definition, state-of-the-art and prospects* my views on satellite oceanography as such were summarised. Those views were shared by a group of influential experts and civil servants and therefore actually reflected the then-accepted national policy on the ways in which satellite oceanography should have been developed.

The late 1970s saw the first reaction of the world oceanographic community to the new era in ocean sampling, an era that had been prepared by the general development of remote sensing and was actually opened up by Seasat – the first satellite totally dedicated to oceanography.

In those years many scientists tried from various standpoints to understand the real meaning for oceanography of new tools for ocean sampling, the true place for remotely sensed data among conventional measurements. Many authors tried to understand those specific features of satellite oceanography that made it a distinct part of remote sensing of the Earth. They discussed the current development of this interdisciplinary branch of science, analysing its first achievements, the obstacles in the way of further development, and put forward their perspectives on the methods and technical facilities of satellite oceanography (Nelepo 1979, 1980; Novogrudskij *et al.* 1978; Fedorov 1977, 1980; Allan 1979; Apel 1977; Nierenberg 1980; Szekielda 1976).

In my article I tried to explain to the Soviet audience (and especially to decision-makers) why satellite oceanography could not be considered merely as the study of the world ocean from space. In my opinion, satellite oceanography at least at that time (and probably not only in the USSR) should have been considered as a multi-disciplinary science and many experts in various branches of science should have been contributing to its development.

1.1.2 Difficulties in development of satellite oceanography

While analysing the development of remote sensing of the Earth from space in the 1970s, Allan (1979) pointed out that the then contribution of satellite-borne data could be estimated as modest. Nierenberg (1980) wrote: 'It is difficult to explain why the development of spaceborne oceanography was so slow, it is difficult to understand why so little attention is paid to this topic (in the world there is only one data acquisition and processing station designed in the interests of oceanography and not a single oceanographic satellite)'.

There were special conditions for the general development of space-based sciences in the world in the 1970s but an analysis of these conditions lies far beyond the scope of this book. Furthermore I will not discuss here the managerial aspects of scientific development (though I consider them as a top priority in many cases). Rather, I would like to list the reasons why the real contribution of satellite information to the understanding of physical, chemical and biological processes in the ocean was relatively small, as explained in Victorov 1982:

1. For the study of many phenomena and processes oceanographic information should be four-dimensional (three spatial coordinates plus time) while remote sensing in principle is sounding of the surface. (Only remote sensing utilising the electromagnetic spectrum will be considered here, though some authors include hydroacoustics in remote sensing.) Knowledge of interconnections between processes occurring at depth and their manifestations at the surface of the ocean is insufficient.

2. Remote sensing methods of obtaining oceanographic parameters are intrinsically indirect. Relationships between the characteristics measured by satellite and the parameters generally accepted by oceanographers are not at all unambiguous. Methods of obtaining truly oceanographic parameters are still to be developed.

3. 'Signals' reaching the satellite from the sea surface are altered in the Earth's atmosphere. The input into the signal from the atmosphere in many cases exceeds the 'useful' signal from the sea. There is no current information about relevant characteristics of the atmosphere, including sea aerosols.

4. The accuracy of oceanographic characteristics determined by satellites is still lower than that of conventional *in situ* measurements. To exclude ambiguity and increase the accuracy of satellite measurements by means of various types of corrections one should carry out synchronous measurements of many parameters of the marine environment and atmosphere. The carrying out of such measurements is a very complicated task, however.

5. Clouds prevent the gathering of regular time series of measurements in very informative bands of the electromagnetic spectrum, i.e., in the visible and infrared (IR). The non-regular character of satellite data prevented many oceanographers from considering satellite data as a self-consistent source of knowledge about the ocean.

6. Deep intrinsic differences between remotely sensed information and conventional oceanographic information (e.g., the well-known problem of the relationship between radiative temperature and thermodynamic temperature) and, in addition, the new form of data presentation (two-dimensional as opposed to traditional 'cross-sections' and 'stations') did not encourage the use of new data. One

must confess that there are no well-developed methods of adaptation of two-dimensional, linear and point measurements. The development of methods for assimilation of satellite data is in its initial stage.

7. Moreover, multichannel satellite sensors with wide swaths started producing such tremendous amounts of data, that 'traditional' oceanographers in traditional oceanographic institutions simply could not handle it.

It soon became evident that the then existing pure oceanographic institutions were not able to process the information flow likely to come from satellites. For example, the transmission rate of the radar on-board satellite Seasat was expected to be 15–24 Mb s^{-1} and 25–30 Kb s^{-1} from the other sensors (Nagler and McCandless 1975). This meant that a few days of operations with this satellite would yield more data than the science of world oceanography had collected since its origin.

The enormous gap between the potential of modern space technology in providing oceanographic (or quasi-oceanographic) data and the real abilities of the oceanographic community to accumulate this information was one more piece of evidence that initially there was no systems analysis approach to the problem of remote sensing of the Earth.

It was pointed out (Victorov 1982) that regular and operational satellite oceanographic information may be obtained only if there existed a whole system consisting of satellites, ground-based systems for data acquisition, processing, analysis and dissemination, and operational systems for sea-truth data collection.

1.1.3 The role of oceanographers in the design and development of satellite oceanographic systems

At the start of satellite oceanography some experience of remote sensing applications in the Earth sciences already existed. This experience and the analysis of difficulties and obstacles in the way of the development of satellite oceanography (given in brief above) provided strong evidence in favour of a systems analysis approach to the design and development of technical means and methods of satellite data collection, processing and use. This approach meant the simultaneous design of technical equipment, including satellite payload, 'ground–satellite–ground' radio links, data acquisition systems, a ground computer centre for satellite and *in situ* data processing, databases of satellite and *in situ* data, an *in situ* data network and packages of methods, algorithms and software for the general and problem-oriented processing of data, and editing and dissemination of output products with user-friendly interfaces.

Figure 1.1 shows the four main logical components comprising this system and modern satellite oceanography (as it was seen in the late 1970s). This figure will be used to illustrate the then proposed role of oceanographers in designing, testing, tuning and running the operational system.

Activities concerned with *satellite payload* were supposed to include:

- working out recommendations of the types of remote sensors and their characteristics;
- investigation of the informational capabilities of the sensors;

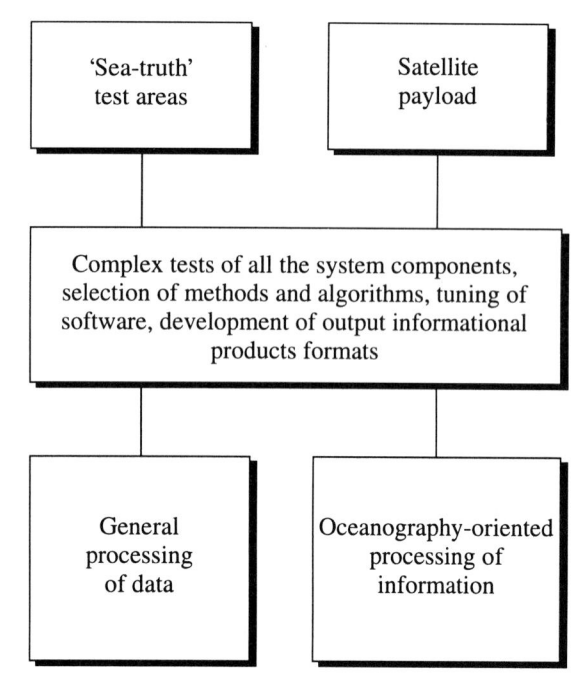

Figure 1.1 The main logical components of an oceanographic satellite system, illustrating the role of oceanographers in its design and development (after Victorov 1982).

- studies of various modes of remote data collection;
- working out recommendations for payload development.

In carrying out these studies an important role was to be played by experts in numerical modelling of various aspects of remote sensing collection. Using the existing and newly developed models of the interaction of electromagnetic waves with the sea surface and models describing the formation of electromagnetic radiation from the sea, one was supposed to estimate parameters of electromagnetic radiation from the sea at the upper boundary of the atmosphere at various physical parameters of the sea surface. Vetlov and Johnson (1978) stated that it could allow one to:

- estimate the range of radiation characteristics to be measured;
- determine the sensitivity of the characteristics to be measured in relation to the range of characteristics of the atmosphere and underlying surface;
- optimise the number of spectral channels in various bands;
- test various methods of solving inverse problems of radiative transfer.

At the final stage of complex tests of the system (the central box in Figure 1.1) the suggested modes of remote data collection should be studied and various options tested. A comparison of remotely sensed data with those collected at the test areas could permit the examination of the informational capabilities of each sensor, of sets of sensors and of the payload plus downlink as a whole.

'*Sea-truth' test areas* were considered as a part of the operational system providing *in situ* data for the calibration of the payload and for tuning the software. The activities under this heading were supposed to be as follows:

- design and set up of a test areas network;
- studies of oceanographic climate in these areas;
- selection of 'test sites' within the test areas;
- optimisation of the network (what sensors to use, where to locate them, etc.).

Oceanographers were meant to work out scientifically based recommendations for the location of test areas (polygons) and dedicated test sites within them taking into account geographical position, hydrological and meteorological climate and the natural variability of various parameters within test areas on various time-scales. Oceanographers were also supposed to participate in the development of methods of data extraction and the complex analysis of information obtained at test areas from sensors located on different carriers – aircrafts, ships, platforms and buoys (Victorov 1980a; Drabkin *et al.* 1982; Ismailov 1980).

Sea-truth information collected at the test areas and stored in databases was meant to be used mainly for the development of a satellite oceanographic system. But these test areas, regularly providing considerable volumes of oceanographic and meteorological data, could have been considered also as self-consistent sources of time series of oceanographic information for conventional studies of regional seas.

Complex Oceanographic Subsatellite Experiments (COSE) were intended to be an essential part of activities at test areas providing additional measurements on non-routine bases (see below).

All the sensors that were to be used for *in situ* measurements at the sea-truth test areas were supposed to have special metrological certificates. In their turn test areas could have been used for the development of metrology of remote sensing applications in oceanography.

General processing of data was supposed to include:

- working out the principles of dividing the world ocean into regions (in rank order of priority), so that these could be used in the general preliminary planning of the work of the satellite payload;
- problem-oriented planning of the schedule for each sensor (or group of sensors);
- working out the structure and content of *in situ* databases.

The first two items were meant to provide measurements at different time intervals and with different spatial resolution of data for different regions of the ocean and coastal zones.

Oceanographers were to have played an important part in the *oceanography-oriented processing of information*, i.e., in:

- the development of methodology, algorithms and operational software for processing the information;
- the running of the 'computer–oceanographer' interactive processing system;
- the development of multiparameter processing (to attempt to obtain synergetic effects);
- the working out of flexible user-oriented formats of output products;
- the editing of output information (using knowledge bases and expert systems).

All the above-mentioned main logical components of the system were supposed to be assembled at the stage of complex tests of the system. The main objective of these tests was to study to what extent the data collected by satellite sensors, downlinked to Earth, processed and analysed in the ground-based computer centre, were comparable to oceanographic information collected by conventional tools and methods. In the course of these investigations possible errors could be detected and the necessary corrections could be introduced in all the links of the 'measurement–communication – processing – analysis – presentation – dissemination' technological chain. Complex tests of the satellite oceanographic system could have been regarded as the actual start of the process of obtaining large-scale oceanographic information on a regular basis.

Thus participation of oceanographers in many activities was supposed to be vital for the design, development and running of the system. Moreover, since carrying out all the above-listed activities was essential for the development of satellite oceanography, a group of these activities appeared to constitute the core of this interdisciplinary branch of science at that period.

I am sure that the actual development of satellite oceanography could have been more successful and could have produced more fruitful results earlier, provided that the above scheme had been realised and the oceanographic community had played its role according to the suggested scenario.

In general the issue of 'satellites and oceanographers' is interesting from both a managerial and a historical point of view and it has only been touched on here. It is important to state that oceanographers were supposed to play an important role in the design and development of satellite oceanographic systems as indicated in the above scheme. This 'ideal' scheme was never realised either in the former USSR or, to my knowledge, elsewhere.

The actual development of satellite oceanography in the world followed another path. Alongside the carrying out of a wide range of methodological studies aimed at the creation of an Earth observation system (including an oceanic component), the 1980s saw the appearance of a quite new phenomenon: *the transition to practical implementations of satellite data in oceanography without waiting for the properly balanced development of all constituent parts of satellite oceanography and in the absence of dedicated truly **oceanographic satellites**, not to mention a satellite oceanographic **system**.*

1.2 Regional and global satellite oceanography

The new phenomenon of practical implementation of satellite data in oceanographic research, first in the form of pilot projects (in the absence of operational or even experimental truly oceanographic satellites), could become a reality due to:

- new knowledge of the actual informational capabilities of existing satellites and sensors initially meant for land applications;
- the coming of new experimental sensors;
- technological development leading to the design of rather simple ground-based (and later ship-based) systems for acquisition of satellite data directly transmitted down to Earth in APT (Automated Picture Transmission) mode, which made this information available for various categories of users;
- progress in the design of computer-based image processing facilities;

- fascinating progress in the development of personal computer hardware and software.

These aspects of the new phenomenon were probably common to many parts of the world oceanographic community. For the former USSR Complex Oceanographic Subsatellite Experiments (COSE) became another and very important constituent part of satellite oceanography. Initially they were meant to provide a practical test of methods of satellite data processing and to check the results of their analysis. Later COSE were used as a tool for regional marine-science-oriented studies of regional seas (see below).

Being 'free' and forced to develop beyond the framework of a non-existent satellite oceanographic system, world satellite oceanography obviously showed a tendency to split into two branches, namely *Global* Satellite Oceanography (GSO) and *Regional* Satellite Oceanography (RSO).

By definition (Victorov 1988a, 1989b, 1990a) regional satellite oceanography (RSO) is a branch of remote sensing, which is based on satellite imagery analysis and deals with processes and phenomena mainly in the seas, coastal zones and separate parts of the ocean, while global satellite oceanography (GSO) deals mainly with processes and phenomena in the world ocean as a whole. Nowadays, RSO is often focused on environmental problems, while GSO concentrates on global climate change and relevant projects.

RSO and GSO differ in their specific features within the whole spectrum of modern satellite oceanographic activities, namely, in:

- the requirements for raw satellite data;
- the configuration of satellite sensors and their modes of operation;
- the algorithms and methods of satellite data processing and analysis;
- the means of sea-truth data collection;
- the configuration of databases;
- the capabilities of links for data transmission;
- users with their requirements and 'standards';
- the formats of value-added output informational products;
- output products dissemination systems.

Let us discuss briefly these specific features and differences between global satellite oceanography and regional satellite oceanography.

1. *Requirements for raw satellite data.* Along with the list of oceanographic parameters to be determined, these include coverage, spatial resolution, frequency of taking images, and delay in data delivery. The fact that these requirements are quite different for RSO and GSO is obvious and Table 1.1 gives the reader a general idea of those differences. We will discuss this issue in more detail later; now it is important only to note that these differences lead to specific features and peculiarities in all the other aspects.

2. *Configuration of satellite sensors and their modes of operation.* This issue includes not only the list of sensors but the structure of the payload (sensors, data storage, communication facilities, etc.). In many cases, to meet the requirements of regional satellite oceanography it is more important to provide a direct data transmission mode rather than to store data on board, while global satellite

Table 1.1 Some requirements to satellite data from global satelite oceanography and regional satellite oceanography (Victorov 1988a)

Data characteristics	Problems of GSO (TOGA Project)	Problems of RSO[a] General	Local
Dimension of studied area	20° latitude band along equator[b]	50–1000 km	1–50 km
Spatial resolution	Averaged 2° latitude by 2° longitude[b]	100–1000 m	20–100 m
Period of observation	10 years[b]	Season–Year	Days–Weeks
Frequency of observation	Not defined	3–24 h	3–6 h
Accuracy	Absolute values (e.g, sea surface temperature 0.3°C, wind 0.25–0.5 m s^{-1})[b]	In some cases qualitative estimates are appropriate	
Means for regular obtaining of *in situ* data	Limited	Existing	

[a] Author's estimate
[b] Report of the TOGA Workshop 1984

oceanography requires just the opposite. It is also important to bear in mind the different requirements of the two while planning operations, especially when the same sensor is used to collect data for RSO and for GSO. For example, in coastal areas you may use fine spatial resolution of a sensor while in the open ocean the data may be averaged. One may also consider the problem of optimisation of payload taking into account the requirements of both RSO and GSO.

3. *Algorithms and methods of satellite data processing and analysis.* These will differ not only because the principal requirements for raw data and output informational products differ but also due to the different amounts of *in situ* data that can be used. The complexity of numerical models and data assimilation schemes, and thus computer facilities, are also different.

4. *The means of sea-truth data collection.* These data are supposed to be used for verification of algorithms and validation of software meant for satellite data processing, for estimation of errors in the determination of oceanographic parameters, and for 'calibration' of satellite data. It is clear that in problems of regional satellite oceanography it is possible to collect more *in situ* data of higher quality for a certain area for a given period and provide their operational usage. The existing meteorological and hydrological networks can be used to collect *in situ* data in coastal areas. In some areas information from the coastal guard can also be made available. Offshore gas and oil platforms are used to collect oceanographic and meteorological data. Availability of *in situ* data helped to obtain more reliable and more accurate satellite information for the problems of regional satellite oceanography.

5. *Configuration of databases.* Databases for the problems of RSO and GSO are different due to the remarkable differences in the spatial and temporal dimensions of datasets and in the types of numerical models used in the two branches of satellite oceanography.

6. *Capabilities of links for data communication and output product dissemination systems.* In many countries it is still a severe problem to provide delivery of satellite data to various categories of users. Many users involved in regional satellite oceanography need high-resolution imagery (gulfs, bays, coastal waters, offshore areas) daily with allowed delivery time of 3 hours, while weather forecast services may need lower spatial resolution data with global coverage twice a day, and climatologists seem not to be concerned about delivery time at all. Thus the total length, structure and capability of links for delivery of satellite data and/or value-added products may differ considerably when we deal with operational RSO and GSO.

7. *Users with their requirements and 'standards'.* For global satellite oceanography a few (often unique) scientific centres are the main users. These centres have high-technology facilities providing processing, analysis and storage of large volumes of data. These centres are networked with national and international satellite data acquisition stations. Regional satellite oceanography is oriented mainly towards users dealing with regional and local problems, and their total number in a certain area may reach a hundred. Their technical facilities may be rather modest (if any), and thus some of them require informational products more or less ready made for their purposes without additional processing. Some of them

require just raw satellite data but pay great attention to the delivery time. Hence regional satellite oceanography has to plan its activities based on, *inter alia*, autonomous satellite data acquisition stations meant for receiving satellite data directly transmitted to Earth and covering the region of interest for a number of users or for a single user. These simplified inexpensive stations should provide processing, storage and (possibly) dissemination of data among the local users. Local administrative boards, institutions, research vessels and fishery ships are also the typical end-users of satellite data on a regional scale. In recent years various boards responsible for monitoring and control of the marine environment have also become users.

Regional satellite oceanography as a branch of science should take care of this kind of user and provide them with methods, software, instructions and manuals matching their requirements and taking into account their current level of expertise, their facilities and human resources.

8. *Formats of value-added output informational products.* If one centralised system provides data and services for many users it should meet their requirements and offer a variety of output informational products. It is obvious that users of different categories require different types of output informational products. GSO users usually have a limited and stable list of formats. Users in regional satellite oceanography are more flexible and may require a variety of formats.

To conclude this comparison of RSO and GSO let us state that *global satellite oceanography* is comprised of the above-listed technological aspects of satellite oceanography applied to general problems of large-scale thermohydrodynamical interaction between ocean and atmosphere, global climate change and weather forecasting. The list of problems covered by GSO includes the monitoring of the dynamics of the quasi-homogeneous layer of the ocean, the description of jet border currents and equatorial zone variability, monitoring of the ice cover edge, and the determination of radiative balance in the 'ocean–atmosphere' system. The core issue of global satellite oceanography is the world climate research program which includes several major oceanographic projects (WOCE, TOGA, JGOFS and others) (see, for example, Berestovskij and Victorov 1981; Kondratyev 1987; Metalnikov and Tolkachev 1985; Nelepo and Korotaev 1985 for an analysis of the satellite component of those programs).

Along with the above-mentioned problems of physical oceanography, another problem of the highest priority for global satellite oceanography is the investigation of the spatial and temporal distribution of the phytoplankton biomass in the upper layer of the world ocean.

Regional satellite oceanography may be characterised by the above-listed technological aspects of satellite oceanography applied to localised problems of studies of oceanographic processes and phenomena in specific conditions of each separate water area (water body). Among the problems covered by RSO are the study of water mass dynamics within the water body, the spread of river discharge, biological productivity in local areas, the monitoring of water pollution, the transfer of pollutants, the interaction between coastal zone and offshore waters, ice cover morphology, etc. There are many international projects dealing with regional seas and coastal zones, and in each of them regional satellite oceanography plays (or should play) an important role. This role will increase with the development of the actual capabilities of RSO.

The natural splitting of satellite oceanography into regional satellite oceanography and global satellite oceanography could probably have been argued for more strictly on a quantitative basis. With this purpose in mind I tried to use the approach developed by Kondratyev and Pokrovsky (1978) and Kondratyev (1983). These authors studied the problem of optimisation of the satellite payload meant for observation of the Earth. They tried to optimise satellite sensors by spectral band and by spatial resolution. All the problems of observation of the Earth from space were divided into four groups (hydrology, geology, forestry and agriculture, oceanography). It is interesting to note that for oceanographical problems the list of requirements for the location and number of spectral intervals for spectral band 0.3–2.6 m km only consists of 102 items. For the systematic description of all the requirements for spectral characteristics the so-called 'data vectors' with 54 dimensions were used. For the description of the requirements for spatial resolution and other parameters of observations 73 'data vectors' with 89 dimensions were used. The authors could not get accurate and unambiguous results. Kondratyev (1983) pointed out that due to the arbitrariness of the requirements for parameters of observations and non-representative data on spectral reflection of various natural objects the results obtained should be considered as merely methodological. While respecting attempts to construct idealised schemes, in my review of satellite oceanography in the former USSR (Victorov 1992a) I had to state that 'though many articles have been published with scientific recommendations on the types of sensors to be launched and their optimal spectral characteristics, in my opinion no significantly new suggestions have ever been put forward. At least, if these suggestions were made they have not been implemented'. Nevertheless, the results obtained by Kondratyev and Pokrovsky (1978) appeared to be interesting from another viewpoint, namely as an additional argument in favour of my concept of regional satellite oceanography.

In fact,

(A) in the course of the optimisation of sensors for four groups of problems (hydrology, geology, forestry and agriculture, oceanography) it appeared that problems of oceanography as a whole could hardly be co-ordinated (in terms of spectral intervals, spatial resolution and other observation parameters) with those of the other three groups, and

(B) some of the problems of oceanography (those I call problems covered by global satellite oceanography) could be fairly well co-ordinated with problems of hydrology and geology, while the rest could be co-ordinated with problems of forestry and agriculture (problems covered by regional satellite oceanography).

This seems to show the internal structure of a set of oceanographic problems and once more leads us to the concept of regional satellite oceanography as a branch of science dealing with a certain number of those problems.

1.3 The concept of regional satellite oceanography

It was pointed out in Section 1.2 that the requirements for satellite information (meaning both the raw satellite data and value-added output satellite-data-based informational products) are actually the basis for the splitting of satellite oceanography into regional and global. Hence it is worthwhile to discuss this point in more detail. Table 1.1 was first published in 1988 and was actually drawn a couple of years before. The 'Regional' column contained the author's estimates of the requirements for satellite information based on the then existing (and known to the author) regional problems. For the 'Global' column the requirements from the Tropical Oceans and Global Atmosphere (TOGA) Programme were cited as a typical example.

To give the reader an idea of the up-to-date formally stated requirements I would like to present Table 1.2 (Report 1995). This table contains requirements worked out by international boards and may be regarded as guidelines for satellite oceanography. The 'Application' column refers to the modules of the proposed Global Ocean Observing System (GOOS) likely to become operational in the 2010s. The following GOOS modules exist on paper:

- (C) climate module;
- (L) living marine resources module;
- (Z) coastal zone module;
- (H) health of the ocean module;
- (S) ocean services module.

It is worthwhile to note that by May 1994 requirements were worked out only for the climate and ocean services modules. The latter is meant for marine forecasts; it should be operational and the listed requirements are actually the values meant to be used in operational numerical models. The second column from the left gives 'regional/global' coverage and may be referred to regional and global satellite oceanography.

Some figures in Table 1.2 are still under discussion and may be regarded as preliminary. Some questions arise when one looks at this table (for example, why are the requirements for sea surface temperature the same for both regional and global coverage when it is doubtful whether the numerical models capable of assimilation of 1 km resolution data exist?). It is clear that the three GOOS modules (L), (Z) and (H) still missing will yield a lot of additional requirements, which will probably make it possible to discuss the updated requirements in more detail in future. But what is important now is that (a) the estimated requirements listed in Table 1.1 are in good or reasonable agreement with those in Table 1.2, and (b) the requirements worked out by international boards of experts and listed in Table 1.2 do actually refer to two different sets of oceanographic problems, i.e., regional and global, which supports my initial idea of distinguishing between regional satellite oceanography and global satellite oceanography.

Moving now to regional satellite oceanography let us remember that practical interest in its development was based on and can be explained by:

- activities in coastal zones all over the world;
- accessibility of satellite data directly transmitted down to Earth;
- abilities of regional institutions for tackling comparatively small amounts of satellite data on a regional level;
- availability of *in situ* data within the region.

Table 1.2 Ocean satellite data requirements (Report 1995)

Parameter	Regional/global	Horizontal resolution	Frequency	Accuracy	Source/application
Wind vector	Global	200 km	12 h	1 m s^{-1}, 10°	GCOS/C
	Global	25 km	6 h	1 m s^{-1}, 10°	WMO/S
	Regional	10 km	6 h	1.5 m s^{-1}, 10°	WMO/S
Sea surface temperature	Global	100 km	5–10 days	0.2 K	GCOS/C
	Global	1 km	6 h	0.1 K	WMO/S
	Regional	1 km	6 h	0.1 K	WMO/S
Ocean wave spectra	Global	10 km	1 h	10°, 0.5 s	WMO/S
Ocean topography	Global	100 km	5 days	3 cm	GCOS/C
Ocean colour	Global	100 km	24 h	—	GCOS/C
Sea ice cover	Regional	10 km	24 h	2%	WMO/S
	Regional	100 km	24 h	1–5	GCOS/C
Sea ice thickness	Regional	25 km	12 h	10%	WMO/S
	Regional	100 km	24 h	10%	GCOS/C
Sea ice edge	Regional	10 km	24 h	2%	WMO/S

These items seem to be common for many countries. While in the former USSR, as mentioned above, Complex Oceanographic Subsatellite Experiments (COSE) were the triggers for the development of regional satellite oceanography. Those COSE used simple satellite data acquisition stations to obtain information for the regions for which large amounts of *a priori* (historical, archived) data already existed. The first oceanography-sounding results obtained helped us to understand the crucial importance of three components needed to carry out regional oceanographic research with involvement of satellite information:

- *own* (independent of unreliable centralised system) source of satellite data;
- *own in situ* datasets;
- knowledge of '*own*' region.

(The last two items were later called the *database* and *knowledge base* components of Geographical Information Systems (GIS).) Understanding of this fact meant one more step towards reinforcing the *regional* accent in satellite oceanography.

A methodology for the study of regional seas, based on complex involvement of all the available satellite data, *in situ* and *a priori* information, was being developed by Victorov (1984a, 1986a). In these papers the basics of 'satellite marine science' or 'regional satellite oceanography' had already been presented. It is important to note that along with practical activities in regional satellite oceanography the methodology of complex satellite monitoring of the seas was being developed in the former USSR (Victorov 1985). The concept of regional satellite oceanography as a science and as a consistuent branch of remote sensing was presented by Victorov (1988a).

A methodological concept of RSO in the context of complex satellite monitoring of the seas was presented in brief by Victorov (1985). This concept formed the basis of complex studies of the Baltic Sea carried out by the Laboratory for Satellite Oceanography and Airborne Methods at the State Oceanographic Institute, St. Petersburg, Russia in the 1980s and at the beginning of the 1990s. These studies will be presented in detail in Chapter 5.

Now let us consider our concept, illustrating each item, where appropriate, with examples relating to the Baltic Sea. The scheme in Figure 1.2 helps to outline the discussion.

The general methodology of regional satellite oceanography can be presented in the form of four modules each consisting of a number of items (operations).

The first module of RSO methodology deals with the working out of general approaches to studies of a given regional sea or water body. It consists of the following operations:

1. analysis of general problems relevant to studies of a certain sea;
2. selection of the principal problems crucial for studies of this sea;
3. setting priorities for these problems.

When starting this work it is natural to use the expertise of oceanographers who professionally investigate the sea in order to obtain their assessment of these three issues.

To illustrate these statements let me remind the reader that several generations of Baltic oceanographers have discovered that the hydrological, hydrochemical and hydrobiological regime of the Baltic Sea is governed by four interconnected issues, namely:

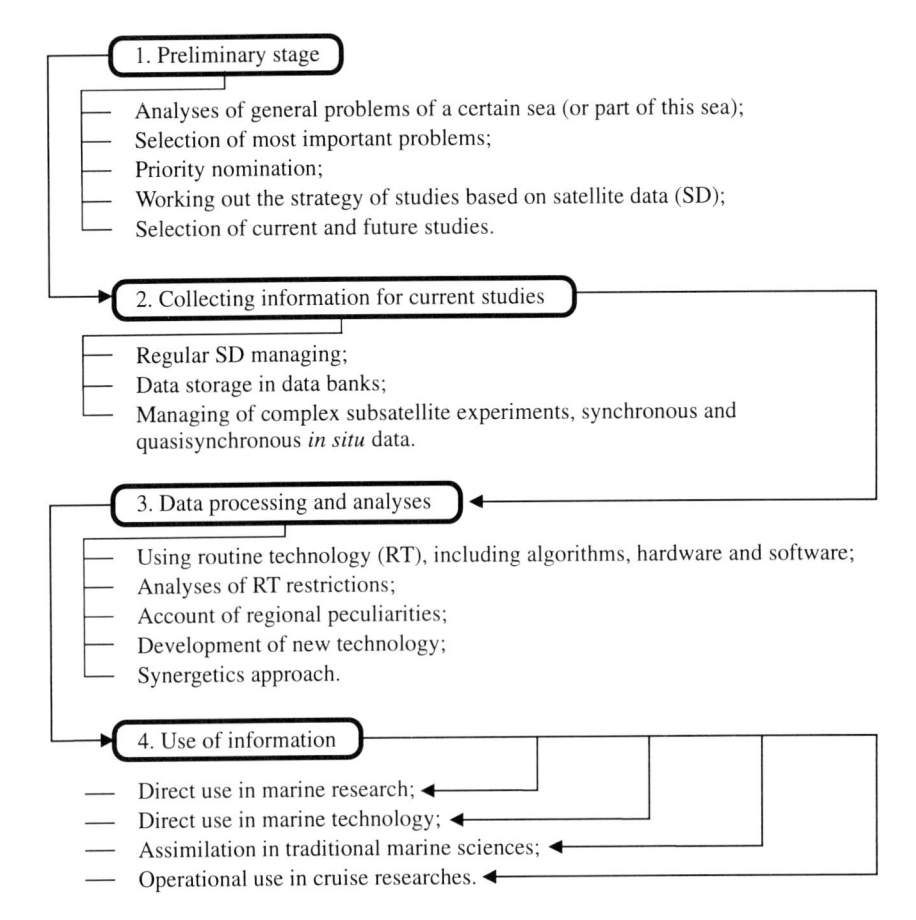

1. Preliminary stage
- Analyses of general problems of a certain sea (or part of this sea);
- Selection of most important problems;
- Priority nomination;
- Working out the strategy of studies based on satellite data (SD);
- Selection of current and future studies.

2. Collecting information for current studies
- Regular SD managing;
- Data storage in data banks;
- Managing of complex subsatellite experiments, synchronous and quasisynchronous *in situ* data.

3. Data processing and analyses
- Using routine technology (RT), including algorithms, hardware and software;
- Analyses of RT restrictions;
- Account of regional peculiarities;
- Development of new technology;
- Synergetics approach.

4. Use of information
- Direct use in marine research;
- Direct use in marine technology;
- Assimilation in traditional marine sciences;
- Operational use in cruise researches.

Figure 1.2 The methodological concept of regional satellite oceanography: sequence of operations (Victorov 1988a).

- water exchange with the North Sea;
- interaction between the water surface and atmosphere;
- river discharge;
- vertical and horizontal mass and energy exchange in the water depth.

Davidan (1981) listed current tasks of the Baltic Sea studies, including:

(A) development of the basic principles of numerical modelling of the sea ecosystem;
(B) development of a primary model of the ecosystem with a hydrodynamical model being its principal constituent part, followed by a set of more sophisticated models;
(C) studies of 'passive tracers' (heat, salinity, turbidity, biological and selected chemical elements and pollutants) transport in the sea;

(D) field studies and the mathematical description of variability and inhomogenities of hydrological and hydrochemical fields and fields of pollutants in the sea;
(E) investigation and mathematical description of biochemical processes, and studies of structural and functional features of biological objects in the marine ecosystyem;
(F) creation of a comprehensive database and directory containing information on physical, chemical and biological fields in the sea and their interaction.

One should note that the expert estimations of different groups of scientists and individual experts may not coincide, particularly in ranging priorities. Sometimes it is difficult to obtain these estimations. The problem of setting priorities is closely connected with the fourth methodological component.

The following items characterise areas of implementation of regional satellite oceanography methods:

4. feasibility analysis of using current and up-coming satellite data for studies of the sea, taking into account the range of problem priorities and the possibility of combining the activities based on remotely sensed data with conventional investigations;
5. decision-making process aimed at working out a general strategy for studies of a certain sea based on the remote sensing technique, selection of problems that can be solved using the available satellite data (current problems) and those problems that require new (not yet available) satellite data (perspective problems).

Victorov (1984a) showed that remotely sensed information may play an essential role in solving problems (C) and (D) in the list of problems of the Baltic Sea studies and also may be useful for problem (E). It is important to note that regional satellite oceanography should have access to the database of characteristics of the sea (under development) and in its turn be regarded as one of the sources of information for the database and the directory (problem (F)).

Among the selected *current* problems, studies of dynamic phenomena – upwellings and eddy structures (using sea surface temperature, turbidity, surface roughness and ice as tracers) – were listed. Spring heating of coastal waters, river discharge spread and frontal zones of various origin (generation, duration, relaxation mechanisms, fine structure) were also on the list.

The second set of methodological components is composed of technological items:

6. Satellite data management. Collection, storage and manipulation of satellite data for a certain region. Efforts aimed at obtaining satellite data on a regular basis.

This is an essential item as unreliable sources of data may totally compromise regional satellite oceanography. Two types of sources of satellite data can be mentioned here – centralised, from data acquisition centres at national level, and autonomous local data acquisition stations. This item will be discussed in more detail in Chapter 2. For now I would like just to note that if you use your own data acquisition facilities, satellite images from 10 orbits can be collected daily (for two satellites of NOAA type, central and side orbits) for the Baltic region.

7. Efforts aimed at obtaining complementary satellite data.

When mentioning this item in the beginning of the 1980s I meant unconventional sources of satellite data. Among them were data that could be obtained on the basis of international exchange, in the framework of joint bilateral projects and also data from experimental satellites. Another source of non-regular satellite images is the photographic satellites. For many years their data have been classified in the former USSR but recently some of them were made available for the oceanographic community (Victorov *et al.*, 1993a; Lukashevich 1994).

8. Management of Complex Oceanographic Subsatellite Experiments (COSE).

These experiments should be considered as part of the more general problem of obtaining *in situ* data as another stream of information flow in regional satellite oceanography. During COSE concentrated efforts are applied in order to collect data of synchronous satellite, airborne and ship measurements and data from atmosphere soundings and thus to obtain information on the sea and atmosphere above the sea. Datasets are meant to develop *regional* methods of satellite information processing.

The managerial aspects and scientific results obtained in a series of COSE carried out in the Baltic Sea in 1982–87 will be presented in Chapters 3 and 5 respectively.

The third module of RSO methodology components covers activities connected with satellite and relevant *in situ* data processing and analysis:

9. Raw satellite data processing and analysis based on existing methods; evaluation of their constraints, analysis of satellite data quality; studies of the applicability of a certain method to a given region.

As one can see, this item contains a wide range of activities some of which will be discussed in Chapter 3. Among our studies of the problems covered under this topic I would like to mention comparative studies of the quality of data transmitted from two satellites of the same type with different times in orbit, and also comparison of thermal infrared data from satellites of 'Meteor-2' type and NOAA. To our knowledge, these studies were carried out for the Baltic Sea region for the first time in 1982 (USSR–GDR 1985).

10. Development of new methods of satellite data processing and analysis taking into account peculiar features of the hydrophysical fields of a given sea and meteorological conditions.

For some purposes the already existing methods and algorithms of data processing could appear inapplicable to a certain region or could need modification. This is the item in which the essential role of regional features in satellite oceanography seems to reveal itself in the most distinctive way. The new or modified methods thus become *regional* methods and can hardly be used in another sea.

It is rather useful to assess the peculiarities of a certain sea before one starts using satellite information in the study of it. For the Baltic Sea the cloud cover is a serious problem, thus restricting the use of satellite data of visible and thermal infrared bands. Cyclones can be traced very often in this region, causing inhomogeneities in water vapour distribution over the sea which may reveal themselves as false thermal or brightness gradients in satellite imagery. Hence the analysis of radiometric data when part of a scene is covered with clouds is not a simple problem. Bychkova *et al.* (1986a) showed that the method of spatial coherent analysis helps in some cases to decrease the total number of 'bad' scenes and to increase the reliability of the results of analysis.

Another peculiar feature of the Baltic Sea is the presence of the so-called 'yellow substance' in water which comes with river discharge. This prevents the use of standard algorithms of satellite data processing developed for oceanic waters and oriented on Coastal Zone Colour Scanner (CZCS) data from the Nimbus-7 satellite. The problem of quantitative determination of chlorophyll and suspended matter in the Baltic using only satellite data appeared to be a difficult one. It is still far from being solved though almost 10 years have passed since the pioneering paper by Kahru (1986).

11. Synergetic approach to data processing and analysis.

Firstly this approach means joint processing of data on a certain oceanographic parameter obtained with various sensors, in various bands of the electromagnetic spectrum, in both passive and active mode. Secondly, various fields (containing two- or three-dimensional patterns) of oceanographic parameters may be 'overlapped' in order to derive new information about the object, phenomenon or process under investigation.

Synergism (synergetics) as applied to remote sensing may be considered as a new advanced methodological approach to data processing and information analysis. It will hopefully allow us to obtain new information on a certain sea or water body as a result of joint analysis of datasets collected with remote sensing and conventional techniques. The result obtained is associated with the non-linear increase in its value as compared to the simple sum of the individual results brought by each individual dataset.

Gidhagen and Hakansson (1987) demonstrated a synergistic approach to the analysis of satellite data collected during the Baltic Sea Patchiness Experiment.

There is a European Project, 'Synergy of Remotely Sensed Data', which started in 1994 and concentrated on vegetation/soil parameters with applications in forest monitoring, land use planning, geology/pedology, and climatology (EARSeL Newsletter 1994). This project is being run by 11 institutions, ITC (The Netherlands) acting as coordinator. To my present knowledge no similar project in oceanography exists.

A pilot project, 'Search for and Studies of Synergism in Remotely Sensed Fields of Oceanographic Parameters', was proposed by the author in 1993 and got a short-term grant from the International Science Foundation. This project was run by the Laboratory for Satellite Oceanography and Airborne Methods, State Oceanographic Institute, St. Petersburg.

The fourth group of methodological components of regional satellite oceanography comprises the resulting steps in studies of regional seas based on satellite information:

12. Working out and testing methods of satellite data and satellite information usage and assimilation in marine sciences.

Besides psychological difficulties of understanding new information and technological obstacles for its use (see Section 1.1) there are a number of methodological problems to be solved. There are three areas where satellite information is being used:

(a) direct use of sets of measurements in current regional oceanographic research as a brand *new* source of knowledge;

(b) assimilation of satellite information, along with traditional data and/or as a mixture of both, in existing models of certain oceanographic and meteorological processes or in coupled ocean–atmosphere models;

(c) operational use of satellite data in the planning of shipborne cruise research.

Fedorov (1980) pointed out that operational guidance of research vessels based on satellite data was one of the first remarkable achievements of satellite oceanography. The direct use of raw satellite data in environmental monitoring of coastal zones and studies of dynamic processes have been demonstated by many authors (see Chapters 4 and 5).

Assimilation of satellite data in regional oceanographic models is also under development in some institutions. Time series of regional satellite data seem to be not durable enough to be used for regional marine climate studies.

13. Direct uses in marine technology.

Offshore industrial operations were probably the first and most remarkable problem area where satellite data served as a reliable source of real-time information on the marine environment. The working out of suitable interfaces between agencies and operators of satellites and this type of user is a challenge for the agencies and their relevant institutions.

When we discuss the problems of *operationalisation* of remote sensing in general, and that of regional satellite oceanography, this category of user should have one of the highest priorities.

The above-listed (and sometimes commented on) items show the sequence of operations that are to be performed when applying satellite information to studies of regional seas or their parts. All the experts and institutions involved do actually perform those operations, sometimes just intuitively, sometimes trying to skip some of the steps. So it seemed worthwhile to present those operations in a systematic way.

The methodology of regional satellite oceanography (Victorov 1984b, 1985) was made a basis for long-term experimental satellite monitoring of the Baltic Sea carried out in the 1980s by the Laboratory for Satellite Oceanography and Airborne Methods, State Oceanographic Institute, St. Petersburg in collaboration with other Russian, German and Lithuanian institutions (see Chapters 3 and 5).

To conclude this discussion of regional satellite oceanography as a science and part of remote sensing, let us consider in brief some basic ideas in the philosophy of science.

What are the features of a mature scientific discipline? Ogurtsov (1988) listed these features as follows:

- the institutionalisation of knowledge;
- the existence of generally accepted ethical regulations of scientific research;
- the formation of the scientific community;
- the existence of specific scientific publications (reviews, textbooks);
- the establishment of certain types of communication between experts;
- the setting up of professional bodies dealing with education and training.

Judging by these requirements, is modern regional satellite oceanography a scientific discipline?

In Section 1.2, regional satellite oceanography was referred to as an interdisciplinary science. But this does not mean that we cannot consider regional satellite oceanography a scientific discipline. Oceanography itself is an interdisciplinary science and is being developed by a union of disciplines which is merely a reflection of the fact that 'many of the basic questions about the ocean are inherently interdisciplinary' (Ocean Science for the Year 2000, 1984).

One of the strongest driving forces of *institutionalisation* is social need. *The Convention on the Law of the Sea*, accepted by many countries in 1982, assigned much of the world ocean to exclusive economic zones where coastal states have jurisdiction over the exploitation of resources (Ocean Science for the Year 2000, 1984). The Law supported and encouraged the coastal countries in their efforts to study the offshore areas and to understand oceanographic phenomena and processes within a few hundred kilometres of their coastline. This actually gave rise to dozens of laboratories and groups in the world dealing with coastal zone studies based on satellite data as a source of information. In many cases the Law caused changes in general policy and practice of oceanographic research in coastal regions; in some cases the coastal states had to use satellite data as the only source of knowledge. So, figuratively speaking the *Law of the Sea* appeared to be the godfather of regional satellite oceanography.

As for *ethical regulation* of scientific research in regional satellite oceanography, general rules accepted in other Earth sciences and other natural sciences are usually applied to RSO. As a rule any problems arising can be solved in a spirit of co-operation. Some problems with copyright, not to mention other sensitive items, still exist in countries with economies in transition.

A *scientific community* has been formed and various forms of *communication* between experts have been established. Many conferences, symposia and seminars on various topics covered by regional satellite oceanography have been held in recent years in many countries.

To my present knowledge there are as yet no textbooks on regional satellite oceanography and no special educational courses, but in recent years the Intergovernmental Oceanographic Commission (of UNESCO) and other international organisations have made some efforts towards strengthening *regional* and *coastal zone* aspects in the existing forms of training experts, including its satellite component. In this context it is worthwhile to mention the computerised course described by Troost et al. (1994).

Is this enough proof that regional satellite oceanography is a scientific discipline? I hope the following chapters of this book will convince the reader that the regional satellite oceanography of today *is* a scientific discipline (though it may not be as mature as it could be).

CHAPTER TWO

Information used in regional satellite oceanography

2.1 Introduction

In the previous chapter some requirements for satellite data and oceanographic products based on satellite information were presented. Now we shall consider whether the satellite sensors provide the required types of data, whether the available satellites and data processing facilities actually produce the required informational products and, finally, whether the value-added products match the stated requirements of the oceanographic community in terms of spatial and temporal resolution and accuracy.

This discussion will be followed by a brief review of satellite programs providing data relevant to regional satellite oceanography. As the information on types of satellites, particular sensors and their technical characteristics referring to operational agencies in the USA, France, Canada, Japan and the European Space Agency (ESA) is widespread and can be easily made available to any interested person through hundreds of publications, numerous conferences and workshops, dozens of booklets, brochures and leaflets, through compact disks, videotapes, etc., while the relevant information on the former USSR (now the Russian and Ukrainian) programs, satellites and their payloads can be picked up only in professional editions with restricted circulation, the author considered it worthwhile to deal in more detail with the activities of the former USSR and its successors.

Regarding the ground segment of regional satellite oceanography, the autonomous low-cost satellite data acquisition stations meant to be installed on the coast and on ships will be mentioned in the context of satellite data availability for regional and local users of various types.

2.2 Matching the requirements(?)

As mentioned above, there are several sets of requirements to ocean satellite data expressed by several international organisations. In Table 1.2. draft ocean satellite data requirements were presented (Report on Satellite Systems and Capabilities 1995) for two GOOS modules: the climate module (C) and the ocean services module (S). Let us discuss whether these requirements are matched by the up-to-date (or planned) satellite missions. Only regional requirements for the GOOS ocean services module will be discussed as they are in line with the concept of regional satellite oceanography. Thus requirements for five parameters can be directly extracted from Table 1.2., namely:

- wind vector;
- sea surface temperature (SST);
- sea ice cover;
- sea ice thickness;
- sea ice edge.

For the sake of consistency the author has added one more parameter: water colour. The reason is that for coastal zone studies (covered by regional satellite oceanography as a scientific discipline) water colour is as important a parameter as SST or sea ice characteristics. Moreover, water colour, in the broad sense of this term, is being determined now and it is probably just a matter of time before the relevant international organisations express their opinion on the requirements referring to this parameter. Meanwhile the author determines his values based on personal experience.

Now six oceanographic parameters will be discussed in terms of (a) feasibility of their determination (as required), and (b) technical effort needed to meet those requirements. It will be up to the reader to say whether these efforts are worth applying. It will be for the decision-makers to say, in terms of costs and benefits (real costs and possible benefits), when and how those efforts will be applied, if any.

The discussion which follows is based on the Report on Satellite Systems and Capabilities (1995) and reflects the joint opinion of members of the Joint CMM-IGOOS-IODE Subgroup on Ocean Satellites and Remote Sensing set up by the Intergovernmental Oceanographic Commission (of UNESCO) and the World Meteorological Organisation.

1 Wind vector

Requirements

Horizontal resolution   10 km
Temporal resolution   6 hours
Accuracy   1.5 m s^{-1} (speed), 10° (direction)

Assessment

Three types of satellite-borne instruments may be used: scatterometer, altimeter and microwave radiometer. None of these instruments matches the requirement for horizontal resolution.

One dual-sided scatterometer requires about 42 hours for 95% global coverage. Hence one needs 6–8 instruments to match the required temporal resolution value. The best demonstrated (for high wind speeds) but not typical accuracy is 1.5 m s^{-1} and about 15°.

Altimeters showed global sampling only and regional applications need development. Two altimeters require about 3–4 days for coverage. To meet the required temporal resolution one needs 12–14 instruments. The best demonstrated accuracy (for high wind speeds but not typical) is about 2 m s^{-1} and no direction measurement is possible.

Microwave radiometers showed 2 m s^{-1} accuracy (no direction measurement). One needs four instruments in orbits to match the requirement for temporal resolution.

It must be noted that if it was stated above that one needed 12–14 instruments, it means not just this number of devices in space, but 12–14 instruments each on a separate satellite, with proper organised intervals between overpasses of each satellite.

2 Sea surface temperature

Requirements

Horizontal resolution	1 km
Temporal resolution	6 hours
Accuracy	0.1 K

Assessment

Two types of infrared radiometers (of AVHRR and ATSR class) may be used, as well as a microwave radiometer.

The horizontal resolution of infrared radiometers is close to the required value, while the microwave radiometer does not match it. To meet the requirement for temporal resolution one needs a set of instruments (satellites) whose number depends on what class of infrared radiometers will be used, as there is a major difference between the AVHRR class and ATSR class. The latter showed accuracy of 0.5–1.0 K with a 500 km swath, while AVHRR showed about 1.6 K. At 1 km resolution there is no improvement by averaging.

The microwave radiometer has near all-weather capability to measure SST with an accuracy exceeding 2.0 K. One needs 4–5 instruments (satellites) to match the required temporal resolution.

Author's additional comments

Based on the experience gained in my laboratory, the above requirement for sea surface temperature may be made less strict for many problems of coastal zone monitoring and management. In my opinion, the values of 24 hours for temporal resolution and 0.5 K for accuracy are quite reasonable and realistic.

Another comment relates to the number of instruments (satellites) actually required to match the, say, 6-hour temporal resolution. In practice the required number of satellites differs for each region in question, depending on the latitude. Hence the term 'latitude band', or something like it, should be introduced to reflect the fact that for the equatorial regions the required number of satellites is maximal, while for higher latitudes it is much less. For example, with two satellites of NOAA type in orbit we managed to receive AVHRR SST data from 12 tracks daily (including central and side overpasses) for the central Baltic Sea using the autonomous data acquisition station located at about 55° North.

3 Sea ice cover

Requirements

Horizontal resolution   10 km
Temporal resolution   24 hours
Accuracy   2%

Assessment

SAR, microwave radiometers, infrared radiometers, visible radiometers and altimeters may be used. All of them match the requirement for horizontal resolution (with the microwave radiometer close to it). One RADARSAT class SAR instrument provides three-day polar coverage in 400 km swathwidth mode, hence one needs three SAR instruments to match the requirement for temporal resolution. The remaining four instruments match this requirement with one instrument in orbit.

SAR matches the required value for accuracy in all weather conditions; visible and infrared radiometers are satisfactory in no-cloud situations only. The microwave radiometer provides about 10% accuracy, and the accuracy for the altimeter needs to be studied.

4 Sea ice thickness

Requirements

Horizontal resolution   25 km
Temporal resolution   12 hours
Accuracy   10%

Assessment

The same set of instruments as for sea ice cover may be used, each of them matching the horizontal resolution requirement (with the microwave radiometer close to it). Again one needs three SAR instruments of RADARSAT class to match the temporal resolution value. All the instruments show accuracies close to that required, with SAR and the microwave radiometer satisfactory in all weather conditions, and the remainder in no-cloud situations.

5 Sea ice edge

Requirements

Horizontal resolution   10 km
Temporal resolution   24 hours
Accuracy   2%

Assessment

The same set of five instruments may be used. The assessment regarding matching the requirements for horizontal and temporal resolution are the same as for sea ice thickness. SAR matches the value for accuracy in all weather conditions, infrared and visible radiometers match it in no-cloud situations, and the microwave radiometer is close to it. Additional studies are needed to assess the accuracy of sea ice edge determination based on satellite altimetric measurements.

6 *Water colour*

Requirements

Horizontal resolution   1 km
Temporal resolution   24 hours
Accuracy   6 levels of concentrations

The last requirement, introduced by the author, needs additional comments. Six (eight at the most in favourable conditions) levels of concentration of one (dominant) or 2–3 (at the most and again in favourable situations) optically active substance(s) are required *in the range of concentrations that is typical for the region in question*. This range is actually typical for a given gulf or bay and may probably differ from another range for a different bay even by an order of magnitude. So, this is a very regionalised and localised requirement but it actually reflects the existing situations in many parts of the regional seas, especially in the sites affected by human development. In general it is not so easy to work out and formally express the requirement for accuracy for this sort of parameter ('ocean colour', 'water colour' or 'water quality', 'bio-optical parameters', etc.). The GCOS community has not expressed the requirement for accuracy of ocean colour determination even for global climate observation; neither has the European team (see below).

Assessment

High-spectral-resolution scanners (of SeaWiFS class) may match the horizontal resolution requirement, and one will need two instruments (satellites) to meet the temporal resolution value. As even the specifications (chlorophyll, sestonic content, attenuation coefficient) are under development it is difficult to make an assessment on matching the yet-to-be-expressed accuracy requirements. In the author's opinion the requirement for accuracy introduced above is flexible and may be used as a tool in practical work with satellite data in regional environmental studies. While there will be no universal algorithms for quantitative determination of unknown substances in the seas, the 'six levels of concentrations' criteria may appear to become a suitable working tool for comparison of different methods and techniques.

Now we pass to another source of satellite data requirements. June 1994 saw the appearance of the document relevant to matters covered by regional satellite oceanography: *Use of Satellite Data for Environmental Purposes in Europe* (Use of Satellite Data 1994). This is a final report produced by Scot Conseil and Smith System Engineering Limited in conjunction with with DGXII-D-4 of the European Commission and presenting the results of a two-year-long study carried out by a group of European experts. This report is based on the data and information requirements declared by pan-national, national and subnational European environmental organisations in terms of spatial and temporal resolution and, where possible, accuracy. These requirements were compared with CEOS members' capabilities up to the year 2008. Report states that 'The process that links the space measurement with the information relevant to the implementation of environmental policy is complex and in many cases not yet well defined or developed. Space data must be input to geophysical and other models of varying degrees of complexity, and combined with other data from different sources, including ground measurements' (Use of Satellite Data 1994).

Along with this kind of obvious statement the report contains some points which are of interest from the viewpoint of nomination of scales (here, in the European context) – the problem that was mentioned in Section 1.2 when regional satellite oceanography was first introduced (Victorov 1988a).

The report gives some terms which 'were adapted and defined for the purpose of this study'.

Application issues are defined as broad sectors of environmental concern, and ten issues are mentioned, of which 'water quality' and 'coastal zones' are of particular interest for us. A set of *information requirements* corresponds to each application issue.

Variables refer to the environmental properties to be determined in order to meet a certain informational requirement. During this study about 300 variables were discussed and about 80 'shortlisted' variables describing 250 scale-dependent measurements were included in the final document. The variables are divided into five *themes*:

- atmosphere;
- ocean;
- coastal zone and big lakes;
- land;
- small lakes, rivers and ground water.

Scale is defined as geographic extent over which information is required. Four scales are nominated, with 'marine' themes covered by only three of them. Some extracts from the report follow:

European scale

- Ocean: refers to the North Sea, Mediterranean Sea and Eastern Atlantic;
- Coastal zones and big lakes: all coastal zones and big lakes in Europe.

National scale (of the order of 500 000 square kilometres)
   No marine or water themes.

Regional or wide area scale (of the order of 100 000 square kilometres)
   Coastal zones and big lakes: Baltic Sea, Black Sea, English Channel and inland lakes as well as coasts.

Local scale (of the order of 10 000 square kilometres)
   Coastal zones and big lakes: selected areas.

(I was very pleased with the fair coincidence of my estimates of 'regional scale' and 'local scale' in the problems of regional satellite oceanography (Victorov 1988a) and those given in Report.)

In the following discussion about whether the satellite data match the expressed requirements, let us restrict ourselves to the regional scale and consider the theme 'coastal zones and big lakes'. (As for this theme on the local scale the report (Use of Satellite Data 1994) expresses for most variables the required spatial resolution of 10–30–50 m with 1 h temporal resolution (with the exception of 24 h frequency for sea surface temperature). Hence the general comment to this section – typical for most of the variables – states that '*At the local level, in general, the scale is too small for space-based measurements, although space-based data may find application in specific circumstances*' (Use of Satellite Data 1994). The report (Use of Satellite Data 1994) lists the following variables for the theme 'coastal zones and big lakes (e.g., Bodensee)':

- surface topography (sea/water level height, rise);
- bottom topography (water depth);
- surface water area extent (lakes);
- waves (height, length, period);
- surface winds (speed, direction, stress);
- currents (speed, direction);
- sea/water surface temperature;
- oil spills (extension, oil type);
- biological activities (chlorophyll, toxic/non-toxic algae bloom, depth distribution of macrophytes, vegetation cover on sea bottom);
- sea water properties (colour, transparency, suspended matter).

These variables are meant to be used in the following 'application issues':

- coastal protection;
- nature and bio-diversity;
- risk management;
- water quality.

Table 2.1 is the author's compressed presentation of requirements for satellite data expressed in the report (Use of Satellite Data 1994).

As a result of an analysis of the correlation between the requirements and the CEOS space segment provision it was stated that for the theme 'coastal zones and big lakes' on a regional scale for only two variables – surface topography and surface water area extent (not included in Table 2.1) – there is 'close compliance between stated requirements and current instruments/missions' (Use of Satellite Data 1994). For the remaining eight variables the conclusion is: '*whilst not closely compliant, stated requirements should benefit from current or planned satellite instruments/missions*' (Use of Satellite Data 1994).

Unfortunately, the report (Use of Satellite Data 1994) presenting two-year-long studies carried out by a large group of experts in most cases does not specify accuracy of measurements, and there is no assessment of the number of instruments (satellites) needed to provide coverage with the requested temporal resolution. Hence we are expected to be satisfied with the qualitative conclusions quoted above.

At the same time for each variable the report (Use of Satellite Data 1994) lists sensor(s) and satellite(s) which are closely compliant with requirements and other significant data sources (operational or planned) in the period 1993–2008.

As a result we have two different sources of analytical assessments on the requirements and the ability of current/planned satellites to match those requirements. Both sources look at the problem from somewhat different positions. In the context of this book (and within the framework of material presented in this section), the

Table 2.1 Requirements for satellite data. Theme 'coastal zones and big lakes', regional scale (Use of Satellite Data 1994)

Variable	Spatial resolution Horizontal	Spatial resolution Vertical	Temporal resolution optimum/minimum	Accuracy	Coverage
Surface topography	100–500 m	—	Yearly	—	Complete
Bottom topography	100–500 m	—	Yearly	—	Complete
Waves	1 km $B = 10$ km	—	Hourly/3 hourly	—	Complete
Surface winds	1 km $B = 10$ km	—	Hourly/3 hourly	—	Complete
Currents	1 km $B = 10$ km	—	Hourly/3 hourly	—	Complete
Sea/water surface temperature	1 km	—	Daily	0.5 K	Complete
Oil spills	20–30 m	—	Hourly (for forecasting)/daily	—	Complete and selected areas for forecasting
Biological activities: Chlorophyll	1 km	5 m	Weekly	—	Complete and selected areas for forecasting
Algae bloom	100 m	—	Hourly (for forecasting)/daily	—	Complete and selected areas for forecasting
Depth distribution of macrophytes	10 m	1 m	Monthly	—	Complete and selected areas for forecasting
Vegetation cover on sea bottom	10 m	1 m	Monthly	—	Complete and selected areas for forecasting
Sea water quality	50–100 m	—	Weekly	—	Complete

WMO/IOC Document (Report on Satellite Systems and Capabilities 1995) is a broader view from the position of a general oceanographic approach and the operational services to be provided, while the European report (Use of Satellite Data 1994) presents a more specified view on environmental issues. To the author's knowledge similar feasibility studies were carried out in the former USSR in the 1970s but their results have not been published in full. Both documents are interesting sources of knowledge and could be considered as complementary to each other. To conclude, it is worthwhile to notice that our concept of regional satellite oceanography is in good agreement with recently issued documents published by authoritative international organisations. But it is more important to note that the existing knowledge encourages researchers to implement the current satellite data in studies of regional seas and their local parts. With this in mind we come to actual sensors and satellites which are relevant to regional satellite oceanography.

2.3 Satellite data for regional oceanographic research

2.3.1 Well-publicised (well-known) sensors and satellites

As stated in the Introduction to this book, it was not the author's intention to include either the basics of remote sensing nor the instrumentation of satellites with descriptions of sensor design and operational modes. Thus in this section only a short list of sensors and satellites relevant to regional satellite oceanography will be presented.

2.3.1.1 Instruments operating in visible and infrared bands

The family of Advanced Very High Resolution Radiometers (AVHRR) consists of AVHRR/2 instruments on current (1984–1995) satellites of the NOAA series, AVHRR/3 scanners on the next (1994–1999) generation satellites and Visible and Infrared Scanning Radiometers (VIRSR) on future (1997–2000) satellites of NOAA type. NOAA satellites operate on polar orbits at an altitude of 850 km with equatorial crossing times of 7:30 a.m. and 1:30 p.m.

Characteristics of the sensors are given below:

	AVHRR/2	AVHRR/3	VIRSR
Spatial resolution, m	1100	1100	1100
Thermal resolution, K	0.12	0.12	0.10
Radiometric resolution, bits	10	10	12
Number of channels	5	6	7
IR calibration	Yes	Yes	Yes
Visible calibration	No	No	Yes
Spectral characteristics (m km):			
Spectral channel 1	0.58–0.68	0.58–0.68	0.605–0.625
2	0.725–1.10	0.72–1.00	0.860–0.880
3	3.55–3.93	1.58–1.64 (day) 3.55–3.93 (night)	1.580–1.640
4	10.3–11.3	10.3–11.3	3.62–3.83
5	11.5–12.5	11.5–12.5	8.40–8.70
6			10.3–11.3
7			11.5–12.5

Since the 1960s data have been directly transmitted to the Earth in APT (Automatic Picture Transmission) mode with spatial resolution of about 4 km and in HRPT (High-Resolution Picture Transmission) mode with resolution of 1.1 km thus providing free access to real-time information from two satellites always in orbit (Sherman 1993).

AVHRR data are the most reliable source of sea surface temperature information. This free source of useful data encouraged many scientific groups throughout the world to start their analysis and implementation. The role that AVHRR on NOAA played in the development of regional satellite oceanography cannot be overestimated.

Since 1985 there was no source of sea colour data in orbit. Thus the oceanographic and remote sensing communities are very interested in the successful launch of the SeaWiFS (Sea-viewing Wide Field-of-View Sensor) instrument on the SeaStar satellite. It will be operated by the private company Orbital Sciences Corporation (OSCO), USA. The instrument will be launched into a Sun-synchronous, near noon descending equatorial crossing orbit with an altitude of 705 km. The spatial resolution will be 1.1 km at nadir in local area coverage (LAC) mode within a 2801 km swathwidth. In global area coverage mode the instrument will have a swathwidth of 1502 km at the equator and a spatial resolution of 4.5 km. (For comparison, the only flown instrument of this class, the Coastal Zone Colour Scanner (CZCS), had a spatial resolution of 823 m in a swathwidth of 1650 km.) Tilting operations will make it possible to observe the sea in one of three positions: nadir, aft ($+20°$) and/or forward ($-20°$). The planned channels of SeaWiFS as compared to CZCS are (intended use is indicated in brackets):

SeaWiFS Wavelength, nm		CZCS
402–422	(Gelbstoffe)	423–443
433–453	(Chlorophyll absorption)	—
480–500	(Pigment concentration)	—
500–520	(Chlorophyll absorption)	510–530
545–565	(Sediments)	540–560
660–680		660–680
745–785	(Atmospheric aerosols)	700–800
845–885		

The relative radiometric accuracy of this instrument is 4%. SeaWiFS is planned to be launched in 1995. The 1.1 km data will be transmitted in real time in a format similar to that of AVHRR HRPT data at 1702.5 MHz. Existing HRPT stations will be able to receive the data provided that a 'decryption' device is purchased from OSCO. Encryption keys will be changed every 14 days. For details of data access and pricing strategy see Lyon and Willard (1993).

In the late 1990s the Earth Observing System (EOS) as part of the US Mission to Planet Earth led by NASA will use polar-orbiting platforms to provide monitoring of the land and the oceans. The candidate sensor EOS Color Instrument (EOS COLOR) is basically the same sensor as SeaWiFS (Sherman 1993).

Another instrument of the future EOS payload is the 36-channel Moderate-Resolution Imaging Spectro-Radiometer (MODIS). Its channels 20 (centre of band 3750 nm/bandwidth 180 nm), 22 (3959/59.4), 23 (4050/60.8), 31 (11 030/500) and 32

(12 020/500) are planned to be used for determination of the sea surface temperature with spatial resolution of 1 km. The same resolution is planned to be achieved for sea colour parameters using channels 8 (412/15) to 16 (869/15) (Sherman 1993).

The four-channel Visible and Thermal Infrared Radiometer (VTIR) (one channel in the visible band and three channels in the thermal IR band) on Japan's Marine Observation Satellite (MOS-1), determines sea surface temperature with 0.5 K radiometric accuracy and 2.7 km spatial resolution within a 1500 km swathwidth. The satellite operates in a Sun-synchronous subrecurrent orbit with an altitude of 909 km, inclination 99°, and descending node local time 10:00–11:00 a.m.

The eight-channel Optical Sensor (OPS) on the Japanese Earth Resources Satellite (JERS-1) has 18×24 m spatial resolution in a 75 km swathwidth. There are three channels in the visible and near-IR band, four in the short-wave IR band and one channel is used to provide stereoscopic images. The satellite operates in a Sun-synchronous subrecurrent orbit with an altitude of 570 km, inclination 98°, descending node local time 10:30–11:00 a.m. and has a design lifetime of two years.

The Advanced Earth Observation Satellite (ADEOS) is a follow-on to the MOS-1 and JERS-1 satellites. It is scheduled for launch in 1996. Its payload includes the Ocean Colour and Temperature Scanner (OCTS), a 12-channel (six visible, two near-IR, middle-IR and three thermal IR) instrument providing imagery with a spatial resolution of 700 m and radiometric accuracy of 0.15 K within a 1400 km swathwidth.

Another instrument on the same satellite, the Advanced Visible and Near-Infrared Radiometer (AVNIR), has five channels (three visible, one near-IR and one panchromatic visible) providing imagery with a spatial resolution of 16 m (8 m panchromatic) in a 75 km swathwidth. The satellite's orbit is Sun-synchronous subrecurrent, altitude 800 km, inclination 98.6°, and local time descending node 10:15–10:45 a.m. Its design lifetime is three years.

The ADEOS satellite has three types of data transmission, including direct transmission of low-rate data from the OCTS instrument for the local users' terminal (Yamamoto 1993).

Many studies of the marine and coastal environment have been carried out using LANDSAT data. The LANDSAT-1, -2, -3 MSS (Multi-Channel Scanner) instrument provided imagery in four visible and near-IR channels with 80 m spatial resolution in 185 km swathwidth. The repetition cycle is 18 days. The satellite, in near-polar Sun-synchronous orbit at a height of 918 km, passes overhead at 10:00 a.m. local solar time.

The LANDSAT-4 Thematic Mapper (TM) has six channels in visible and near-IR bands providing imagery with 30 m spatial resolution, and a thermal IR channel with 120 m resolution. The repetition cycle is 16 days.

The High Resolution Visible instrument on the French SPOT satellite has three channels in multi-spectral mode (0.50–0.59; 0.61–0.68; 0.79–0.89 m km) with a spatial resolution of 20 m and one channel in panchromatic mode (0.51–0.73 m km) with a 10 m resolution and repeat coverage of several days (Cracknell and Hayes 1991).

The future European Polar Platform will have the Medium Resolution Imaging Spectrometer (MERIS) with 288 channels in the range from 400 to 1050 nm, 2.5 nm apart. The spatial resolution will be 1 km over the ocean and 0.24 km over the land and coastal zones (Guide to Satellite Remote Sensing of the Marine Environment 1992).

2.3.1.2 Instruments operating in the microwave band

The Synthetic Aperture Radar (SAR) on ERS-1 with a wavelength of 5.6 cm has a spatial resolution of 20 m within a swathwidth of 100 km. The SAR swath covers the range from 20° to 26° incidence angle from a 780 km height Sun-synchronous polar orbit.

A similar SAR was launched on ERS-2 in 1995.

Canada's RADARSAT (Radar Satellite) will carry an SAR operating at 5.6 cm wavelength. The orbit will be Sun-synchronous at an altitude of 800 km, and an inclination of 99°. Repeat coverage is every three days with a 6:00 local time descending node equatorial crossing. A seven-mode operational configuration is planned including:

- a four-look processing mode providing a 28 m spatial resolution within a 130 km swathwidth;
- a high-resolution mode with 10 m spatial resolution (one look) with a swathwidth of 50 km;
- a SCANSAR mode providing a 100 m spatial resolution (six look) over a swathwidth of 500 km.

The incidence angle can be programmed in the range 10° to 60°. The design lifetime is five years without servicing and 10 years with servicing. The SAR can be operated up to 28 minutes in sunlight during each orbit. A network of ground acquisition and processing facilities and dedicated communication links will make it possible to deliver information to end-users within three hours of acquisition (Sherman 1993).

Among the space-based scatterometers, in the context of regional satellite oceanography, the next-generation NASA scatterometer to be flown on Japan's Advanced Earth Observing System (ADEOS) satellite is worth mentioning. This instrument (designated as NSCATT) will operate at 2.1 cm wavelength and provide surface wind vectors at a spatial resolution of 50 km with an accuracy of 2 m s^{-1} or 12%, whichever is larger, for velocity and 20° for direction. The planned swathwidth is 1500 km with a 175 km nadir gap, which will provide a 79% global wind coverage daily and a 95% coverage every 2 days (Sherman 1993).

We conclude this section with some remarks on the selection approach and data availability from the sensors.

Characteristics of some satellite sensors which, in the author's opinion, are relevant to studies and observations covered by regional satellite oceanography were given above, and deal with current and planned activities of agencies and companies of the USA, Canada, France, Japan and the European Space Agency (ESA). Selection was made bearing in mind the following points (for current sources of data):

- well-proven use in problems covered by regional satellite oceanography;
- proven reliability and quality of data:
- availability of long time series;
- easier (as compared to other similar sources) access to data (price policy was not taken into account);

and (for planned sources of data):

- scheduled launch prior to the year 2000.

Data from the sensors mentioned in this section are available either from operating agencies or the appointed organisations, or through the national contact points. Cracknell and Hayes (1991) provide a useful list of relevant addresses.

2.3.2 Sensors and satellites of the former USSR relevant to regional satellite oceanography

2.3.2.1 General remarks

In order to help the reader to gain a clearer view of the present situation and to estimate the possible activity of Russia and Ukraine on the international arena, where so many agencies and organisations are currently being involved in the research and business activities covered by the term 'regional satellite oceanography', and in order to illustrate current abilities and future perspectives, some milestones of Soviet satellite oceanography will be mentioned and some historical facts will also be included in the text, along with a presentation of the technical characteristics of current and planned sensors and satellites.

A brief history of satellite oceanography in the former USSR was presented by Victorov (1992a). In that paper satellites meant for (and those used for) oceanographic research and their payloads were reviewed. Satellite oceanography in the former USSR as a scientific and technological phenomenon was analysed. The results obtained by 1992 were compared with the objectives proclaimed in 1977. A description of the then existing schools and teams involved in this problem area were also presented.

My paper was a response to Professor A. P. Cracknell's kind offer to arrange a set of presentations by leading Russian experts on various aspects of remote sensing in the former USSR at the 18th Annual Conference of the Remote Sensing Society (Dundee, Scotland, 15–17 September, 1992) in order to fill certain informational gaps existing at that time. Invited lectures dealing with remote sensing data applications in geology, meteorology, ice research, agrometeorology and oceanography were given (Remote Sensing from Research to Operation 1992).

Accepting this offer to make a sort of review of space-based oceanographic research in the former USSR I knew very well that it would not be very easy to give a balanced judgement of the situation in this branch of remote sensing in my country. Still I thought the time had come to sum up what had been done in this field of science and the way it had been done in the former USSR. Three years ago I wrote:

> This work seems to be still more timely in the light of the fact that remote sensing will further develop on an utterly different logistic and financial basis. In this new situation some scientific schools and research groups will just exist no more, others will be transformed; many scientists will have to abandon research. Changes in state policies bring about a change in correlation between fundamental and applied research, including its military aspect. Quite a few scientific schools and industrial works which had formerly worked in close cooperation found themselves in different independent states within the Commonwealth of Independent States (CIS) or outside the CIS. It cannot but tell on their work and will do so in the future.
>
> These arguments seemed to me sound enough to try to sum up an important phase in remote sensing as applied to marine science in the former USSR. However it is

somewhat too early for a comprehensive survey on or a review of this subject for some reasons:

first – research in remote sensing has always been scattered in governmental agencies and industrial ministries which makes it difficult to have an integral (complete) picture;

secondly – some research in such a sensitive field of science as oceanography was classified and some of the results have not yet been published. They may never have a chance to be published, just because of the lack of interest in those problems nowadays and

thirdly – adequate evaluation of the past is possible only after the elapse of some period of time.

This presentation (Victorov 1992a) was meant to be the first step, just an attempt to make an outline of a future more detailed and comprehensive paper. Three years have passed since the publication of this paper. Unfortunately some of my forecasts on the fate of scientific schools and research groups in the new economic situation came true, and it actually proved difficult or impossible to save the co-operation between science teams banned by national borders. To my knowledge no attempt to present an analysis of remote sensing developments in the former USSR has been made so far. I leave this task to future historians of science and come back to sensors and satellites which formed the basis of satellite oceanography in this country.

2.3.2.2 How it started

Nelepo *et al.* (1982) stated that the first experiments in satellite remote sensing of the ocean and atmosphere had been started in the USSR on board satellites KOSMOS-149 (Space Arrow) and KOSMOS-243 followed by experiments on board the satellites of the KOSMOS and METEOR series (Vinogradov and Kondratyev 1971; Basharinov *et al.* 1974; Space Arrow 1974). But the formal start for development of a national satellite oceanographic system was made on 5 May 1977 when the highest authorities of the former USSR issued a decree on spaceborne observation of the Earth. It was declared that a national system for studies of natural resources should be created, consisting of a photographic subsystem (RESURS-F), an operational subsystem for the investigation of terrestial natural resources (RESURS-O) and an operational subsystem for studies of the world ocean (OKEAN-O). (RESURS is the Russian word for 'resource' and OKEAN for 'ocean'; the letter 'O' stands for operational, and 'F' for photographic (*Foto* in Russian)).

The main objectives of the system were formulated and the final technical characteristics of the output informational products were specified. For the oceanographic subsystem the satellite payload parameters and characteristics of the output products were similar to those of Seasat and Nimbus-7. Those parameters were to have been achieved at the final stage as a result of the development of a series of prototypes and experimental satellites. It is interesting to note that the module for atmospheric sounding had to be installed on board the oceanographic satellite to provide the necessary correction for the atmosphere of all the measured oceanographic parameters. It was supposed that the comprehensive payload on board a heavy satellite should gain a set of valuable informational products regularly for all branches of the national economy, for proper management of marine resources, for monitoring the marine environment, for transport operations in the polar regions,

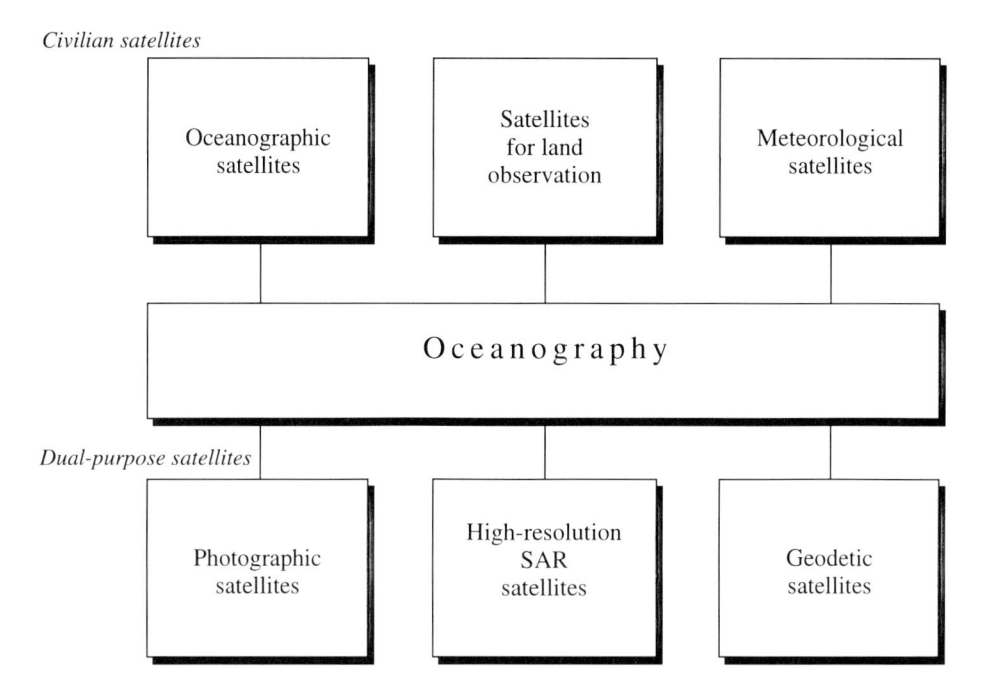

Figure 2.1 General scope of Russian satellites relevant to oceanography (Victorov 1994b).

as well as for the navy. Figure 2.1 shows these and other sources of satellite data relevant to oceanography (Victorov 1994b).

This encouraging programme had good financial support and attracted the attention of many experts among the remote sensing community in the USSR. A great deal of financial resources within the framework of the oceanic subprogramme was given to Ukrainian organisations: the Marine Hydrophysical Institute (Sevastopol, Crimea) and the Southern Machinebuilding Enterprise (Dnepropetrovsk) (Nelepo et al. 1982, 1983, 1986).

Two experimental oceanographic satellites with similar payloads were launched on 12 November 1979 (KOSMOS-1076) and on 23 January 1980 (KOSMOS-1151). These satellites had quasi-polar circular orbits with an inclination of 82.5° and a mean height of 650 km. The payload consisted of (Nelepo et al. 1982):

- a non-scanning spectrometer in the visible band (455–675 nm) with six channels 3–8 nm wide, and spatial resolution of about 20 km;
- a multi-channel non-scanning IR-radiometer (9.04–18.4 m km) with 10 channels 0.135–0.325 m km wide, spatial resolution about 25 km;
- a multi-channel microwave non-scanning radiometer (0.8; 1.35; 3.2 and 8.5 cm) with spatial resolution from 18 km to 85 km;
- a system for collecting information from drifters.

The experiments with these two satellites which had been carried out in the Atlantic for about two years helped to understand the real problems and difficulties in creating the operational satellite oceanographic system.

General outlines and some parts of the paper by Victorov (1992a) were used to prepare the lecture on Russian oceanographic satellites (Victorov et al. 1993a) in which the programme of the Russian Federation on satellite oceanography was considered as a set of oceanographic or ocean-related satellite subprogrammes. It will be suitable for the reader to follow this sequence of subprogrammes. Sections 2.3.2.3–2.3.2.8 and the relevant tables are based on Victorov et al. (1993a).

2.3.2.3 Medium-resolution radar satellite subprogramme

This subprogramme was started by launching the satellite KOSMOS-1500 on 28 September 1983 with a side-looking real aperture radar system (Kalmykov et al. 1984) operating at a wavelength of 3.2 cm. Its real aperture antenna was 11 m long. The swath was 475 km with a ground resolution of 1 (3) km. The first figure stands for the spatial resolution of radar imagery transmitted to dedicated acquisition centres, while the figure in brackets shows the resolution of radar imagery transmitted directly to simplified data acquisition stations (APT mode).

These parameters, and the fact that it was possible to receive the data in APT mode using rather simple receiving stations commonly used in the USSR for receiving imagery from meteorological satellites, made possible the wide use of new information for such routine and useful applications as ice cover mapping in the night period (which is very important for polar regions) and in cloud conditions. Other oceanographic and meteorological applications of radar imagery included:

- mapping the wind at the sea surface (scatterometer mode);
- tracing tropical and polar cyclones;
- tracing atmospheric fronts;
- tracing atmospheric internal waves;
- studies of the fine structure of major oceanic currents and tracing their seasonal variability;
- studies of eddies and internal oceanic waves.

A lot of articles have been published in scientific magazines after the side-looking radar images appeared. A comprehensive presentation of the new instrument for oceanographic and atmospheric research was given in *Radar Studies of the Earth from Space* (1990). This book contains many images (though very poorly printed) of various natural phenomena in the world's oceans and in the atmosphere as recorded in radar imagery.

Radar imagery of the KOSMOS-1500 type was a success among the world scientific community and users, including governmental agencies, because at that time there were no other instruments in orbit with such an equilibrium between the swath and ground resolution. A series of satellites of that type was launched: KOSMOS-1602 (28 September 1984), KOSMOS-1766 (29 July 1986) and OKEAN (5 July 1988). They were also equipped with a scanning microwave radiometer and routine scanners of the visible and near IR bands. A system for data collection from drifters was also installed. Payload characteristics of this series of satellites are presented in Table 2.2.

Radar images and either microwave or visible images could be transmitted directly to the Earth in a single frame. Figure 2.2 shows an example of these frames.

Unfortunately the radiophysical payload on board this type of satellite was not reliable: some malfunctions occurred soon after the launch, or there was trouble during the planned period of operation. For example the sixth satellite in this series,

Table 2.2 Characteristics of sensors flown on the OKEAN satellite

Instrument	Spectral band	Ground resolution (m)	Swathwidth (km)
Side-looking radar (SLR)	3.2 cm	1200 × 1500	450
Scanning microwave radiometer (RM-08)	0.8 cm	15 000 × 20 000	550
Scanners:			
MSU-M	0.5–0.6 m km 0.6–0.7 0.7–0.8 0.8–1.1	1900	1900
MSU-S	0.6–0.7 0.8–1.1	370	1100
MSU-SK	0.8–1.1 m km	500	1150
Non-scanning microwave polarimeter	3 cm	12 along orbit 6 perpendicular to orbit	
Data collection and relay system from DCPs		Data collection and location of DCPs	

launched on 4 June 1991, started experiencing its mission degradation by September 1991. Except for data collection missions and imagery missions, which have been normal, the satellite has an anomaly in its side-looking radar and scanning microwave radiometer performance. It is planned to launch in the near future a subsequent spacecraft of this type. The overall programme will continue with at least two more satellites of OKEAN type (WMO 1992).

The seventh satellite of this series was launched in October 1994.

2.3.2.4 High-resolution SAR satellite subprogramme

Impressive images of the sea surface, similar to those from Seasat, were obtained from the satellite KOSMOS-1870. According to Salganik *et al.* (1990):

> As far as it was known to the authors, for the first time the idea to use synthetic aperture radar (SAR) on board spacecraft was put forward and the peculiarities of such systems were considered by them in 1962, while the first article published abroad appeared in 1967 ... In the 1960s–1970s the prototypes of satellite SAR had been made ... but ... only in 1987 was it decided to launch the satellite KOSMOS-1870 with SAR ... By this time in the USA SAR systems had been already launched on board 'Seasat-A' (1978), SIR-A (1981), SIR-B (1984) with parameters close to those of SAR KOSMOS-1870.

There is no explanation yet why this SAR was actually kept stored in the backyard of one of the launching pads at the Tyuratam launching site for a period of several years.

The KOSMOS-1870 satellite had an orbit of 250–280 km height with inclination of 71.9°. Its operational swath was 25–30 km within a 200-km-wide area on both sides of nadir. The reported ground resolution was 25–30 m (for ALMAZ, the

Figure 2.2 Full-resolution frame transmitted from the OKEAN-1 satellite on 3 September 1988 (main acquisition Centre/Moscow). A large part of the Baltic Sea is covered. The left part is the image from the microwave radiometer RM-08 (no useful information for regional satellite oceanography). The right part is the image from the side-looking radar. Sea surface roughness anomalies can be seen. The bright spots on the coast are the big cities of Stockholm, Riga and Turku.

second satellite in this pair, the ground resolution is estimated to be 10–15 m). The operational wavelength is 10 cm. An original optical data processing system was used in the ground facilities; quantitative information did not seem to be available. For selected examples of KOSMOS-1870 SAR imagery see Chelomei et al. (1990).

The very narrow swath of SAR imagery means that it could be most successfully used to solve the problems of regional satellite oceanography (Victorov 1988a, 1990a) where detailed information for local areas is needed.

Figure 2.3 shows an ALMAZ SAR image of the Neva Bay in the eastern part of the Gulf of Finland, the Baltic Sea. The satellite environmental monitoring of this aquatoria will be discussed in detail in Chapter 5. The city of St. Petersburg is located in the right part of the image; the island of Kotlin with the town of Kronstadt is in the middle of the Bay; the flood barrier connecting the island with the northern and the southern coasts of the Bay is under construction. (More examples of ALMAZ and KOSMOS-1870 radar imagery will be given in Chapter 5.)

The launch schedule of the next satellite of this type has not been settled, though some sources give the planned date as 1996–1997 for ALMAZ-1B, the next satellite in the series (Use of Satellite Data 1994).

2.3.2.5 RESURS satellite subprogramme

The lack of narrow-band visible data (similar to CZCS) forced oceanographers in the former USSR to use available data in broad spectral bands from the satellites initially meant for land surface studies. Such images were provided from the scanners MSU-S (medium spatial resolution), MSU-SK (conical scanning, medium ground resolution) or the 'push-broom' sensor MSU-E (high ground resolution) on board satellites of the RESURS series.

For various applications of this imagery, including oceanography, see *The Earth's Nature from Space* (1984), the first ever Soviet album–monograph on remote sensing. (Chapter 9 of this book is devoted to oceanographic applications of visible and IR bands satellite imagery.)

Some good examples of using imagery of the visible band from the METEOR series of meteorological satellite given in this book, actually referred to the METEOR–PRIRODA section of this series. ('Priroda' means 'nature' in Russian, and those experimental satellites – in a sense – were predecessors to the RESURS series of satellites meant for studies of natural resources.)

As for the METEOR-2 and -3 data the author cannot recommend these data for application in the problems covered by regional satellite oceanography. The spatial resolution of data in the visible band (about 2 km) is worse than the resolution provided by other available sources. The data of the infrared band meant for determination of sea surface temperature with a spatial resolution of 8 km actually appear with 24 km resolution (3 pixels are averaged) with a delay of at least several days. This information has no chance of competing with AVHRR infrared data from the NOAA series of satellites directly transmitted to the Earth with a spatial resolution of about 4 km (APT mode) and 1.1 km (HRPT mode). For this reason alone AVHRR data were and are widely used in my country. Besides, we have studied other characteristics of METEOR infrared band data (see Chapter 3) and came to the definite conclusion not to use them in regional investigations of the seas.

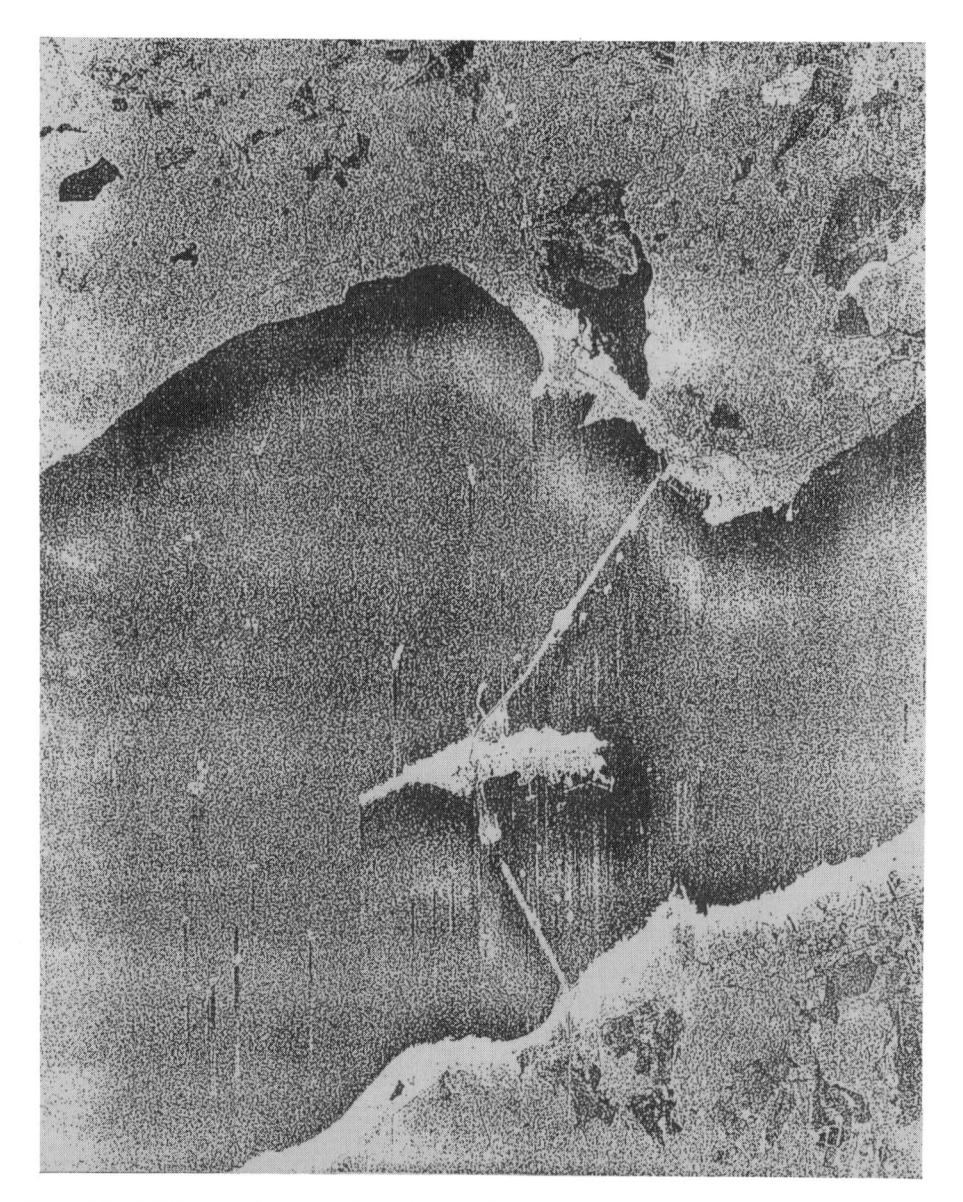

Figure 2.3 SAR image of the Neva Bay taken by the ALMAZ satellite on 28 May 1991, 05:21 a.m. GMT, polarisation HH.

Table 2.3 summarises RESURS-01 baseline specifications. (For the sake of consistency and also bearing in mind that certain developments of the payload are being planned, I include Tables 2.4–2.6 showing characteristics of the METEOR-2 and -3 series of satellites. It is worth noting that the bottom of Table 2.5 gives the technical description of direct transmission of 'high' resolution and 'low' resolution data with recommended diameters of ground antenna.

The recent RESURS No. 2 (or KOSMOS-1939) launched 20 April 1988 has remained an operational satellite far beyond its planned lifetime. In spite of the fact

Table 2.3 RESURS-01 current and projected* baseline characteristics

Instrument	Spectral band	Spatial resolution (m)	Swathwidth (km)
Optico-electronic high-resolution sensor (MSU-E)	0.5–0.6 m km 0.6–0.7 m km 0.8–0.9 m km	45 × 30	45 within 600–700 km coverage (80 in 'double' instrument mode). Repeat cycle 18 days. Tilt angles ±32° in steps of 2°
Five-channel medium-resolution scanner (MSU-S)	0.5–0.6 m km 0.6–0.7 m km 0.7–0.8 m km 0.8–1.1 m km 10.4–12.4 m km	170 170 170 170 600	600 Repeat cycle 3–5 days
* High-resolution scanner (MSU-V2)	0.45–0.49 m km 0.53–0.59 m km 0.65–0.69 m km 0.70–0.74 m km 0.83–0.87 m km 1.55–1.75 m km 2.10–2.35 m km 10.5–12.5 m km	15 15 15 15 15 15 15 45	200
* Medium-resolution scanner (MSU-SK-M)	0.5–0.6 m km 0.6–0.7 m km 0.7–0.8 m km 0.8–1.1 m km 1.4–1.7 m km 3.1–4.2 m km 10.2–12.6 m km	500	1500

Table 2.4 Characteristics of sensors flown on METEOR-2 and METEOR-3 satellites and output data. (METEOR-3 data are given in brackets)

Instrument	Spectral band (m km)	Swathwidth at altitude 900 km (1200)	Spatial resolution (km)	Output products
Scanning TV system for direct relay of cloud and underlying surface imagery	0.5–0.7	2100 (2600)	2 (1 × 2)	Individual images, photomosaics of images from 2–3 passes over receiving station within 200 km in radius
Scanning TV system with on-board data recorder to provide global coverage	0.5–0.7	2400 (3100)	1 (0.78 × 1.4)	Individual images, global photomosaics of images of various regions of the globe (2–3 times daily), cloud-free photomosaics of Arctic and Antarctic Oceans once in 5 days
Scanning IR radiometer for global coverage with direct broadcast capabilities	8–12 (10.5–12.5)	2600 (3100)	8 (3 × 3)	Global photomosaics; individual images; digital SST and top cloud heights charts, tropical cyclones coordinates, cloud amount data
Scanning eight-channel IR radiometer for atmospheric thermal sounding	11.1–18.7 (9.65–18.7)	1000 (2000)	32 × 32	SATEM messages with angular atmospheric thermal sounding data (total ozone content)
NASA's Total Ozone Mapping Spectrometer 6 channels (TOMS)	312.5–380.0 nm	(2900)	(63 × 63)	Global ozone charts
Microwave scanner, LR	1.5; 0.86; 0.32 cm	(1500)	80; 50; 20	Integral humidity, rain rate, snow ice delineation, sea roughness
Radiative–metric complex	0.15–90 MeV			Data on radiation fluxes

INFORMATION USED IN REGIONAL SATELLITE OCEANOGRAPHY 43

Table 2.5 Basic characteristics of METEOR-3M

Item	
Orbit	Circular, near-polar, Sun-synchronous
Altitude	800–900 km
Inclination	98°
Payload mass	800–1000 kg
Stabilisation	Three-axis, with error not worse than 10′, angular velocity not more than 10^{-3} deg s^{-1}
Power	Not less than 1000 W
Launch vehicle	ZENIT
Lifetime	Not less than three years
Scheduled launch	1996

Data transmission
Three-channel digital link at 1.7 GHz:
- transmission of stored data, 15 Mbit s^{-1} data rate; (05 m)
- direct transmission of HR data, 960 kbit s^{-1} data rate; (03 m)
- direct transmission of LR data, 96 kbit s^{-1} data rate; (01.5 m)

that the spacecraft had substantially exceeded its in-orbit designed lifetime (12 months), imagery in all three spectral channels (high-resolution scanner MSU-E) and in five channels of the medium-resolution scanner MSU-S showed normally good quality in 1992. However, due to problems with the energy supply, imagery missions were performed with certain constraints in time.

Standard image data on magnetic tape and film are available to the user community upon request. The RESURS programme will continue at least until the end of this decade.

2.3.2.6 *Geostationary Operational Meteorological Satellite (GOMS)*

Of some interest for oceanography is the future GOMS satellite to be placed in a geostationary orbit. It was announced in the 1970s for launching in the framework of the WMO activities of the former USSR. According to WMO (1992), 'The current phase of the program includes the manufacture of GOMS No. 1, procurement of the second spacecraft and the implementation of the ground segment. Because of budget constraints started last year, most of the industrial contracts were negotiated with delay and, therefore, it has not been possible to keep the original schedule. However, the integration of this satellite is continuing and the work has also progressed on implementation of the ground segment. It is scheduled to finish integration tests and arrange for shipping the GOMS No. 1 to the launch site by the end of this year' (WMO 1992).

The planned launch of three axis-stabilised spacecraft GOMS to be stationed over 76 E will undoubtedly represent a major change and advance from the current national environmental satellite observing system. On-board instruments of GOMS will provide continuous observations of the Earth disk in visible and IR spectral

Table 2.6 Summary of candidate instruments to be flown on METEOR-3M

Instrument	Applications	Spectral range	Ground resolution	Swathwidth
Multichannel visible and IR radiometer	Cloud, ice, snow imaging; SST, NDVI	0.5–0.7 m km 1.5–1.75 m km 3.55–3.93 m km 8.25–14.5 m km 10.3–11.3 m km 11.5–12.5 m km	1.5 × 2.0 km	3100 km
Microwave scanner, LR	Integral humidity, water content, rain rate, snow and ice boundaries, sea-surface roughness	1.5 cm (V, H) 0.86 cm (V, H) 0.32 cm (H)	80 km 55 km 20 km	1500 km
Scanning spectrometer–interferometer	Vertical temperature and humidity soundings	15 m km absorption band, spectral resolution 0.5 cm^{-1}	20 × 20 km (31 elements) per line	2500 km
Scanning microwave sounder	Vertical temperature soundings (cloud contaminated FOV)	10–12 channels in 52–56 GHz band (H)	50 km	1500 km
ScaRaB	Radiation budget measurements	0.2–4.0 m km 0.2–50 m km 0.5–0.7 m km 10.5–12.5 m km	60 km	3000 km
Active cavity irradiance monitor	Total solar irradiance, solar constant	0.001–1.000 m km	Solar	Printing
Side-looking radar	Operational sea ice and snow cover mapping	3 cm	200 × 400 m	Two symmetric 400–500 km swathwidths

Table 2.7 Basic characteristics of GOMS

Spacecraft characteristics	
Satellite mass	2400 kg
Payload mass	800 kg
Stabilisation	Three-axis
Power	1500 W (per day)
Instrument configuration	Combined visible and IR imager, independent radiation/magnetometric system, data collection and relay complex
Lifetime	Not less than three years

Instrument	Spectral band	Coverage (km)	Spatial resolution (km)
Scanning TV imager (8000 lines per frame)	0.46–0.7 m km	Full disk	1.25
Scanning IR radiometer (1400 lines per frame)	10.5–12.5 m km (6.0–7.0 m km)	Full disk	6.5
Radio complex for data collection, transmission and relay		Main and regional centres, DCPs, APT stations in the radiovisibility zone (75N–75S)	
Radiation/magnetometric	0.02–600 MeV		
Monitoring complex	3.0–8.0 KeV 100–1300 A		

bands. The scheduled mission also includes collection and relay of hydrometeorological and other environmental data from DCPs, information exchange between main and regional centres, including data from polar-orbiters and end-products (Karpov 1991). The following services will be available to various users:

- Imaging-GOMS will acquire images of the full Earth disk in two spectral channels (in three, beginning as from GOMS No. 2), up to 48 times per day. Images will be preprocessed in the main centre before distribution to users.
- Analogue image dissemination – preprocessed data will be retransmitted from the main centre via spacecraft to national and foreign user stations.
- Data collection and relay – environmental data from the national and foreign DCPs will be collected and retransmitted to regional centres and users.
- Space environment monitoring – various parameters of magnetic and radiation fields in space at the geostationary orbital altitude will be measured and relayed to the regional centres.
- Meteorological data dissemination – image fragments, charts and other meteorological products in alphanumerical form will be retransmitted from regional centres via spacecraft to national and foreign users.

Table 2.7 presents the basic characteristics of GOMS. To my knowledge there is no practical experience in using data from geostationary satellites in regional satellite oceanography in this country.

2.3.2.7   Heavy oceanographic satellite subprogramme

Within the framework of CIS co-operation a heavy oceanic satellite with a set of active and passive sensors in visible, IR and microwave bands is being developed. Possible characteristics of its payload are listed in Table 2.8.

The current status of the subprogramme is uncertain. This satellite, to be designated OKEAN-O, will actually be the one whose future appearance had been announced in 1977. Hence it will be interesting to compare its payload with that proposed in 1977. But let us wait till the launch.

Meanwhile it is worthwhile noting that, judging by Table 2.8, there is no high spectral resolution scanner for ocean colour studies and no imaging spectrometer on the list.

Some general remarks on Tables 2.2–2.8   The characteristics of payload presented in these tables may differ slightly from the characteristics of payload configurations and individual sensors given in other sources (for example, Kramer 1994). Moreover one may note that even the designation of satellite series and individual satellites was

Table 2.8 Summary of candidate instruments for the heavy oceanographic satellite OKEAN-0

Instrument	Characteristics
Multichannel high-resolution scanner MSU-V	Swathwidth 180 km Resolution 50–250 m 8 channels: 0.46–12.6 m km
Multichannel high-resolution scanner MSU-E	Swathwidth 300 km (80 km) Resolution 30 m 3 channels: 0.5–0.9 m km
Multichannel medium-resolution scanner MSU-SK	Swathwidth 600 km Resolution 175–820 m 5 channels: 0.5–12.6 m km
Side-looking radars (left and right)	Swathwidth: 2×450 km Resolution 1.8×2.2 km Wavelength 3.1 cm
Multichannel microwave scanner DELTA-2	Swathwidth 900 km Resolution 15–90 km 8 channels: 0.8–4.3 cm
Spectroradiometer with polarisation TRASSER	Spectral resolution 1 nm in 0.43–0.80 m km band
Non-scanning microwave radiometer R-600	Resolution 130 km Wavelength 6 cm
Non-scanning microwave radiometer R-225	Resolution 130 km Wavelength 2.25 cm
Multichannel low-resolution scanner MSU-M	Swathwidth 1900 km Resolution 2 km 4 channels: 0.5–1.1 m km
Data collection system from DCPs	Swathwidth 1600 km Transmission rate 500 bit s^{-1}

very complicated in the former USSR, and one and the same satellite could well be shown under different names in different sources. Various modifications of sensors of the same type, and various configurations of payloads could also be designated as new items, which also led to misunderstandings. For this reason, even the national sources (and individual authors) sometimes give somewhat contradicting information on the type and version of sensors flown aboard a certain satellite. Data presented here are generally based on information formally submitted to WMO by the operating agency.

The proposed launch schedule for satellites operated by the Federal Agency for Hydrometeorology and Monitoring of the Environment is presented in Table 2.9.

Table 2.9 needs some comments. OKEAN-01 satellite No. 7 was not launched in 1992. Of the five proposed satellites, only one – METEOR-2 No. 21 – was launched during the whole of 1993 (Zaitsev 1994). This was the period that could be called 'hard times' for Russian satellite oceanography (Victorov 1994a,b).

In 1994, METEOR-3 No. 7 was launched on 25 January.

After a very long pause a bunch of satellites was launched in Russia within a three-week period. OKEAN-01 satellite No. 7 was launched on 11 October 1994 (two years delay) in the following configuration: side-looking radar, MSU-M and MSU-S scanners and IR scanning radiometer.

GOMS 'Electro' was launched on 31 October 1994 (more than 10 years delay). And finally RESURS-01 No. 3 was launched on 4 November 1994 (two years delay) in the payload configuration: MSU-E, MSU-SK and IR scanning radiometer. So the whole operational programme (OKEAN-01 and RESURS-01) seems to be shifted by at least two years. But does it mean that the 'hard times' are over for Russian satellite oceanography? I am afraid not. While it is true that the new satellites with slightly modernised payload did replace the dead satellites, neither the ground segment nor the services were affected. The planned further commercialisation of the satellite data dissemination scheme will ban scientific applications of satellite information.

2.3.2.8 Photographic satellites

It is necessary to mention the data from photographic satellite systems run in the former USSR and nowadays in Russia. Hundreds of photographic satellites have been launched. Their high spatial resolution (tens of metres and metres) led to the

Table 2.9 The proposed launch schedule for Russian satellites operated by ROSKOMGIDROMET (Victorov et al. 1993a)

Satellite	1992	1993	1994	1995	1996	1997	1998	1999	2000
METEOR-2	—	No. 21	—	—	—	—	—	—	—
METEOR-3	—	No. 7	No. 8	—	—	—	—	—	—
GOMS	—	No. 1	—	No. 2	—	No. 3	—	No. 4	—
METEOR-3M	—	—	—	—	No. 1	—	No. 2	—	No. 3
RESURS-01	—	No. 3	No. 4	—	No. 5	—	No. 6	—	No. 7
OKEAN-01	No. 7	No. 8	—	—	—	—	—	—	—

fact that these images were classified and thus have not been accessible for the common national oceanographic community. Recently these restrictions were made less severe. Though in practice these images may be made available only some months after they were actually taken, the high quality of satellite photographic imagery makes it possible to use them (as additional satellite data) when planning the regional investigation of coastal zones, for environmental monitoring of marine ecosystems, etc. (see, for example, Victorov 1991b).

Satellites of the RESURS-F1 series can operate in active mode for 14 days and in stand-by mode for 11 days, while satellites of the RESURS-F2 series can operate in active mode for 30 days with no stand-by mode. Satellites are launched into non-Sun-synchronous polar orbit with an inclination of 82.3°. After the end of a mission films are returned to the Earth in descent capsules. Some characteristics of satellites and their payloads are presented in Tables 2.10 and 2.11.

'Spin-off' satellites which can be regarded as complementary to RESURS-F satellites are equiped with doubled cameras KFA-3000, providing 3 m spatial resolution from an altitude of 275 km. Lukashevich (1994) stated that an additional elliptical orbit with a minimum altitude of 180 km is being planned for these satellites. Zaitsev (1994) mentioned several photo-reconnaissance satellites with elliptical orbits and perigee of about 180 km launched in 1993.

Some examples of satellite photograhic imagery will be presented in Chapter 5.

2.3.2.9 Russian geodetic satellites

This section is based on our original presentation at the Bergen conference (Victorov *et al.* 1993c). Altimetric data from the former Soviet Union geodetic satellites of the GEOIK series originally were primarily meant for geodetic applications. Only recently have some extractions from over 20 million measurements been made available for oceanographic research.

Table 2.10 Russian photographic satellites. Geometric characteristics of photo information (Lukashevich 1994). (System characteristics after modernisation are shown in brackets)

| Characteristics | Photo cameras ||||
	KFA-200	KFA-1000	MK-4	KFA-3000
Flight altitude H, km	275 (235)	275 (235)	240	275
Focal length, mm	200	1000	300	3000
Picture format, cm	18 × 18	30 × 30	18 × 18	30 × 30
Film length, m	250	580	500	600
Swath	0.9 H	2 × 0.3 H (3 × 0.3 H)	0.6 H	2 × 0.1 H
Spatial resolution, m,				
one spectral band	25–30 (23–25)	— (4–6)	10–12 (7–10)	3
spectrozonal survey	—	8–10 (6–8)	12–14 (8–11)	—

Table 2.11 Russian photographic satellites. Spectral characteristics of on-board photo instruments (Lukashevich 1994)

Satellite	Photo camera	Spectral band (nm)
RESURS-F1	KFA-200	510–600
		700–840
		600–700
	KFA-1000	570–800
RESURS-F2	MK-4	640–690
		810–860
		515–565
		460–510
		610–750
		435–680
RESURS-F1 M	KFA-200	600–700
	KFA-1000	570–800
RESURS-F2 M	MK-4 M	640–690
		520–560
		800–870
		610–760

Satellites of GEOIK type have an orbit inclination of 73.6°, an altitude of 1500 km and a period of 116 min. The radar altimeter basic parameters are as follows:

nominal accuracy (m)	0.6
pulse duration (ns)	150
carrying frequency (GHz)	0.7
pulse repetition rate (Hz)	670
beam width (degrees)	2
ground resolution (km)	15
accumulation interval (s)	1

The programme has run now and may be continued with more precise altimeters on board future satellites. On my initiative – as a by-product of this geodetic programme – the pilot project Satellite Altimetry Applications in Oceanography is being carried out with the following objectives:

- design of experimental technology which could provide data selection, extraction and archiving in problem-oriented databases;
- development of methods of data processing which could produce sets of satellite measurements of sea level in a given region on the arcs and on a grid;
- development of optimal procedures for retrieval of oceanographically sounding information.

Preliminary results of a case study of the North Atlantic seas showed that, based on GEOIK satellites data, the root mean square sea surface height anomalies of 10 m could be computed for areas of 400 × 400 km averaged over three-month periods. This result (rather poor as compared with the current results of the TOPEX/POSEIDON mission) was discussed at the international conference on 'Satellite Altimetry and the Oceans' (Toulouse, France, 29 November–3 December 1993), where our experimental technology of Russian satellite altimetric data processing was presented (Victorov et al. 1993d).

2.3.3 Concluding remarks

To conclude this discussion of the sensors and satellites of the former USSR the author would like to mention a peculiar feature characteristic of a society with an economy in transition as reflected in activities associated with satellite-based remote sensing observation of the Earth. I mean some ethical and legal aspects of satellite remote sensing.

In recent years with the development of a market-oriented economy more and more organisations are being involved in the area of satellite data tackling. Agencies and their organisations dealing with satellite platform design and manufacturing, sensor design and manufacturing, data transmission and relay, data acquisition and storage, raw data processing, extractions of information of various levels and problem-oriented data processing and interpretation, assume that they are the owners of satellite data/information/output products and in this capacity tend to present themselves to the emerging market. Based on this assumption they also start commercial activities in the international market. New 'independent' organisations spring up covering each minor step of data tackling. Some awkward and embarrassing situations occur in the relations between national organisations involved in the many-staged technological process of satellite data collection and processing. One can give examples of such contradictions and foresee more of them in the future. The reason for it is the lack of legal regulations or inadequate legal regulations in this area.

We conclude this section with some general remarks. The current and planned satellites and their payloads have been presented with some comments on usage of the data from those sources in the problems of regional satellite oceanography. One should clearly understand that the full set of users' requirements will never be met because of natural basic physical constraints (one cannot detect a small object from a far distance especially at night), atmospheric conditions (clouds and fog, aerosols act as obstacles and filters or cause distortions in imagery) and cost limitations (from the viewpoint of common sense one cannot launch more satellites and spend more money for observation of the Earth – however important – than a reasonable budget can afford).

God was clever to create things in such a way that all the secrets of Nature cannot be discovered and comprehensive knowledge will never be obtained. So a reasonable compromise between the desired requirements and the amount and quality of the available satellite datasets, a proper balance in using both remotely sensed data (from satellites and aircraft) and conventional *in situ* measurements (which also need to be improved and persistently developed), extensive use of local experts and their knowledge of local marine environment – these are the plausible ways of solving current problems of regional satellite oceanography in the context of the ever-important tasks of complex monitoring of the regional seas and their coastal zones.

This means that there will always be room for flexible programming of observations of regional seas and local aquatoria, with satellite data playing their role both in providing valuable information and in harmonisation of data from various sources.

2.4 Autonomous satellite data acquisition systems

While discussing methodological aspects of regional satellite oceanography in Chapter 1, we mentioned as an essential item 'Satellite data management. Collection, storage and manipulation of satellite data for a certain region. Efforts aimed at obtaining satellite data on a regular basis'.

There are two principal sources of satellite data for a regional user, namely, the centralised source, when data from a single source are disseminated to all the regional users by mail, wires or with radio links, and autonomous sources owned, maintained and operated by regional users for their own purposes.

Figure 2.4 illustrates two types of satellite data supply for regional and local users.

Each type of supply has its advantages and disadvantages. The potential advantages of a centralised system are:

1. high-technology facilities capable of receiving high-resolution raw satellite data transmitted with a high rate (these facilities may be unique and very expensive);
2. high quality of data acquisition;
3. good conditions for data storage;
4. provision of informational services (catalogues, browsers, easy access for users via communication links, etc.);
5. provision of raw data and processed data of various levels (radiometric correction, atmospheric correction, geo-referencing, filtering, etc.);
6. capability to deliver data and informational products to the users in due time.

The crucial module of a centralised system is the data dissemination module. If it is not effective, the whole system does not meet users' requirements. One of the sensitive points is the availability of communication links. Even good imagery will not reach the user in due time in the absence of communication lines or if the lines are not capable of transfering large amounts of data or if the lines are too expensive for the user. Even if the dissemination module is not effective, some categories of users will purchase the centralised system's data provided that the other potential features of this system listed above also exist. But if this is not the case, users will turn to other ways of collecting data – they will purchase autonomous satellite data receiving equipment (or design and manufacture home-made facilities). This is what happened in the former USSR with many users among the marine community in respect of satellite data on sea surface temperature.

Even nowadays the level of operational satellite data handling and processing is still far below modern requirements, not to mention the situation 15 years ago. The

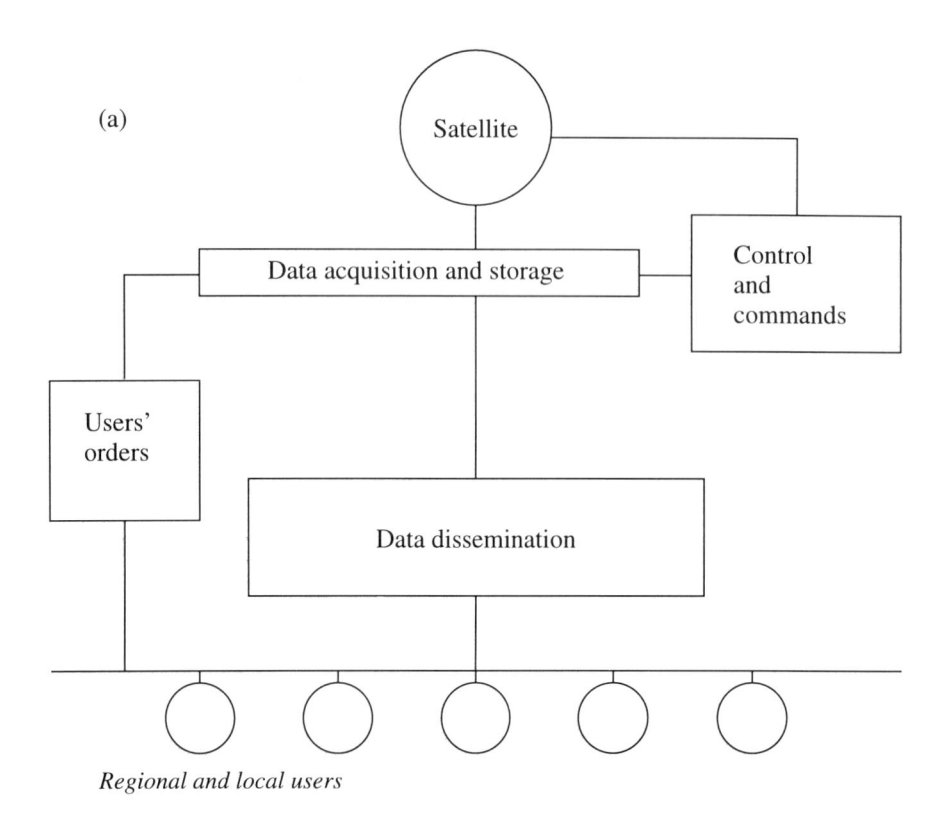

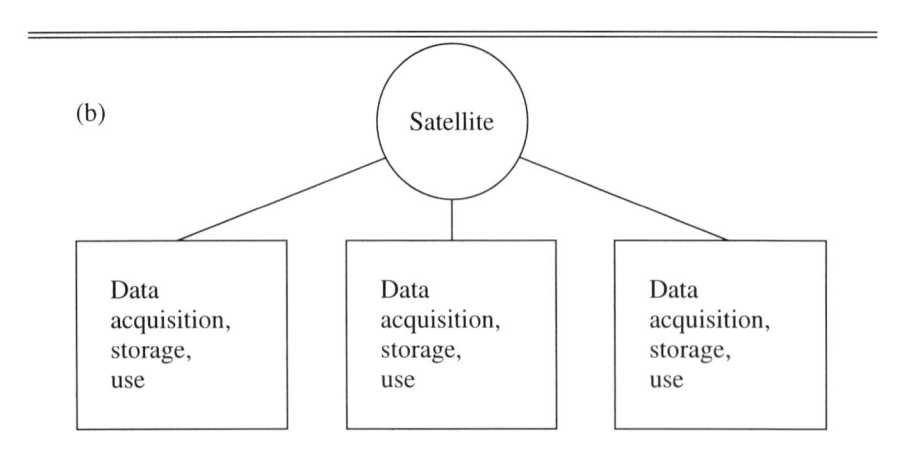

Figure 2.4 Two types of satellite data supply for regional and local users: (a) centralised system; (b) direct supply.

Russian Agency for Hydrometeorology and Environmental Monitoring – the agency operating METEOR, RESURS, OKEAN and GOMS satellites – stated in a document of the Agency Board held on 10 November 1993 that

> Due to insufficient finance:
>
> - the maintenance of ground-based data acquisition and processing equipment whose planned operational period was over, is insufficient;
> - the development of satellite payload and the ground-based segment is not intensive enough, which results in still more lag behind the world level and in producing information in which nobody is interested;
> - there are no real capabilities to create an up-to-date archive of digital satellite information. (Victorov 1994b).

The ground-based segment has always been the weakest section of the former Soviet Union Earth observing system (insufficient acquisition capability, 'low technologies' used for data tackling, poor communication lines, etc.). Lack of up-to-date services (browsers, CD ROM catalogues, 'quick-look' data analysis, easy access to some types of data from external terminals, user-oriented value-added products on sale, flexible prices, etc.) which is the fact in the Russia of today is no longer admissible in the present world market situation (Victorov 1994a).

Going back some 15 years, the oceanographers involved in remote sensing activities (designated by the term 'regional satellite oceanography') realised that the centralised system of satellite data supply was not (and in the years to come would not be) able to meet even rather modest requirements of regional and local users. One could do nothing regarding high-resolution imagery in the visible band; due to technological and economic reasons acquisition of those data could be performed only by unique centres. But as for satellite data in the infrared band, the oceanographic community in the USSR realised the reliability and high quality of the AVHRR SST data transmitted in APT mode from operational satellites of the NOAA series. Probably the first attempt to digitise this information was made by the late V. Vinogradov of the State Oceanographic Institute, St. Petersburg (then Leningrad). Before 1974 there was no experience of regular implementation of satellite data in producing SST charts for various regions of the world ocean. The first regular charts of SST in the Soviet Union were produced based on METEOR and NOAA-2 and -3 satellites data. METEOR charts had a scale of 1 : 45 000 000 with spatial resolution of 250 km (Bychkova *et al.* 1988).

In the Laboratory for Satellite Oceanography at the State Oceanographic Institute we started the programme with the aim of designing and manufacturing hardware and software enabling us to obtain AVHRR NOAA data on a regular basis. The level of technical expertise, available equipment, and finance enabled us to design a series of prototypes and manufacture a small series of autonomous portable acquisition systems capable of:

- receiving AVHRR NOAA data in APT mode (analogue signal);
- digitising these data;
- conversion of digits into radiative temperature data using on-board calibration information presented in the NOAA manual (Lauritson *et al.* 1979);
- georeferencing digital data using orbit predicting information;

- displaying SST charts on black-and-white monitor;
- producing hard copies (some levels of greyscale were formed using routine line-printer characters).

This programme was carried out in co-operation with the Leningrad Polytechnic Institute (now St. Petersburg Technical University) and the Odessa Polytechnic Institute.

Since 1981 – on an experimental basis – the prototype and modernised versions have been working in St. Petersburg producing SST charts for the Baltic and the White Seas, and also for the Ladoga and Onega Lakes (Victorov et al. 1982b; Temporary Manuals on Acquisition 1985).

As one of the objectives of this programme was to install the satellite data acquisition and processing system on board research vessels of the State Oceanographic Institute, we had to implement the design on the mainframe computers Minsk-32 and EC-1022 which had already been installed on those vessels. Later versions used home-made microcomputers. (Personal computers were not available in the Soviet Union for oceanographers until the late 1980s.) The bottle-neck problem was the satellite real-time data loading into 'low-rate' computers and data storage.

Autonomous satellite data acquisition and processing systems worked on board the research vessels *Georgiy Ushakov* and *Musson* in the 1980s in the Atlantic Ocean providing charts of SST to support current oceanographic cruises and also – on an experimental basis – were transmitted via facsimile radio link to Soviet fisheries operating in the neighbouring region of the Atlantic. For technical details of these systems and algorithms of data processing see Bychkova et al. (1988b), Victorov et al. (1982b), Bychkova et al. (1986b), and Temporal Manuals on Acquisition (1985). The general algorithm of satellite data processing and presentation of SST charts used in our autonomous system PARSEK was meant for regional oceanographic research; it produced a chart in operational mode from each available orbit.

Similar autonomous systems were designed in the other research organisations in the former Soviet Union. The system DISTERM designed in the Institute for Automation and Problems of Control (Vladivostok) used an algorithm enabling it to obtain averaged 10-day charts of SST from the NOAA and METEOR series of satellites (Ivanov et al., 1983). The NOAA data-based system designed in the Marine Hydrophysical Institute (Sevastopol) used problem-oriented flexible algorithms for temporal and spatial averaging of SST data (Terekhin et al. 1983). For a review of shipborne autonomous systems for acquisition and processing of satellite remote sensing data in the former USSR see Ivanov et al. (1986).

While oceanographers in the former USSR were busy designing their home-made autonomous systems meant to help them in collecting satellite data from the NOAA series of satellites and thus providing independent sources of data for regional oceanographic studies, a number of these systems were commercially available elsewhere. Some fishery regional administrations in the Soviet Union managed to purchase these systems. The autonomous system JAA-2N manufactured in Japan was used in the framework of a pilot programme of satellite monitoring of the Baltic Sea carried out by the State Oceanographic Institute in the 1980s (courtesy of the VNIIRO Research Institute for Fishery and Oceanography, Kaliningrad). This system was equipped with the image processing module JCV-5 and for many oceanographers it was the first personal experience of manipulating the satellite-borne SST images in false colours on the colour monitor.

Nowadays there are a large number of proprietary 'turn-key' autonomous satellite data acquisition and processing systems available on the market. Some of them provide facilities for tackling AVHRR NOAA data transmitted in HRPT mode with a spatial resolution of 1.1 km. It is far beyond the scope of this book to describe these systems but the reader should know that these systems can be regarded as a reliable source of satellite data for many types of problems covered by regional satellite oceanography. Their advantages are:

- easy access to data for the regional/local user;
- flexibility of planning one's simultaneous collection of satellite and *in situ* (subsatellite) datasets;
- 'zero delay' of satellite data delivery.

Besides, there is the 'human factor' – operators of an autonomous system are often the oceanographers who have a strong motivation to collect as much satellite data as possible for their own scientific research or for local applications with which the operators are familiar.

These attractive features of autonomous ways of satellite data collection have been recognised by the oceanographic community. The tremendous and generous input of the NOAA satellites into regional oceanographic and marine environmental research was highly appreciated worldwide. Regular conferences of NOAA users also help users to exchange their experience and encourage them in their efforts.

Direct transmission of radar imagery from satellites of KOSMOS-1500 type in APT mode was the next step towards meeting the requirements of the regional and local users. Direct transmission of satellite data and wide use of autonomous data acquisition systems are of great importance especially for large countries with poor communication services. (For example, even in 1994 it was practically impossible to transmit a high-resolution satellite image in the visible band from Moscow to St. Petersburg at a distance of 700 km at a price reasonable for a typical local user.) Thus radar imagery from these satellites was a success among the local users in the USSR and also abroad.

This development in satellite technique, coupled with the availability of sophisticated ground-based (and ship-based) autonomous data acquisition and processing systems, which was highly appreciated by the world oceanographic community, led to a tendency to provide direct data transmission from oceanographic satellites of the next generation (e.g., the SeaWiFS SeaStar ocean colour data).

Moreover, following this tendency Avanesov (1994) suggested installing identical modules for data compression and storage on board satellites of LANDSAT, SPOT and RESURS-O types to provide a 'regional service for personal users' by means of direct transmission down to Earth of selected frames of high-resolution imagery in the visible band. In his opinion this could help to meet the requirements of users (who are interested in the data for an area within a radius of about 500 km of their location) and thus to overcome a certain contradiction between the centralised way of satellite data acquisition and decentralisation of data processing by a multitude of users. This contradiction cannot be compensated by the ground communication links because of their high cost (Avanesov 1994).

It is possible that the years to come will witness a competition between centralised and autonomous methods of satellite data acquisition and processing. But most probably both methods will be in use, complementing each other, with a balance between them varying over time for each country and problem area.

CHAPTER THREE

Methodological aspects of regional satellite oceanography

3.1 Introduction

Sometimes one may have the impression that in recent years interest in the problem of satellite data calibration, in tuning algorithms of data processing using *in situ* data, has decreased. Problems of collecting subsatellite 'sea-truth' data in the mid-1980s drew the attention of many experts and organisations involved in remote sensing of the Earth in general, and satellite oceanography in particular. The keen interest in this matter could be traced in the large number of scientific publications in the world science literature. This 'information explosion' could be explained from the viewpoint of a general assessment of the state-of-the-art of Earth observation from space at that period: as mentioned earlier, on the one hand, the amount of satellite data was increasing and, on the other, a lot of methodological and technological problems of data processing and information presentation and use have not yet been solved.

Here the term 'data' means the 'raw data' or 'primary data', that is the data collected by satellite sensors and transmitted from satellite down to Earth either by radio link or physically in the form of photographic film or magnetic tape. Sometimes this term is also used to describe the transitional (intermediate) products based on raw data and useful for further processing. The term 'information' stands for the end product, the result of the entire process of data handling which may include joint processing of satellite data and data from external sources.

The lack of, the non-representative character of, or the inconsistency of subsatellite 'sea-truth' data has become 'the actual break in further development of methods of satellite oceanography' (Nelepo *et al.* 1983), preventing proper transformation of satellite data into oceanographic information or bringing less valuable information.

At first sight, judging by the number of publications on this topic, nowadays this problem seems to be in the shade. Pure scientific oceanographic and environmental issues are dominant in the present publications, reflecting the fact that satellite oceanography has become a mature discipline. Still, the keen professional reader will often find in almost every publication either solid 'sea-truth' support of remotely sensed data, or some uncertainties in the results caused by the lack of or inadequacy of 'sea-truth' data.

If one analyses the problem of collecting 'sea-truth' data in more depth it becomes clear that this problem was in the past and is nowadays within the scope of activities related to regional satellite oceanography. A special programme of subsatellite synchronous measurements of oceanographic and meteorological parameters has been organised during the test flight of the Seasat satellite launched in 1978 (lifetime 105 days). Preliminary results of the test flight were published in a set of eight articles in *Science* magazine in June 1979; these are the source of the following brief review of subsatellite activities aimed at collecting 'sea-truth' data during the test flights of Seasat (Victorov 1982).

'Sea-truth' datasets of two types have been collected. First, in August–September 1978 in the Gulf of Alaska a special subsatellite 'polygon' was set up with buoys, three ships and four aircraft involved. Shipborne measurements included standard meteorological observations, and surface wind and wave measurements 10 minutes before and after the satellite pass. Airborne measurements included wind and air temperature at the altitude of flight, sea surface temperature and wave measurements. There were 60 passes of Seasat over this test area. Secondly, 'sea-truth' data collected in the course of an 'independent' international experiment on sea–atmosphere interaction in the eastern part of the Atlantic Ocean in July–September 1978 could be made available. In this test area 14 research vessels and three aircraft were working; 36 buoys were used and more than 900 radiosounds were flown. There were 200 overpasses of Seasat in this region. In addition, subsatellite measurements were carried out from oil and gas platforms in the North Sea and also from a British buoy and weather vessels in the eastern Atlantic (Allan 1979). The large amount of satellite data permitted an assessment to be made of the reliability and quality of various types of 'sea-truth' shipborne data and to rank these data according to the type of vessel. The general conclusion was drawn that the management of subsatellite oceanographic measurements needs improvement, and a thorough selection of 'sea-truth' data is necessary (Victorov 1982).

During the flights of the experimental oceanographic satellites KOSMOS-1076 and KOSMOS-1151 in 1979–80, relevant shipborne measurements were carried out in the Atlantic Ocean (Nelepo *et al.* 1983).

That was in the past; this was the history of satellite oceanography. Nowadays, in the course of preparation for the launch of the SeaWiFS sensor on board the SeaStar satellite, a series of airborne measurements of the optical characteristics of the seas was carried out using a prototype of the satellite instrument. A special programme to collect shipborne data to be used as coefficients in various water optical models with an emphasis on the SeaWiFS data is currently being carried out in the Baltic Sea as a co-operative effort of the Institute for Remote Sensing Applications of the Joint Centre of the European Communities (Ispra, Italy), the Institute of Oceanology of the Polish Academy of Sciences (Sopot, Poland) and the Institute for Baltic Sea Research (Rostock-Warnemunde, Germany) (Sturm 1994; Hoepffner *et al.* 1994).

The above examples indicate that collection of 'sea-truth' data is actually on the agenda nowadays as it was 20 years ago. The difference is that the generalities are more or less clear and now there is no need to collect subsatellite data to prove the basics of a certain method of data extraction; rather, nowadays we need relevant *in situ* data to account for the regional and local peculiarities of the oceanographic and meteorological conditions of taking satellite-based measurements.

Summarising the results of a four-day workshop specifically addressing the problems of sea surface temperature retrieval from satellites held at the University of Rhode Island in September 1982, with experienced users of satellite data from all major US satellite oceanographic research groups and several European groups participating, Byrne (1983) wrote that one of the recommendations was: 'Designated areas of the ocean should be identified for ongoing calibration/verification exercises for both operational and experimental satellite sensors and systems'.

Thus he touched on another aspect of the problem: selected areas of the ocean, test areas and test sites to be used in sea-truth data collection exercises. In the late 1970s and the beginning of the 1980s particular attention was paid to the setting up of a network of subsatellite test areas (polygons) in the former USSR. This problem was discussed by Garelik *et al.* (1977), Belyaev *et al.* (1977) and Drabkin *et al.* (1982); the theoretical aspects of the design of subsatellite sea-truth networks were considered by Nelepo *et al.* (1976), Dotsenko (1980), Dotsenko *et al.* (1981), Nikanorov and Svetlitskij (1986) and Bondarenko *et al.* (1986).

Description of some field work at these test areas will be presented in brief in Section 3.3.

3.2 Definitions relevant to collection of subsatellite data

Let me remind the reader that in Figure 1.1 (Chapter 1) four main logical components comprising a satellite oceanographic system were shown. Along with satellite payload, general and problem-oriented subsystems of data processing, *sea-truth test areas* were indicated as one of the essential subsystems. All the system components were to be tested during the *complex tests of the system* consisting of in-flight tests of satellite payload and the ground-based segment of the system with sea-truth test areas being used to collect the necessary subsatellite *in situ* data. Some terms will be used in the following text.

- Scene – satellite image or its part which is being considered, processed and analysed as a separate item.

- *In situ* data – the data used to analyse the scene and collected by means of aircraft, ships and other mobile platforms, and stationary measuring equipment located on the sea surface, above the sea surface and at depth. *In situ* data may be collected simultaneously with the satellite overpass (synchronous data) or almost simultaneously (quasi-synchronous data).

- *A priori* data – the data used to analyse the scene and collected before the satellite overpass.

- *A priori* information – the processed and analysed *a priori* data. Books describing some local features of oceanographic or meteorological fields or some natural phenomena related to the scene, directories and records, old maps, etc., containing information on the scene, may all be regarded as *a priori* data and information. In the modern GIS context they may be considered as a *knowledge base*.

- Test area – part of a sea where *in situ* data are being collected (sometimes the word 'polygon' is used to designate the same item).

- Test site – a small part of a test area with some peculiarities: either specific measuring devices are located within test areas, or it is characterised by a specific spatial distribution or local anomaly of a field of oceanographic parameter(s).
- COSE (Complex Oceanographic Subsatellite Experiment) – a system of measurements of *in situ* parameters of the sea and atmosphere, preprogrammed in time and organised in space, by means of mobile and stationary platforms operating at various heights above the sea surface. Passive COSE are strictly preprogrammed in time and space, while conducting active COSE one uses current operational data to manipulate the mobile platforms in order to match the targets of the experiment.

Subsatellite *in situ* data may be collected at the test areas and in the course of COSE. COSE may be carried out at the test areas or elsewhere. The principal objectives of COSE are as follows:

- collecting *in situ* data to be used in the routine calibration of satellite sensors;
- collecting complementary data in the course of an 'additional observation' survey (when analysts need additional data for a part of a scene where they detected an anomaly but want to avoid 'false alarm' events);
- collecting data to be used in finding a relationship between various phenomena and processes at depth and their manifestations at the sea surface.

Why are these subsatellite oceanographic experiments called complex? Because in carrying them out, the satellite, airborne, shipborne and other types of measurements are being performed at the same period and in the same area; because both the sea and the atmosphere above the sea are being observed; because a set of oceanographic parameters are being determined from a satellite; and because for each parameter determined from a satellite a group of relevant *in situ* parameters is being measured.

How do COSE relate to test areas? The latter are mainly associated with stationary measuring equipment and facilities located in a certain part of the sea at least during a period of some years, while COSE is a more flexible and more operational form of collecting *in situ* data. But mobile platforms can also be used as part of permanent facilities at the test area. Regular routine measurements of *in situ* parameters may produce long time series which is an advantage of test areas.

How do COSE relate to in-flight tests of satellite payload and complex tests of the system? During flight tests the payload should be tested quickly and in an unambiguous way. Hence the set of *in situ* parameters to be measured should be minimised. The list of COSE objectives is broader; collecting *in situ* data for calibration of satellite payload is only one of its aims, among others. Still the technological procedures of *in situ* data collection should be standardised for various phases of satellite operation: during its in-flight tests, during the satellite lifetime and at the periods of COSE performance.

In terms of costs and benefits COSE are more flexible and they do not require investment in stationary facilities (as permanent test areas or test sites do). Some economical aspects of running permanent test sites were considered by Drabkin and Allabert (1982).

3.3 Some field activities at the test areas in the selected seas

In the 1980s the oceanographers in the former Soviet Union still believed that the test areas in the seas would become a constituent part of a satellite oceanographic system. This idea stimulated activities in some seas aimed at creating the scientific basis for proper selection of those areas. In the European part of the former USSR three seas could be used to set up a subsystem of regular subsatellite measurements of *in situ* parameters and arrange a network of test areas, namely the Baltic Sea, the Black Sea and the Caspian Sea.

The geographic location of these seas permitted the collection of *in situ* sea-truth data within the range of 40–60° North and within 20–50° East. A pair of seas (either the Baltic and the Black or the Black and the Caspian) could be observed from the same orbit by means of the MSU-S scanner with a swathwidth of about 1000 km^2 or the AVHRR scanner with a larger swath.

Thus within minutes of the flight time two independent sets of subsatellite measurements could be provided to calibrate the satellite sensors or tune the algorithms of satellite data processing. Figure 3.1 shows the location of the Baltic Sea, the Black Sea (with the adjacent small Azov Sea) and the Caspian Sea with swaths of some sensors flown on satellites of the former USSR.

To compare the satellite measurements with *in situ* sea-truth measurements one should select the test areas with a homogeneous distribution of the oceanographic parameter in question. It is possible that a test area homogeneous with respect to one parameter is not homogeneous with respect to another oceanographic parameter. Hence a set of test areas may be required to calibrate all the satellite sensors. Moreover the spatial distribution of various parameters within a certain test area may vary with time; it may depend on the season, synoptical conditions, etc. This means that the 'oceanographic climate' within the preselected test areas should be studied.

This sort of study has been carried out in the Baltic Sea and the Caspian Sea with respect to sea surface temperature (SST) using airborne IR radiometers (Allabert *et al.* 1982). Our approach differed from the then widely accepted one based on calibration of a satellite image (two-dimensional field) using a set of point *in situ* measurements. We suggested an approach based on comparison of a satellite-borne image (two-dimensional field) with an airborne two-dimensional field obtained by means of low-flying aircraft. This approach is based on the assumption of a 'frozen' field of oceanographic parameters. As one of the main objectives of test areas is to provide sea-truth subsatellite data comparable with satellite data, the essential problem which arises is the selection of test areas (sites) within which the field is homogeneous. Spatial resolution of satellite sensors varies from metres to tens of kilometres, and this is also the spatial range of homogeneity of oceanographic parameters to be provided at the test areas.

At a first guess the determination of homogeneity of the SST field may be limited to the estimation of the mean value of the measured parameter, its dispersion and the amplitude of the measured values (the difference between the maximum and minimum values). The critical (threshold) value of the dispersion may be nominated, so that if the measured value of dispersion is more than the threshold dispersion, this set of data is considered to be inhomogeneous. Such an area cannot be used as a test area. Allabert *et al.* (1981) considered various variants of the experimental determination of homogeneity of the preselected test area and various solutions of

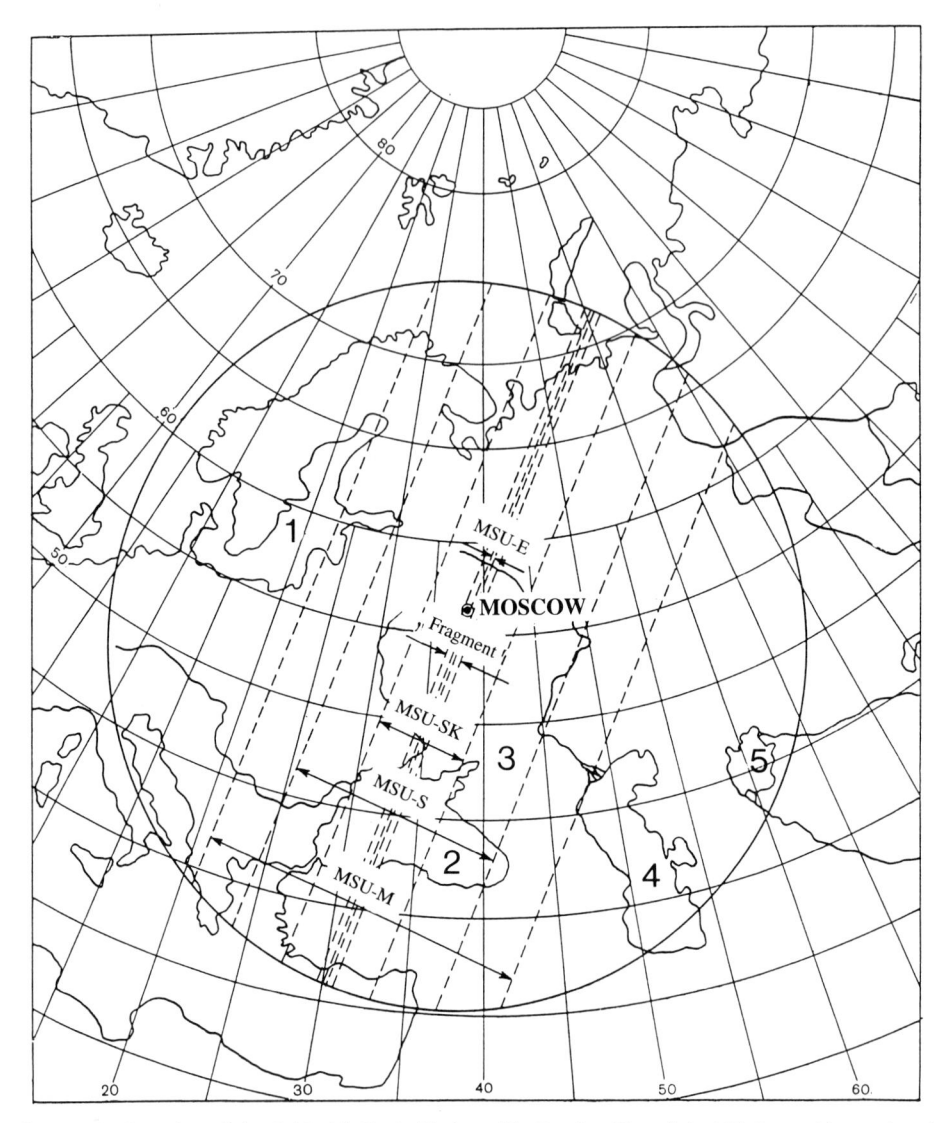

Figure 3.1 Location of the Baltic (1), Black (2), Azov (3), Caspian (4) and Aral (5) Seas with swaths of the MSU-E, MSU-S, MSU-SK, 'Fragment' and MSU-M sensors. (To clarify the drawing the actual direction of swaths is turned.)

this problem based on airborne measurements. They also showed that the assumption of a 'frozen field' of a hydrophysical parameter could be proved in the course of its measurement from aircraft.

This approach was used to analyse sets of airborne SST data collected in the Baltic Sea on a number of 80–100 km-long flight routes located perpendicular to the eastern coast. Spectral analysis of data was also applied to study inhomogeneities. As an example of the results obtained, based on the spectral analysis of airborne data collected in June and September 1980, spatial inhomogeneities of 11–15–16 km have been recorded in most of the spectra. This value gives the scale of 'patchiness' of the SST field in the Baltic for the period of observations.

The same 'threshold dispersion' approach was used to study the SST field at the preselected test site in the Caspian Sea. The spatial and temporal variability of the SST field within a square of 18×18 km^2 has been studied using airborne IR radiometers. Fourteen series of airborne measurements have been carried out at this test site in different seasons in 1980–1981. The flights were performed on a 1 km grid. The SST field dispersion for each series of measurements was calculated, and appeared to be less than or equal to the dispersion of the radiometer instrumental errors. This indicated that the SST field was homogeneous within the test site. During each series of airborne measurements, the SST within the test site did not change by more than 0.5°C. Moreover, the temperature field was homogeneous also in the third dimension – in depth. *In situ* shipborne data showed that on 8 and 9 August 1980 the depth of the upper homogeneous layer with a temperature of 25°C was 20 m.

In the course of airborne studies of the SST field at the preselected test areas in the Baltic Sea and in the Caspian Sea some areas had been selected as the sites where representative sea-truth subsatellite data could be collected. The proposed approach was proved, and it could be used also to select test areas meant for calibration of other satellite sensors measuring various oceanographic parameters. Some selected test areas in the Caspian and the Baltic Seas which were thoroughly investigated in the 1980s could be used in future for calibration of sensors on board the satellites planned to be launched in the late 1990s.

The eastern part of the Gulf of Finland was also suggested for international co-operation as the test area for satellite sensors in the visible band (Victorov and Sukhacheva 1994), based on the available *in situ* data (Usanov *et al.* 1994a) and knowledge of the local distribution patterns of the suspended sediments obtained in the course of experimental satellite monitoring of this area which has been carried out since the 1970s (Victorov *et al.* 1989a; Victorov 1991b; Sukhacheva and Victorov 1994). For satellite imagery of this area see Chapter 5.

Victorov *et al.* (1986b) presented the results of complex studies of the Kurshi Bay of the Baltic Sea carried out in 1981–1985 in which satellite images, airborne data, shipborne and coastal *in situ* measurements were used. Kurshi Bay is a remarkable area separated from the Baltic Proper by a narrow (about 1 km) sand spit. The colour and the optical properties of the bay waters differ from those of the sea (see satellite image in Figure 3.2).

The bay is actually the shallow estuarine area of the Nemunas River with high concentrations of suspended sediments which are being regularly measured by the Klaipeda marine laboratory. A well-shaped area with solid boundaries, the Kurshi Bay waters are always in sharp contrast (regarding the temperature and colour) with the neighbouring waters of the Baltic Proper. Thus the Kurshi Bay was suggested as a natural test area for satellite data calibration (Victorov *et al.* 1986b).

Some activities at the test areas in the Black Sea were presented by Nelepo *et al.* (1986). In the adjacent small Azov Sea the test area was selected (Dyadiunov *et al.* 1986) in the vicinity of the Arabat spit for calibration of satellite sensors of the visible band. In May–July 1983 an airborne survey of the sea was performed, and three test sites with homogeneous colour characteristics were identified as the most stable and representative. A set of isolated water bodies of various colours at the spit were proposed as the densitometric grey scale of natural origin. Detailed studies of these test sites were carried out using helicopter-borne optical measurements and water sampling. Figure 3.3 shows the location of the test area and three test sites.

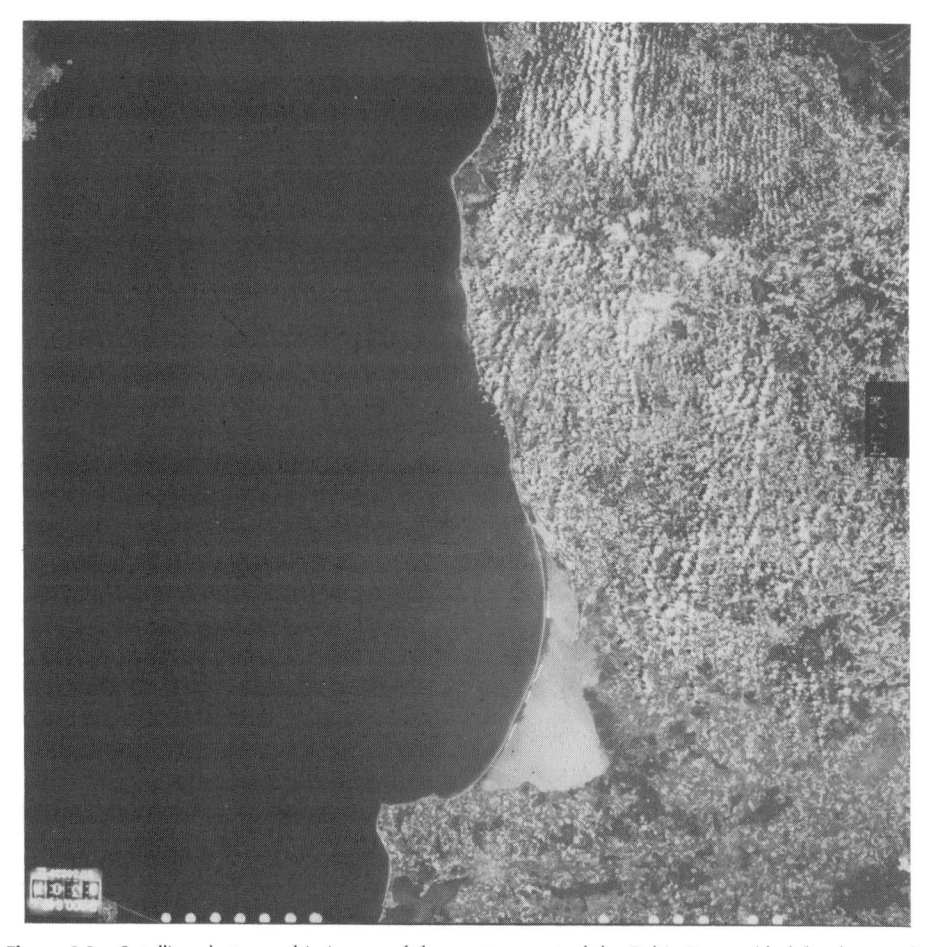

Figure 3.2 Satellite photographic image of the eastern part of the Baltic Proper (dark background) and the Kurshi Bay (bright triangular object in the middle of the scene). 27 May 1987.

Four hydrochemical parameters and the concentration of suspended matter were measured at the networked stations. Mean values and dispersion were calculated for each test site and distribution maps for each parameter were obtained (Dyadiunov *et al.* 1986).

In the 1970s the problem of monitoring the marine environment was put on the agenda in the former USSR. Monitoring of *oil slicks* on the sea surface was to be carried out using remote sensing technique. Pilot projects, field work and laboratory experiments were performed in the Laboratory for Satellite Oceanography and Airborne Methods, State Oceanographic Institute, Leningrad, USSR (now St. Petersburg, Russia).

Special methods for airborne control of the sea surface have been developed, and a set of sensors has been designed and manufactured. These sensors have been installed on board the specially equipped long-range low-height-flying aircraft IL-14. An airborne subsystem of the optical band included a lidar, an ultraviolet

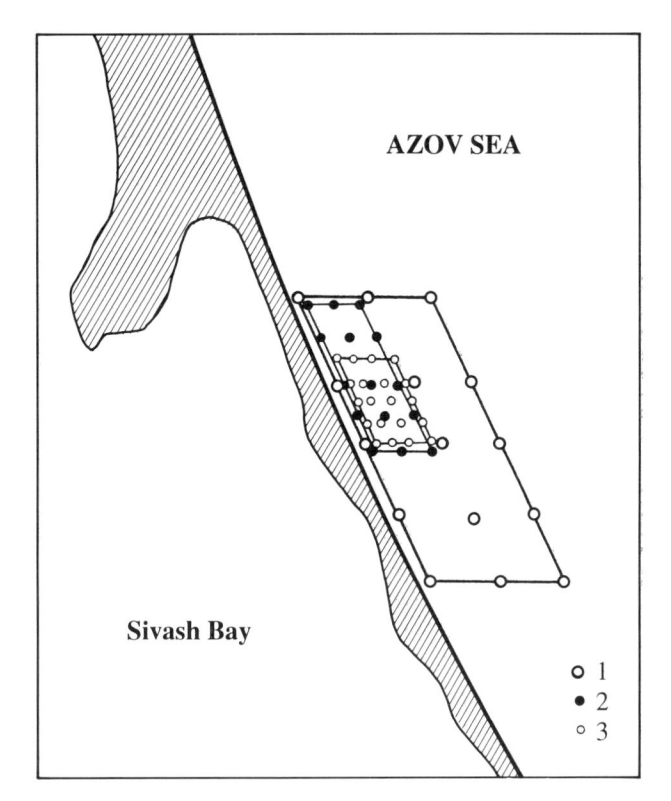

Figure 3.3 Test area in the Azov Sea meant for calibration of satellite sensors of the visible band. For the sensors with different spatial resolutions the test sites with areas of 72 km² (1), 23 km² (2) and 12 km² (3) were selected (Dyadiunov et al. 1986).

polarimeter, a trajectory infrared radiometer and a television camera. Standard aerial reconnaissance film cameras were also included. The aircraft payload was described by Victorov (1980b). A lidar with a laser generating beam with a wavelength of 10.63 m km was capable of detecting oil slicks 0.1–2.0 m km thick (Osadchij 1983). Its design was based on theoretical calculations (Gurevich and Shifrin 1982) and laboratory simulations (Osadchij and Shifrin 1979). The characteristics of the IR radiometer will be given in Section 3.4. The original TV equipment is described by Voitsekhovskij et al. (1986).

The unique aircraft and its experimental equipment was used to test the methodology of airborne monitoring of oil slicks and to work out procedures and write manuals for routine operational monitoring of the seas in the former USSR. The Caspian Sea was selected as the test area, with marine oil platform clusters being used as test sites.

A programme of experimental airborne environmental monitoring of the Caspian Sea was worked out. It consisted of four seasonal (winter, spring, summer, autumn) aircraft surveys of all the major oil-producing offshore regions each year with correlated *in situ* sampling of oil slicks. Standard navigational schemes of flights were used. Current meteorological data from the local network were collected as well.

In 1975–1990 the programme was carried out in semi-operational mode. To run the routine work a separate unit was set up which was responsible for maintenance of equipment, performing regular seasonal flights over the Caspian Sea and analysis of data. Output products consisted of processed aerial photographic films and paper material (charts of oil slicks based on quick-look analysis of films and data of manual observations), magnetic tapes and paper tapes containing original records of sensors. Thus a unique comprehensive archive of oil-slick pollution airborne data for the Caspian Sea was created (Victorov 1994c; Victorov and Shaposhnikova 1995).

To conclude let us summarise the results obtained in this problem area – one of the essential parts of the proposed satellite oceanographic system. The activities under the heading 'sea-truth test areas' were supposed to include the design and setting up of a test area network, studies of oceanographic climate at these areas, selection of test sites within the test areas, and optimisation of the network (Victorov 1982). The actual results obtained were as follows:

> 'The network of 'sea-truth' data collection areas (polygons) has not been set up. Though some efforts have been made and the areas in the Baltic Sea, the Ladoga Lake and in the Caspian and the Black Sea have been selected. Temporal and spatial characteristics of the sea-surface temperature and other parameters have been studied. Data on the seasonal and synoptic scale variability of those parameters made it possible to work out the recommendations on the possible use of the selected test areas as the components of the oceanographic satellite system. The difficulties that we came across included: very poor management, lack of reliable sensors for measurements of *in situ* oceanographic and meteorological parameters and lack of communication links' (Victorov 1992a).

Though the system was not created, the data collected at the proposed sea-truth test areas could be considered as a useful contribution to the oceanography of regional seas. Nowadays in the post-cold-war era it seems plausible to use these data in the framework of international co-operation to help in the calibration of oceanographic satellite sensors (no matter what country they belong to) at some of the test areas selected many years ago. The historical *in situ* sea-truth data collected at these test areas are a valuable source of background information and thus could become an essential part of knowledge bases in the relevant Geographical Information Systems (GIS).

Nowadays the current problem in the former USSR is to save databases and to convert them into real GIS (Victorov 1994c) thus providing easy access to the data for the international oceanographic and remote sensing communities.

3.4 Complex Oceanographic Subsatellite Experiments (COSE)

3.4.1 Objectives of COSE carried out in the Baltic Sea in the 1980s

In 1982–1985 in the Baltic Proper three COSE have been carried out as a co-operative venture of the former USSR and the former German Democratic Republic (GDR). The partners in this co-operation in the field of regional satellite oceanography were the Institute for Sea Research, Warnemunde (Principal Investigator Dr. H.-J. Brosin) and the State Oceanographic Institute, Leningrad (Principal Investigator Dr. S. V. Victorov), with the State Centre for Natural Resources Research, Moscow, the Atlantic Research Institute for Fishery and Oceanography, Kaliningrad, Klaipeda Marine Laboratory, the Lindenberg Aerological Observatory and Meteorological Observatory, Potsdam having also been involved. A general description of the first COSE was given only in Russian (USSR–GDR 1985). Much later a review of the three COSE appeared as part of the five-year Joint Report (in Russian and German, with restricted circulation). Some extracts from these sources are to follow. The main formal objectives of the first experiment were:

- field test of the scheme of interaction between satellite, airborne and shipborne measuring devices in the framework of COSE in order to develop operational methods of sea-truth data collection;
- collection of systematic data of synchronous and quasi-synchronous satellite, airborne and shipborne observations and measurements meant for the development of recommendations and manuals on the determination of sea surface temperature and suspended matter in the Baltic, based on satellite data;
- carrying out a methodical comparison of satellite images of the visible band from an experimental satellite of the METEOR series and airborne images taken by means of the multichannel photographic camera MKF-6M. (The METEOR No. 18 or METEOR-PRIRODA satellite payload included the experimental high-resolution scanner 'Fragment' manufactured by Karl Zeiss, Jena; the camera MKF-6 was also made in GDR.)

We also wanted to collect data for comparison of AVHRR NOAA imagery and METEOR-2 SST charts, and also to compare the informational capabilities of two satellites of the METEOR-2 type with different lifetimes in orbit.

For the second and third COSE, new objectives were put forward (see below).

3.4.2 Programme of the first COSE

Satellite data and equipment installed on aircraft and research vessel Satellite data included:

- SST images (films) and digital charts (20 km spatial resolution) from the satellites METEOR-2 No. 5, 7, 8.
- Images in the visible band from METEOR No. 30 (medium and high spatial resolution) and METEOR No. 31 (medium resolution).
- SST data from NOAA satellites No. 6 and 7 (additional information).

For characteristics of satellites and their payload see Chapter 2.

Airborne data consisted of:

- Charts of SST obtained by means of the IR non-scanning radiometer MIR-3, measurements of air temperature and moisture; data of visual observations of the sea surface from low-altitude aircraft IL-14 (USSR).
- Photographic images taken by means of the multichannel camera MKF-6 and standard camera AFA-TE-200; data of visual observation of the sea surface and clouds from high-altitude aircraft AN-30 (USSR).

In the context of COSE some parts of the airborne data were considered as *in situ* data.

Shipborne data collected from the GDR Academy of Sciences research vessel *Alexander von Humboldt* – according to the main objectives of the first COSE and in

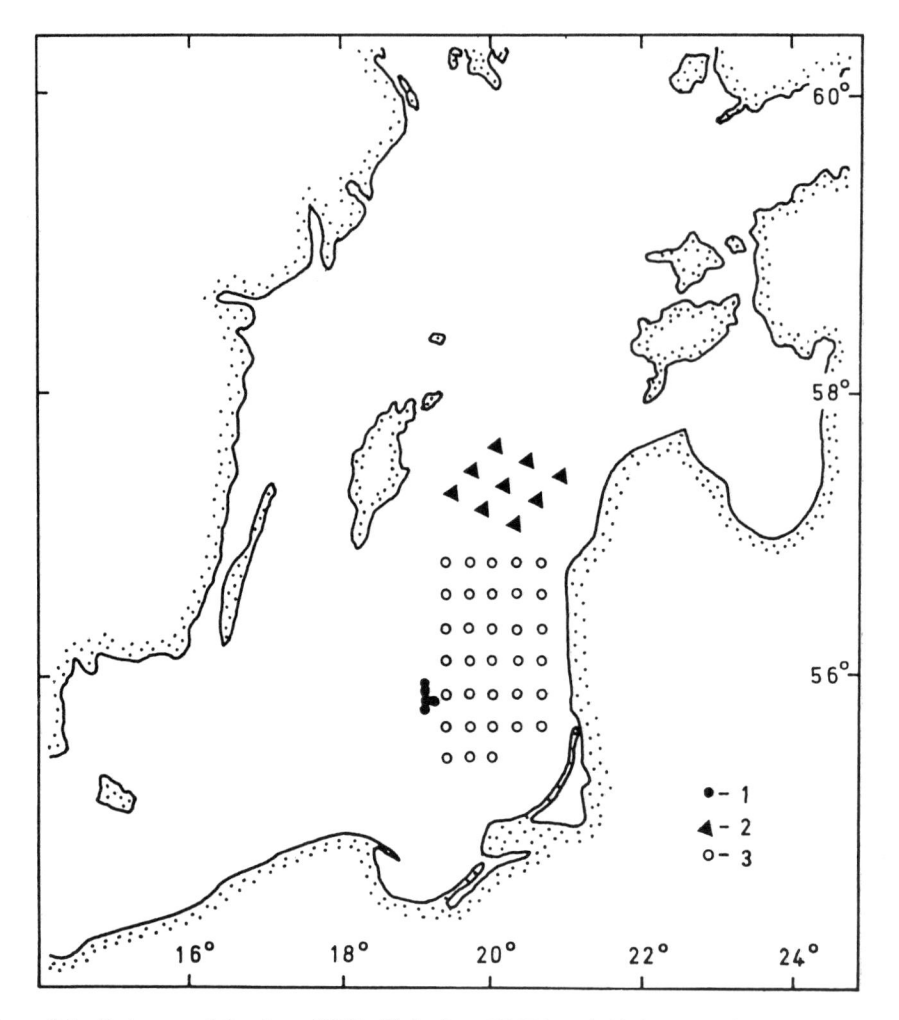

Figure 3.4 Test areas of the three COSE: (1) the first COSE (1982), (2) the second COSE (1983), (3) the third COSE (1985).

agreement with the internationally accepted recommendations (Sorensen 1979) – included the measurements of:

- concentration of seston and chlorophyll-a or phaeopigment;
- spectral flux and upwelling and downwelling irradiation in the 0.4–1.0 m km band;
- spectral brightness of the sea;
- light attenuation coefficient in the 0.38–0.725 m km range;
- light scattering coefficient at 0.633 m km;
- in-water irradiation at 0.435 or 0.545 m km;
- spectral transmission of the atmosphere;
- radiative temperature of the sea surface;
- vertical distribution of water temperature;

and routine oceanographic and meteorological measurements and observations. SST was measured by means of the same IR radiometer MIR-3 which was used on board aircraft IL-14.

All the available data from the ships of opportunity were collected and stored. Data of aerological measurements from balloons flown in the region of COSE were also collected, and data of standard oceanographic and meteorological observations carried out at the network of coastal and island stations were collected as well.

Geographical location of the three COSE The location of the test area of the first COSE was selected, taking into account the following circumstances:

1. the amount of oceanographic *a priori* data available for the area;
2. mutual interest from both participating sides in the study of the area;
3. accessibility of the area for the aircraft based in mainland airports of the USSR; and
4. the relatively short distance from the harbour of Warnemunde.

For the first COSE the two sides selected the region in the vicinity of station 9A (international Baltic network number) with geographical coordinates: 56° 06′ North, 19° 10′ East. This station is located in the central part of the Baltic Sea (Baltic Proper), in the southern part of the Gotland Deep. Figure 3.4 shows the location of the three COSE.

This station was used in the research programme of the International Baltic Year 1969–1970 and the programme of the international experiment BOSEX (1977); it is included in the routine oceanographic programmes of many institutions.

Operational structure of the first COSE A general methodical approach was based on the principle of independent ship movement along the fixed route with strictly fixed time intervals of movement in each part of the route between the local stations. When planning the schedule of the first COSE we took into consideration the spatial resolution of satellite sensors, the expected temporal and spatial variability of the oceanographic parameters to be studied, and also the real technical characteristics of the research vessel (cruise speed, time consumption for each type of measurement, special manoeuvres to match the requirements of scientists, etc.). These items and the orbit inclination of METEOR satellites led to the design of the scheme of COSE shown in Figure 3.5. The diagonal of the triangle coincided with the satellite ground track.

The ship was the most 'conservative' element in the COSE programme. All the measurements were managed in a way that each day the ship was moving along the fixed route shown in Figure 3.5; at each moment the location of the ship was known within 250 m accuracy (which sounds rather poor nowadays). Daily in the morning the ship started from point 9A, it took 8 hours to reach point 9 (with four optical stations and nine oceanographic stations in between). It took 5 hours to go from point 9 to point 15 (with two optical stations at points 12 and 15, and six oceanographic stations at points 10 to 15). In 3 hours the ship returned in point 9A. Hence the total operational time of the ship was 16 hours daily.

The aircraft AN-30 could work at the test area for about 3 hours, and aircraft IL-14 for up to 5 hours. There were two modes of AN-30 operation: in the 'morning/evening' mode (at high values of the Sun's zenith angle), and in the 'noon'

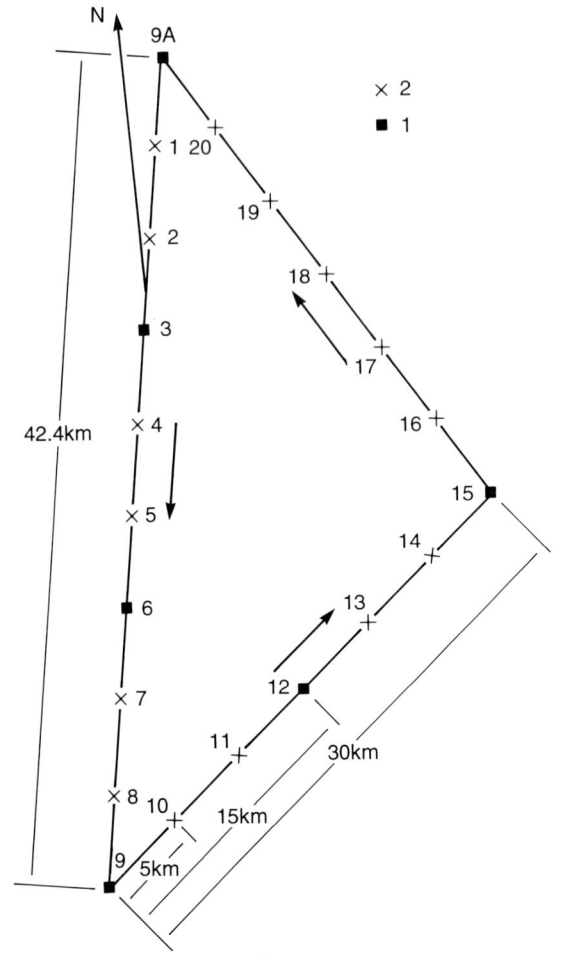

Figure 3.5 General scheme of the first COSE and location of combined optical/oceanographic (1) and oceanographic (2) stations. Arrows show the direction of the ship movement (USSR–GDR 1985).

mode (at low zenith angles). Flight height was from 1 km to 6 km with flight routes along and across the ship route.

The aircraft IL-14 was operated on a special grid of flight routes covering the whole test area (Figure 3.6). SST data were recorded continuously from the flight height of 300 m. The grid was designed to provide maximum coverage of the test area, synchronicity of airborne and satellite measurements, and also overlapping of some flight routes with the ship route. Figure 3.6 gives an example of the actual flight route and the ship route on 19 April 1982. Note that the last long aircraft route (No. 6) crosses the test area along a diagonal which nearly coincides with the ship route. Several times the selected test site was flown over at heights of 100 m, 200 m, 300 m, 400 m, up to 1 km, to estimate the correction of the SST data for the atmosphere.

Management of the COSE Figure 3.7 presents the general diagram of management of COSE. The variant shown refers to the third COSE when in course of the experi-

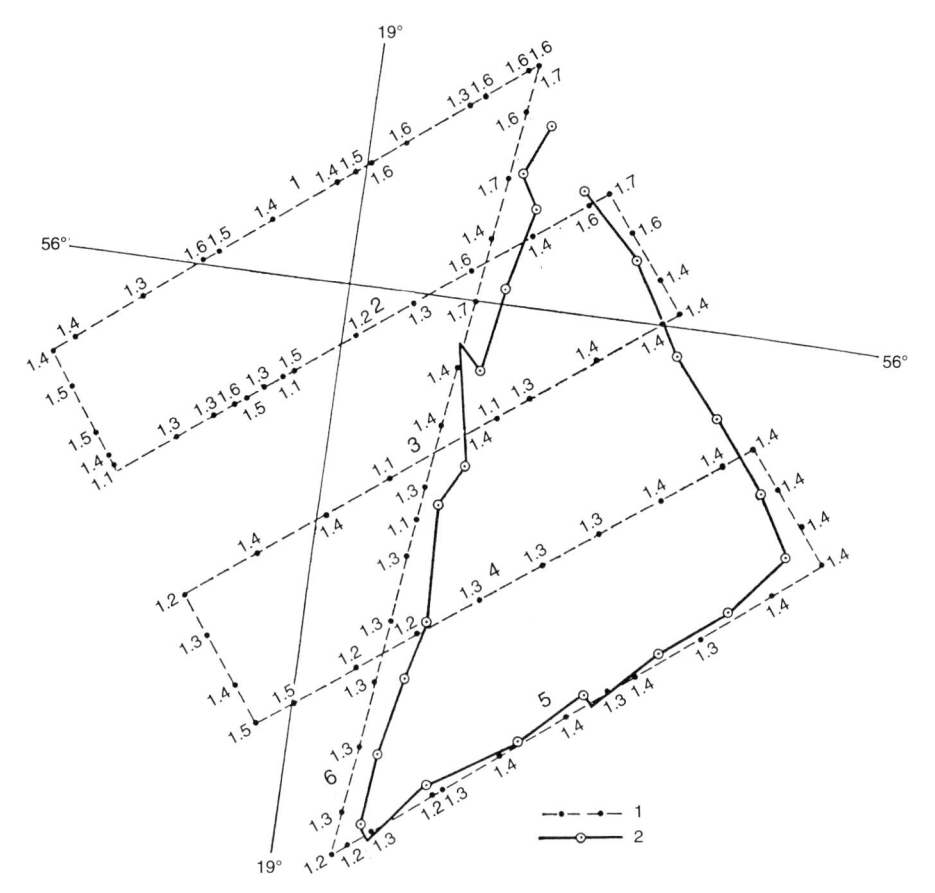

Figure 3.6 The actual routes of aircraft IL-14 (1) and research vessel *Alexander von Humboldt* at the test area of the first COSE on 19 April 1982. Figures show the temperature in °C. The long parts of the aircraft route are numbered from 1 to 5 (USSR–GDR 1985).

ment the SST data from NOAA satellites were collected using the autonomous data acquisition station located on the mainland in the village of Lesnoe and operated by members of the Steering Board. Two different airports (Palanga and Vilnius) were used for the aircraft.

All the stationary modules in the diagram were connected with telephone lines providing voice communication only. There was no direct communication between the two aircraft, between the Steering Board and aircraft, or between the ship and aircraft. Voice communication could be made available between the Steering Board and the ship only once a day at a fixed time.

Of special interest, both in their scientific and managerial aspects, was the performance of a complicated manoeuvre, with the ship in the course of the first COSE meant to provide collection of *in situ* data within the narrow (85 km) swath of the high-resolution (80 m) scanner 'Fragment' on board METEOR No. 30. It crossed the COSE test area only twice – on 23 and 24 April. The updated forecast with the predicted location of ground coverage was received by the Steering Board on 22 April. The relevant calculations were made and the command was given for the ship

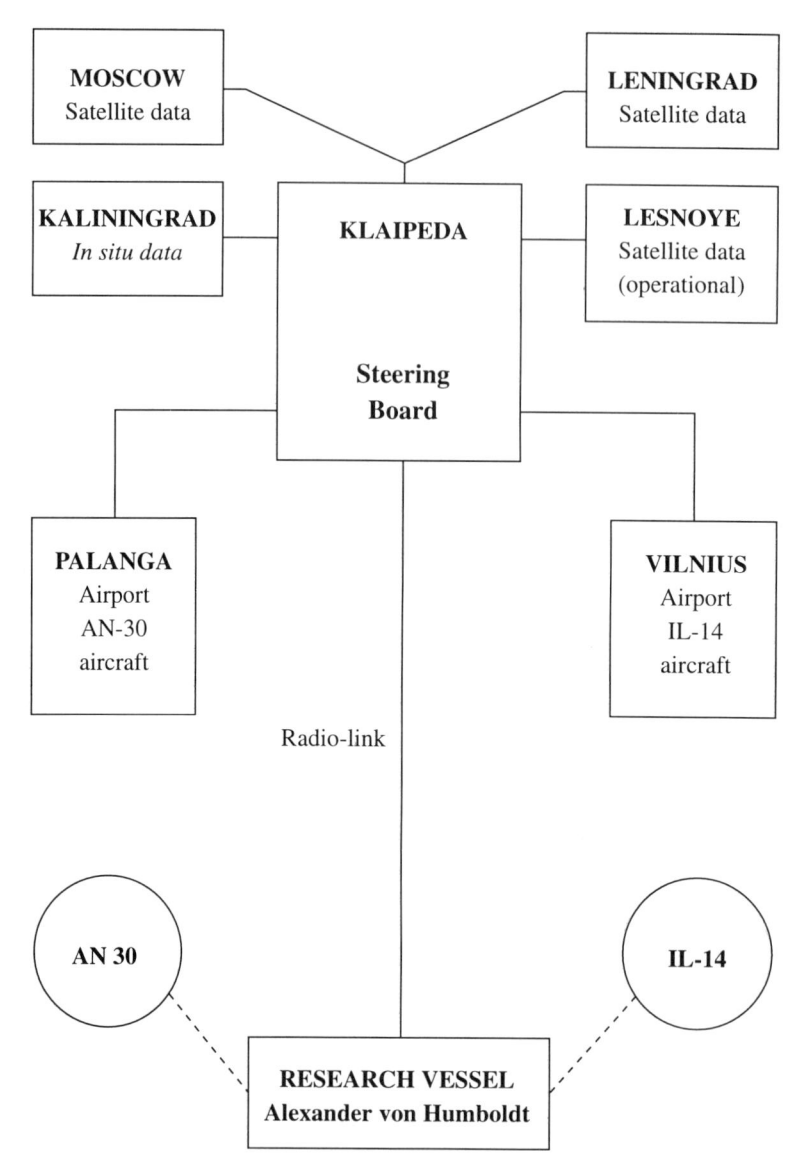

Figure 3.7 Diagram of management of the third COSE.

to move from the initial position (point 1 in Figure 3.8) to point 2 on 23 April and to point 3 on 24 April. This provided the location of the ship within both swaths (also shown in Figure 3.8). The next day the ship moved to the present position.

3.4.3 Development of COSE programme

The first COSE was briefly described by Berestovskij *et al.* (1983, 1984). Some results and findings will be reviewed in Section 3.4.4. The programme of the second

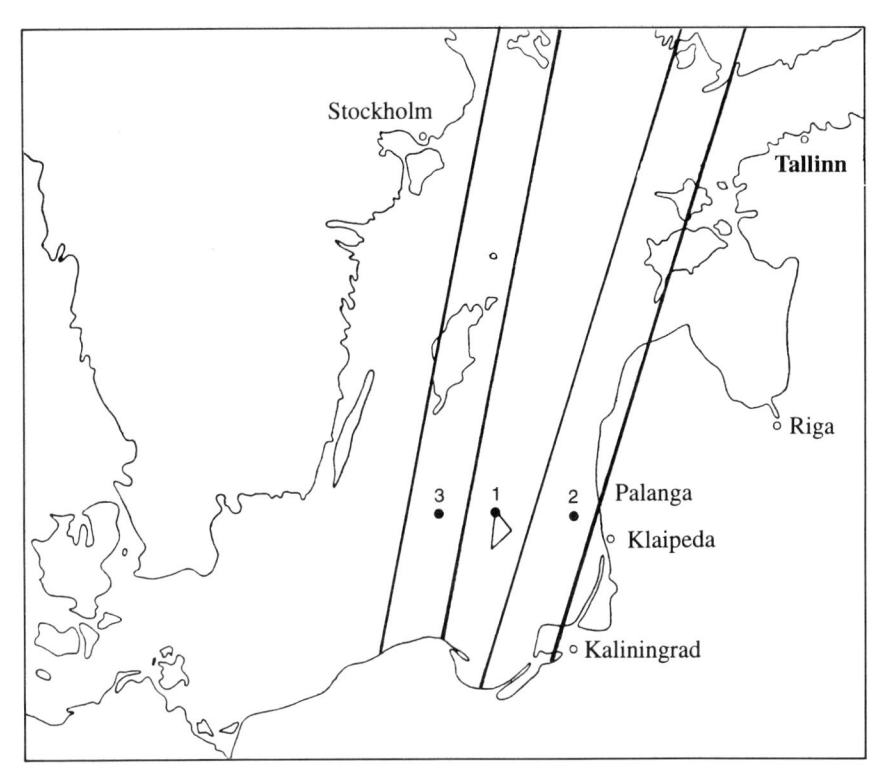

Figure 3.8 Location of the first COSE test area (1) and positions of the research vessel on 23 April 1982 (2) and 24 April 1982 (3) with the swaths of the 'Fragment' scanner (USSR–GDR 1985).

COSE (May–June 1983) was modified by expanding the test area and increasing the operational phase of the COSE. A spectrophotometer was used on board aircraft AN-30. A second research vessel, *Rudolf Samoilovich* (USSR), also participated.

The successful manoeuvring of the research vessel during the first COSE in order to match the satellite swath encouraged us to plan one of the next COSE as an 'actively controlled' or rather 'actively managed' experiment with not merely the satellite swath but the operational oceanographic information extracted from satellite data being the driving force. The idea was not a new one. For example, Stewart (1984) wrote about satellites which '... transmit data to ships at sea, and these help guide ships into regions where measurements should be made'. He explained that during the 'Ring' experiment in the Gulf Stream area in 1982 satellite images were received and processed on the mainland and the results in the form of simplified drawings (charts) were sent by facsimile radio-link to ships at sea participating in the experiment (Stewart 1984). Earlier this kind of exercise was highly appreciated by Fedorov (1980) who considered the use of satellite data for the guidance of research vessels to the interesting oceanographic phenomena as one of five main achievements of satellite oceanography.

(The idea of trying to guide the research vessel to an interesting oceanographic object in the framework of a COSE, when the cruise programme of the vessel was rather flexible – contrary to conventional cruise programmes of the former USSR research vessels with their strictly fixed (in time and space) 'no-matter-what-happens' schedules – perhaps became particularly relevant to the author who, in the late 1960s and early 1970s, was involved in remote sensing applications to the study of the lunar surface and in remote control and guidance of another movable platform with a scientific payload – the lunar vehicles LUNOKHOD-1 and LUNOKHOD-2 (Kocharov *et al.* 1972, 1975). It is curious that one of the lunar vehicles was operated in the 'marine' environment – in the Sea of Rains on the Moon, and another investigated the transition zone between the lunar 'sea' and lunar 'mainland' – which is reminiscent of the 'coastal zone' in oceanography.)

As was mentioned above we could not transmit facsimile messages, hence the Steering Board had to receive and analyse the satellite image to the extent that could enable us to detect and identify a certain oceanographic phenomenon, make a forecast of its motion and give a command to the ship either guiding her to a point where further detailed measurements were supposed to be carried out or suggesting a cross-section meant to detect and localise the phenomenon by means of conventional *in situ* measurements of three-dimensional oceanographic parameters. Figure 3.9 shows details of this 'guidance'.

Note that to work out the command for the vessel some *a priori* knowledge must be applied (to identify the 'object'-anomaly-phenomenon), and current meteorological data must be used (to predict the motion of the 'object'). Another point to note is that when the command is received the ship will be carrying out some other measurements to solve a current problem, and (if these measurements have to be interrupted) some decision-making mechanism should be activated to weight the value of the current problem and the potential value of the new problem. Provided that the choice was made in favour of the new problem, on the way to the test area the researchers should work out the new survey plan.

The above concept was used as a basis to prepare the third COSE in May 1985 as an actively managed one. For the test area of the third COSE (Figure 3.4) the eastern part of the Baltic Proper near the Klaipeda Strait was selected. When in the AVHRR image a 'mushroom-like' structure was detected and identified as a dynamical dipole worth studying in detail using shipborne measurements, the scenario shown in Figure 3.9 was used to guide the German research vessel *Alexander von Humboldt* to this phenomenon. It resulted in the first ever detailed investigation in the Baltic Sea of the three-dimensional structure of this eddy and its dynamics. Oceanographic description and analysis of this phenomenon will be presented in Chapter 5.

3.4.4 Some findings of the COSE

What was actually studied in course of the three COSE carried out in the Baltic? What were the major findings? Along with obviously routine collection of the sea-truth *in situ* data which were used to tune algorithms of satellite data processing and to check their oceanographic interpretation, some important results have been obtained which might be of general interest. Among them were the following oceanographic and methodological matters:

- the so-called 'allowed interval of non-synchronicity' between measurements carried out from various platforms – satellites, aircraft and ships – was suggested, developed and experimentally tested in the course of COSE (this item will be presented in Section 3.5);

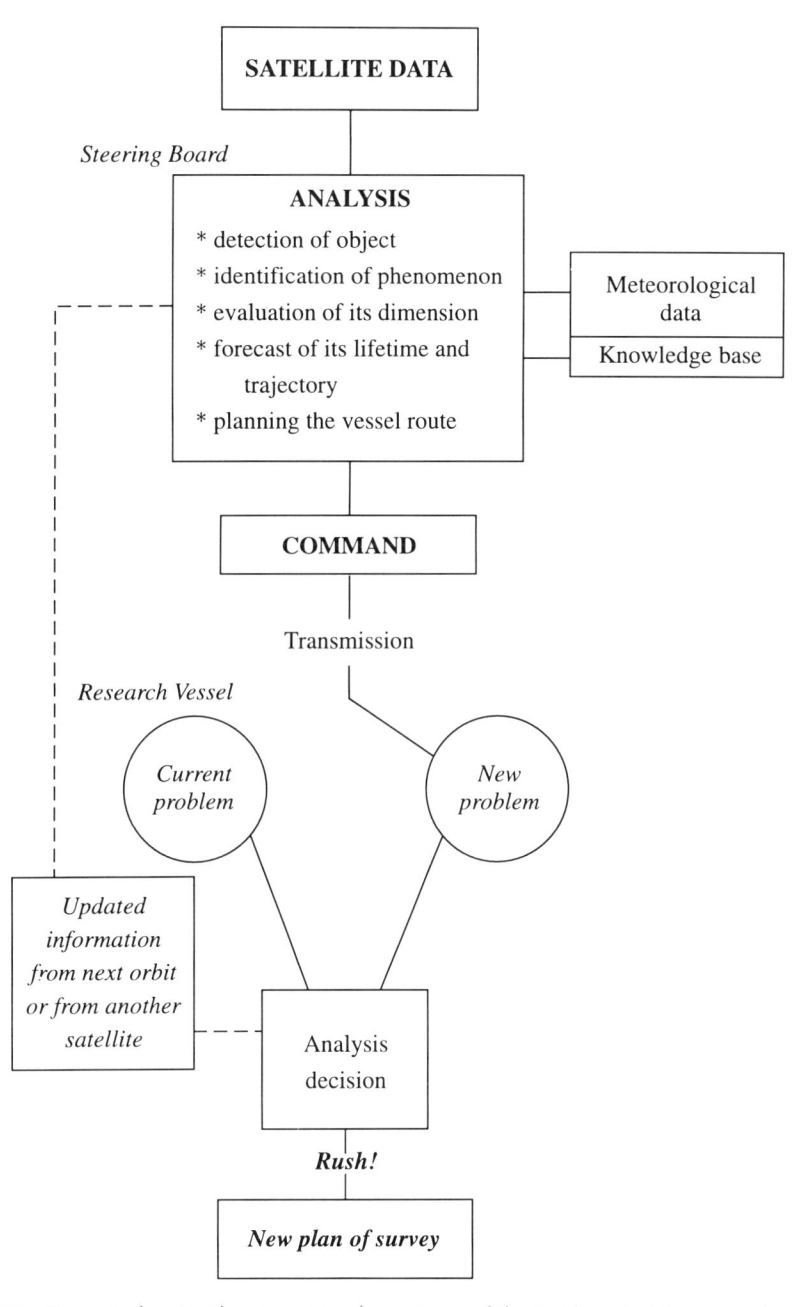

Figure 3.9 Diagram showing the sequence of operations of the 'guidance' of the research vessel to an oceanographic phenomenon.

- complex study of a 'mushroom-like' eddy structure (to be presented in Chapter 5);
- the study of the relationship between the internal structure of the deep water mass and the manifestations of that structure on the sea surface (to be discussed in Chapter 5 as part of the presentation on eddy structures);

76   REGIONAL SATELLITE OCEANOGRAPHY

- the comparison of informational capabilities of the two satellites of METEOR-2 type with different lifetimes in orbit.

The latter item will be presented in brief below.

A comparison of informational capabilities of the two satellites of METEOR-2 type   With the development of satellite technology the possibility of acquiring quasi-synchronous images of the same water area taken from different satellites is becoming a reality. Comparing the data from different satellites one can carry out an extra validation of satellite information or obtain some knowledge of the payload in-flight performance, for example, to single out sensors (or channels) with high levels of noise.

It is convenient to compare the data from different satellites using the theory of planning an experiment at the specially selected test areas. Such a comparison was carried out during the first COSE in the Baltic Sea in April 1982 (USSR–GDR 1985). The comparison was made as applied to the METEOR-2 No. 5 and No. 7 IR sensors with the Gulf of Riga being the test area. Charts of the SST in cloud-free situations were selected for the analysis. The time interval between the successive passes of two satellites over the test area was 1 hour.

The analysis demonstrated that the sum total of the SST data on each chart could be divided into two groups – the coastal group (block 1) and the central group (block 2) – with 21 measurements in each block (Figure 3.10).

These groups of data were significantly different in the average values of SST resulting from the warming up of the coastal zone in the daytime and also due to possible partial encroaching of the warmer coast into the field of view of the satellite

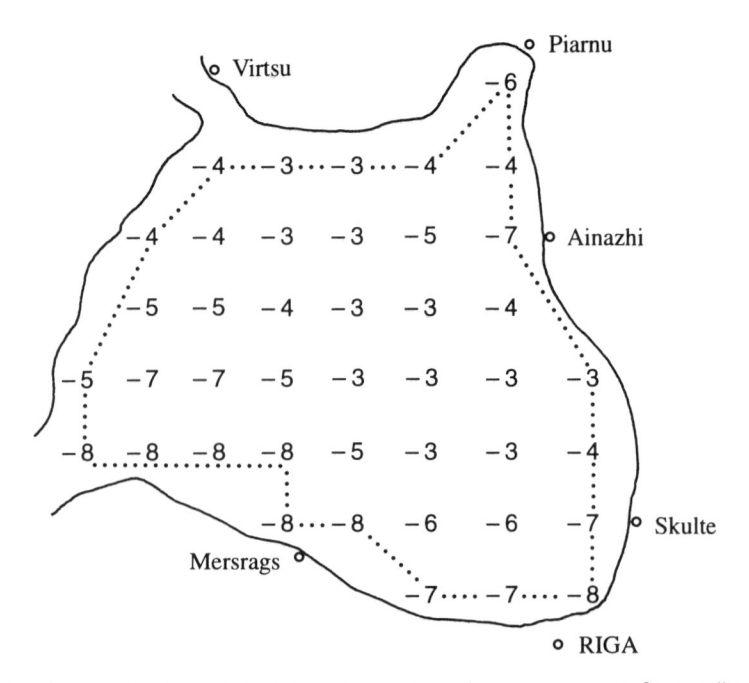

Figure 3.10   Chart of SST of the Gulf of Riga. Figures show the temperature in °C. Satellite METEOR-2 No. 7, 24 April 1982, 1:00 a.m. GMT. Dots show the coastal zone (USSR–GDR 1985).

radiometer. The data processing was carried out according to the 'plan of a randomised block experiment'. It was considered that the SST data met the assumptions of the dispersion analysis for random values. The assumption of the homogeneity of dispersions was proved using the Kochren criteria (see Tables 3.1 and 3.2).

To prove the comparability of the two satellites' data it was necessary to compare the residual dispersions. The calculated statistics

$$F = \frac{S_3^{2\prime}}{S_3^{2\prime\prime}} = 4.3$$

were compared to the critical value of the F-distribution with (3;3) degrees of freedom. For the value of significance 0.05, the critical value of F was 9.28. This meant that the calculated value of F was less than the critical value, which made these two satellites' data comparable with each other.

On the whole the SST values from METEOR-2 No. 7 were higher than those obtained from METEOR-2 No. 5. Both testify that the sea surface was at its warmest on 25 April; colder water was recorded on 19 and 20 April. These results correspond to the observations *in situ* carried out at the coastal hydrometeorological stations. The correction for the atmosphere during the first COSE calculated from subsatellite observations was 4–7°C for METEOR-2 No. 5 and 3–5°C for METEOR-2 No. 7. The correction for the atmosphere carried out in Moscow at the Main Centre for satellite data acquisition using monthly averaged climatic data amounted to 4°C for both satellites. That meant that it was possible to use this method for METEOR-2 No. 7 (with less than a year in orbit). The SST data from METEOR-2 (with more than two years in orbit by that time) needed a more serious

Table 3.1 SST values for the Gulf of Riga (°C), METEOR-2 satellites

Satellite no.	Block	Days of April 1982			
		19	20	23	25
5	1	−3.0	−3.9	0.4	1.2
	2	−4.9	−4.0	−1.4	−0.7
7	1	−1.0	−2.0	1.5	1.7
	2	−2.5	−2.0	−0.2	−0.25

Table 3.2 The results of the dispersion analysis of the SST data for the Gulf of Riga, METEOR-2 satellites No. 5 and No. 7

Source of variability	Sum of squares (K²)		Number of degrees of freedom	Estimation of dispersion (K²)	
	No. 5	No. 7		No. 5	No. 7
Blocks	4	3	1	4	3
Days	30	13	3	10	4.3
Residue	1	4	3	0.3	1.3
Sum	35	20	7	5	3

approach. It could be due to the possible change of spectral characteristics of the sensors or the decline of sensitivity of the instrument. It was evident that one should take into account the change of the instrument response during exploitation. (A lot of other questions arise: on the artificial ageing of the instrument at the laboratory, methods of in-flight tests, etc.)

The patterns of the SST isotherms plotted on the basis of the data from the two satellites were also different (see Figures 3.11 and 3.12), which to some degree might have been connected with the high level of noise in the METEOR-2 sensor.

As a result of this study we came to the conclusion that it was expedient to use the data from the IR band from the METEOR type of satellites for the qualitative analysis of the oceanographic processes on a large scale or elsewhere but not in the problems covered by regional satellite oceanography. Our experience of tackling the METEOR SST data in connection with the first COSE taught us that it was not

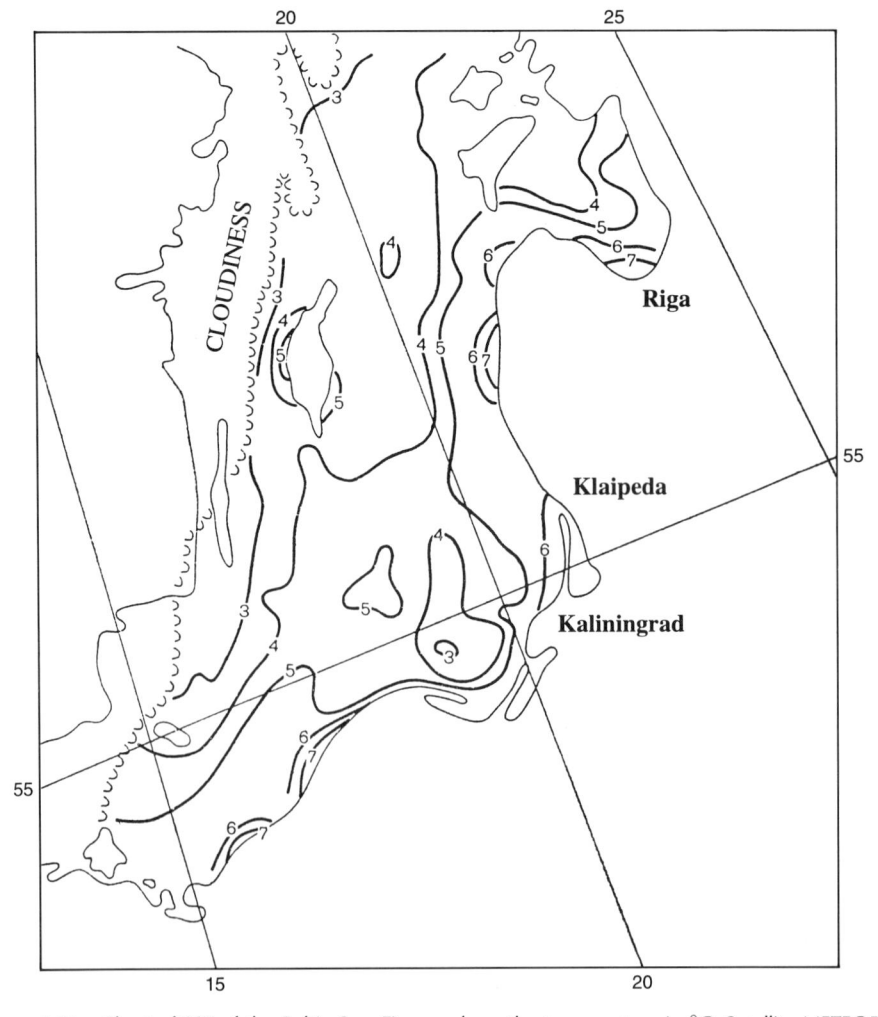

Figure 3.11 Chart of SST of the Baltic Sea. Figures show the temperature in °C. Satellite METEOR-2 No. 5, 23 April 1982, 11:00 a.m. GMT (USSR–GDR 1985).

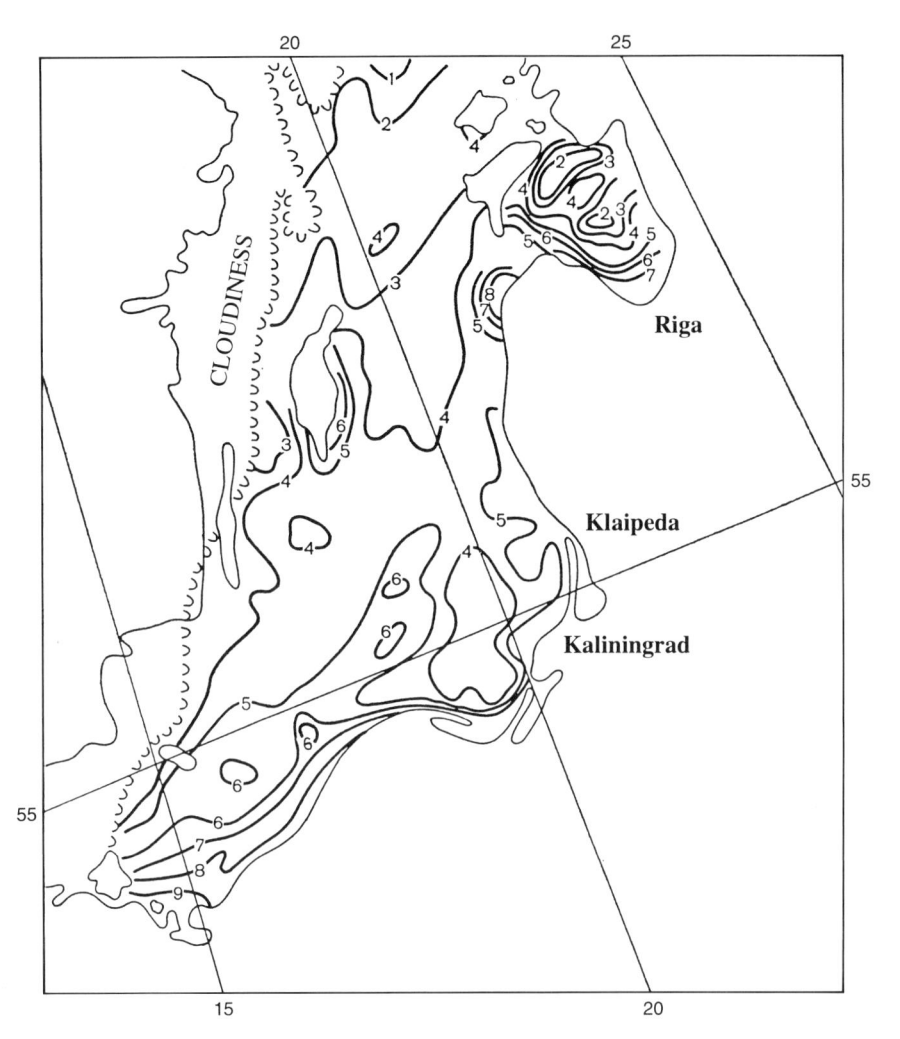

Figure 3.12 Chart of SST of the Baltic Sea. Figures show the temperature in °C. Satellite METEOR-2 No. 7, 23 April 1982, 12:00 a.m. GMT (USSR–GDR 1985).

possible to rely on the data in operational regional satellite oceanography, as it brought ambiguous results in terms of the fine structure of the SST field. It was not wise to use these data for coastal zone studies as the ground resolution was actually about 20 km and, besides, near coastal pixels contaminated with higher temperatures were to be excluded as well. The satellite data could reach the user only in a week's time (mail service), and certainly one could not expect to get 8–10 images of the test area daily from the METEOR services.

So we came to a definite conclusion to focus mainly on AVHRR NOAA SST data. The results of the first COSE backed our position set forward in February 1981 when we held the Inter-ministerial Workshop on Satellite Oceanography in Leningrad. Experts and decision-makers from 24 organisations participated, representing the State Committee for Hydrometeorology, Academy of Sciences, Ministry of Fishery, Ministry of Higher Education and the navy. The aim of the workshop was to exchange experience and to make an attempt to co-ordinate the efforts in satellite data acquisition and processing to the benefit of oceanography. The recommendation from the workshop was to focus on autonomous satellite data acquisition systems with NOAA AVHRR data being the main source of SST data.

Complex Oceanographic Subsatellite Experiments carried out in the Baltic Sea in the 1980s were among the cornerstones of regional satellite oceanography in the former USSR. They yield a lot of satellite and *in situ* data (which otherwise could not have been collected) which were used in methodical and pure scientific exercises. Practical experience was gained by scientists and technical personnel in carrying out multilayer satellite–aircraft–ship measurements in passive mode and also in active mode with direct guidance of the ship to the oceanographic object. COSE encouraged us to perform experimental long-term satellite monitoring of the Baltic Sea in the second half of the 1980s.

3.5 The study of 'allowed interval of non-synchronicity' (AINS) between oceanographic observations at different levels

3.5.1 Methodical approach

No interpretation or use of remote sensing data is feasible nowadays without some *in situ* information. Subsatellite sea-truth measurements are used for evaluation of the accuracy and calibration of satellite data, and in some cases for direct calibration of satellite payload. But there inevitably arise methodical aspects of the problem of conducting subsatellite measurements associated with the accuracy of sea-truth measurements themselves, with the accuracy of fixing the geographical position of different subsatellite platforms (vessels, aircraft, etc.), and with non-synchronicity between satellite measurements and sea-truth measurements. The problem of allowed disagreement in time in the 'satellite–*in situ* measurements' is one of the most consequential ones that arise during subsatellite observations and collecting *in situ* information, including special Complex Oceanographic Subsatellite Experiments (COSE).

There is no universally accepted opinion about the evaluation of the allowed interval of non-synchronicity (AINS); neither is there a unified terminology. However, Zlobin *et al.* (1983) subdivided all the satellite activities into synchronous, quasi-synchronous and non-synchronous. They called 'synchronous' the activities conducted on the day when the image was taken during the period of $t + \Delta t$, where t is the time of image taking. The activities also conducted during the period of $t + \Delta t$ not on the day of image taking but during several days (ΔX) before and after the image was taken, are called 'quasi-synchronous'. Zlobin *et al.* (1983) call 'non-synchronous' the activities conducted at any time within 24 hours within the limits of a certain period, during which synchronous and quasi-synchronous activities were conducted one or more times. The authors' evaluation of Δt, ΔX are indeterminate, they depend on the type of natural objects, conditions of the survey, etc. According to Zlobin *et al.* (1983), one can take, for example, $\Delta t = 30$–60 min, $\Delta X = 24$ hours for the study of vegetation.

These definitions of synchronicity and non-synchronicity, though they may be usable in geology or in the study of soil and vegetation, seem to make no sense in oceanography.

Dotsenko (1980) offered an analytical equation for the determination of the optimum interval between remote and *in situ* measurements, based on the minimisation of the inaccuracy in measurement. According to this equation, the estimation of t required the knowledge of the characteristic time and space scales of the field, the speed of the field movement with regard to the fixed point, and a number of geometric characteristics of the relative position in the satellite–*in situ* measurement system. It was difficult to get a good idea of the characteristics of the field under study before COSE. In this way evaluation of the space scales and the speed of the field transfer can be used to support aircraft or satellite data which are to a considerable extent liable to changes connected with atmospheric fluctuations and cloud conditions and must be calibrated using *in situ* data.

The objects of oceanographic studies – hydrophysical fields – are more liable to time changes than those of closely related geophysical disciplines. For that reason it is so important to have objective criteria of synchronicity of the satellite and the sea-level measurements. Synchronicity in conducting complex oceanographic subsatellite experiments can be guaranteed only at a few points of the scene, where *in situ* measurements are done exactly at the moment of image taking. One must bear in mind though that to do a comparative analysis of satellite information covering an area of up to 3000 km long it is necessary to have a wide enough range of subsatellite observations to get statistically significant estimations. The number of carriers – ships and aircraft – at the oceanographers' disposal is limited within reasonable bounds. That means that some parts of the subsatellite observations under analysis will need to be non-synchronous. The task is to determine the AINS of measurements within the limits of which it is expedient to use different time data of the multilevel measurements within the ship–aircraft–satellite system. It stands to reason that the estimation of this value will be different for different oceanographic parameters. This section deals with the field of the sea surface temperature, but the methods described below can be used for the AINS evaluation for other hydrophysical parameters.

The usage of carriers of different level in COSE is functionally dependent. Aircraft with IR radiometers are able to measure the radiative temperature on the surface SST, i.e., the same physical value as the one that can be taken by satellite. It must be noted though that methodical inaccuracies of the IR radiometry caused by the atmosphere influence IR radiation and the background cloud radiation makes determination of the real value of SST impossible. The accuracy of such measurements does not exceed 0.1 K (K here stands for degrees Kelvin).

Standard hydrological measurements from ships make it possible to measure the temperature of the surface layer of the sea (the so-called 'thermodynamical temperature') with an accuracy of 0.01 K, but the latter is not physically equivalent to the radiative temperature measured by means of remote sensing methods. The existence of the skin layer on the ocean–atmosphere boundary causes a typical difference of some tenths of a degree in the values of two types of temperature. So it is expedient while conducting COSE to use remote and *in situ* data jointly, turning to account the advantages of carriers of different levels in full.

Considerable difference in the values of non-synchronicity can be found in various publications. In addition, these data are based on expert evaluations of the variability of the hydrophysical fields in the work area, without using objective methods of analysis. In this way Gonzalez *et al.* (1979) considered synchronous data with regard to the Seasat satellite, measurements from ships, taken 10 minutes before and after the moment of the satellite pass, and measurements of the buoys taken 30 minutes before and after the satellite pass. The non-synchronicity of the airborne measurements at the oceanographic test area was 1 hour 15 minutes on average and 3 hours 24 minutes at maximum with respect to the time of the satellite pass.

Subsatellite observations on the ships of opportunities in the Arabian Sea during the MONEX-1979 experiment were carried out with non-synchronicity of measurements in the satellite–ship system of 2–6 hours (Pathak 1982).

As was mentioned by Champagne et al. (1982a), at the oceanographic test area in the front region near the island of Malta the measurements taken by the ship at the front cross-sections 6 hours before and after the moment of the satellite pass were regarded as quasi-synchronous. The authors of this paper pointed to considerable diurnal sea surface temperature variation, significant for the measurements conducted during 12 hours.

Legeckis and Bane (1983) described satellite and aircraft measurements of the Gulf Stream surface temperatures. Non-synchronicity of aircraft observations with regard to the time of the satellite overpass was from 6 to 10 hours. The position of the zone of maximum gradient according to the aircraft data was shifted 4–6 km relative to the satellite data. It was caused by the advection of surface water and must be taken into account in AINS estimation.

At the State Oceanographic Institute, Leningrad the methods of AINS evaluation in SST measurements were worked out within the framework of the first and second USSR–GDR COSE on the Baltic Sea (Bychkova and Victorov 1985). The test area location was shown in Figure 3.4. During the experiments the data from all the levels were analysed: the measurements taken by research vessels and the ships of opportunity, as well as airborne and satellite data. Hydrometeorological observations on the standard networked stations of the Baltic Sea were also used.

3.5.2 AINS in the satellite–ship system

In carrying out the comparative analysis of information in the satellite–system ship we will consider the time interval during which the autocorrelation function of the time series of the sea surface temperature measurements at a given point will drop e times to be the allowed interval of non-synchronicity. To estimate the AINS one can use regular sea surface temperature measurements taken either from an anchored buoy or from a ship located in one and the same point during the whole period of the subsatellite experiment. While using the data from the ships of opportunity, collected from large areas such as from the whole sea, it is expedient to use in the analysis the data of regular measurements taken at several points. It will allow us, within one oceanographic object, to single out the zones where it is possible to use this or that value of AINS. The normalised autocorrelation function $R(t)$ was estimated according to the equation:

$$R(t) = \frac{1}{n-t} \sum_{k=1}^{n-t} \frac{(T_k - \bar{T})(T_{k+t} - \bar{T})^2}{S}, \tag{3.1}$$

where $\bar{T}$ is the average value of the sea surface temperature at the point obtained from the selection of n measurements; S^2 is the time series dispersion, and T_k and

T_{k+t} are the instantaneous values of the sea surface temperature at the point at time moments k and $k + t$, respectively. The size of the selection necessary for obtaining statistically reliable results is determined by proceeding from the condition of minimum error m in evaluating the normalised autocorrelation function. Assuming that $m = 5\%$ one can, with a routine approximate formula linking m with the size of the selection n and the value of $R(t)$

$$m = \frac{(1-R)^2}{n-1},$$

derive $n = 300$.

During the second USSR–GDR COSE in the Baltic Sea hourly sea surface temperature measurements were performed at the point with coordinates of 55°16′ North, 16°00′ East. These data were used for the calculation of $R(t)$; the number of measurements was 350, and $S^2 = 99$ K^2. Figure 3.13 illustrates the results of this estimation of $R(t)$. The analytically obtained relation can be written as $R(t) = e^{-0.04t}$, i.e., the allowed interval of non-synchronicity in measurements is 25 hours. Similar data were obtained for the analysis of sea surface temperature observations at the point with coordinates 57.8° North, 21.6° East. Therefore during the experiment all the ship measurements taken during the same 24-hour period could be used to compare with the satellite data. It must be noted that during the period mentioned above the AINS between satellite and ship data appeared to be considerable, which can be explained by comparatively low variability of synoptic conditions; in other cases the AINS can be much shorter (about several hours).

While planning subsatellite cruise research it is desirable to estimate AINS in advance, using standard observations of the sea surface temperature on light vessels, anchored buoys and other platforms located not far from the place of the cruise research. In the course of the experiment this preliminary estimation must be made more precise. The AINS estimation obtained in the State Oceanographic Institute

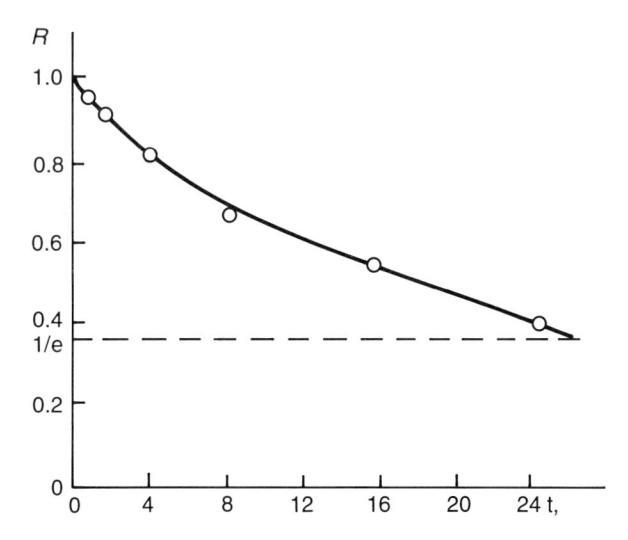

Figure 3.13 Normalised autocorrelation function of SST field based on point shipborne measurements during the second COSE (Bychkova and Victorov 1985); time in units of hours.

(St. Petersburg) during the second USSR–GDR COSE made possible the objective selection of ship observations used for correction of satellite data for the atmosphere. The measurements from ships taken during the 12-hour interval from the satellite pass were used for the analysis. Both the data from the ships operating in the experiment and the data from the ships of opportunity were turned to account. Correction for the atmosphere of the satellite values of the radiative temperature (T_r) was done according to the regression equation:

$$T_r(x) = aT(x) + T, \tag{3.2}$$

where $T_r(x)$ is the value of T_r obtained from the satellite at point x; T is the value of the sea surface temperature at point x, reconstructed from the satellite data; and a and T are the parameters obtained by the least squares technique. Figure 3.14 shows the SST chart based on the data from the METEOR satellite taken at 15 hours (Moscow time) on 1 June 1983, corrected using equation (3.2). The series of subsatellite observations consisted of 20 terms (see Figure 3.15).

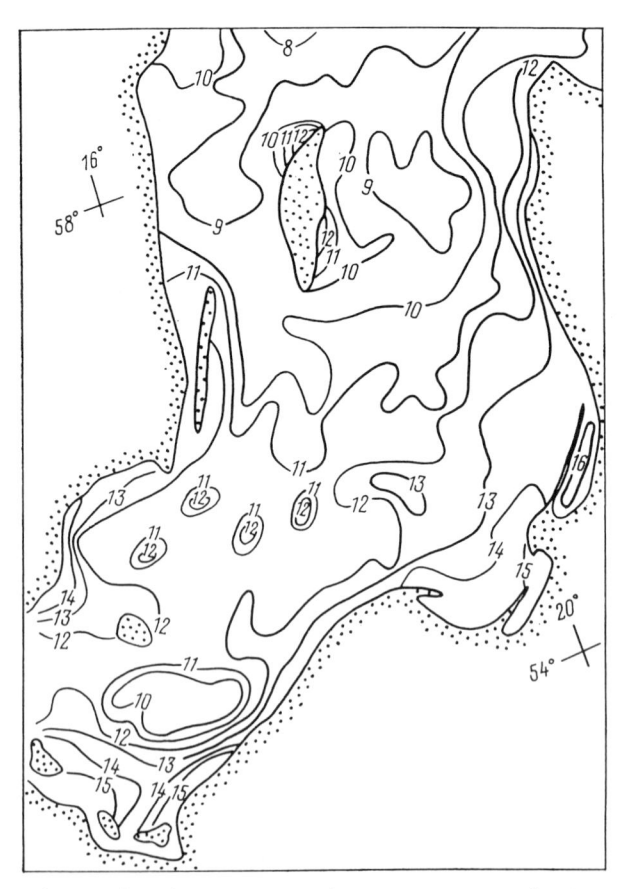

Figure 3.14 Chart of SST in the Baltic Sea. Figures show temperatures in °C. Correction for the atmosphere was made using shipborne *in situ* data. METEOR-2 data, 1 June 1983, 15:22 Moscow time (Bychkova and Victorov 1985).

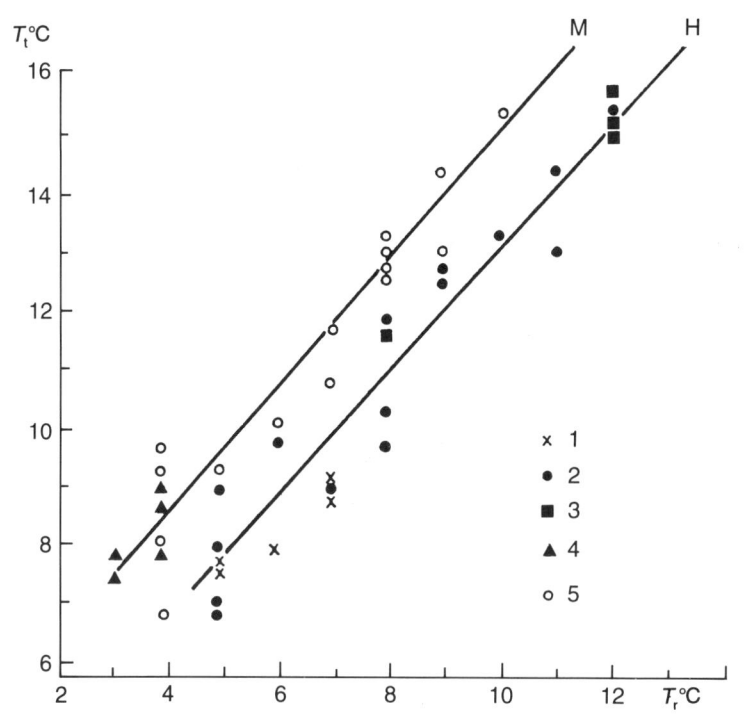

Figure 3.15 The regression between radiative temperature T_r and thermodynamic temperature T_t. N designates NOAA-8 data taken on 1 June 1983 at 15:52 Moscow time, M designates METEOR-2 data taken on 1 June 1983 at 15:22 Moscow time. (1), (4) data from research vessel *Alexander von Humboldt*, (2), (5) data from ships of opportunity, (3) data from local coastal hydrometeorological stations (Bychkova and Victorov 1985).

The dispersion of measurements calculated by means of a routine technique using *in situ* shipborne measurements amounted to 0.7 K. It proved to be impossible to use the *in situ* data collected at local coastal hydrometeorological stations for correction of the METEOR data because the *in situ* observations were obtained in a narrow heated coastal zone while satellite information was presented in the form of a 20 km spacing digital grid. For the NOAA satellite with the high-resolution IR radiometer (4 km in APT mode) the results of observations from hydrometeorological stations were used together with the research vessels data in deriving the regression equation of type (3.2). The dispersion of measurements amounted to 0.9 K.

3.5.3 AINS in the satellite–aircraft system

The technique of AINS estimation described above for the satellite–ship system cannot be applied to aircraft measurements because it is impossible to acquire from an aircraft a durable time series at one point. Besides, the AINS value obtained with shipborne *in situ* observations cannot be extended to aircraft observations on account of considerably greater dependence of the sea surface radiative temperature (compared to the thermodynamic temperature of the upper layer) on diurnal variability of the Sun's radiation, the variation in cloud conditions, wind speed and other parameters, and heat exchange in the sea–atmosphere boundary layer.

Bearing in mind that aircraft surveys of oceanographic test areas could take several hours it is necessary to evaluate the significance of diurnal variation of radiative temperature for airborne measurements.

For airborne measurements at the COSE test area let us define the value of the AINS as the time interval during which the change of average radiative temperature at the test area is statistically insignificant. This will allow us to get statistically reliable estimations for the satellite information on the test area using the airborne measurements taken during the period equal to the AINS from the time of the satellite pass. These estimations are obtained by means of dispersion analysis of the data collected at the test area in the course of two or more flights performed within 24 hours on the same routes. Table 3.3 shows the results of such a two-factor dispersion analysis based on the observations done on 12 June 1983 at the test area of the second COSE. On this day two flights (factor F) were made: at the period from 11 hours 14 min, up to 12 hours 42 min, and from 13 hours 32 min, up to 14 hours. During both flights a grid consisting of five long routes (factor R) crossing the test area was used. The estimation of the total dispersion and its components was calculated according to the formula:

$$S^2 = \frac{1}{rv-1} \sum_{i=1}^{r} \sum_{j=1}^{v} (T_{ij} - \bar{T})^2 = \frac{Q}{rv-1}, \tag{3.3}$$

where Q is the total sum of the squares of deviations of single observations from the total average.

$$S_1^2 = \frac{v}{r-1} \sum_{i=1}^{r} (\bar{T}_i - \bar{T})^2 = \frac{Q_1}{r-1}, \tag{3.4}$$

where Q_1 is the sum of the squares of differences between the average value on the flights ($\bar{T}_i$) and the total average; $v = 5$ is the number of levels of the factor 'Routes' (R); $r = 2$ is the number of levels of the factor 'Flights' (F).

$$S_2^2 = \frac{r}{v-1} \sum_{j=1}^{v} (\bar{T}_j - \bar{T})^2 = \frac{Q_2}{v-1}, \tag{3.5}$$

where Q_2 is the sum of the squares of differences between the average on the routes ($\bar{T}_j$) and the total average.

Table 3.3 The results of a two-factor dispersion analysis of airborne measurements of the sea surface temperature on 12 June 1982 (Bychkova and Victorov 1985)

Dispersion components	Sum of squares (K^2)	Number of degrees of freedom	Estimation of dispersion (K^2)
Between averages along lines (factor 'Flights')	0.041	1	$S_1^2 = 0.041$
Between averages along columns (factor 'Routes')	0.175	4	$S_2^2 = 0.044$
Residual	0.074	4	$S_3^2 = 0.0185$
Total	0.29	9	$S^2 = 0.03$

$$S_3^2 = \frac{1}{(r-1)(v-1)} \sum_{i=1}^{r} \sum_{r=1}^{v} (\bar{T}_{ij} - \bar{T}_i - \bar{T}_j + T)^2 = \frac{Q_3}{(r-1)(v-1)}, \qquad (3.6)$$

where Q_3 is the residual sum of the squares representing the influence of the factors which failed to be taken into account.

To determine the significance of the F and R factors it is necessary to compare the factor dispersions with the residual dispersion. The statistics are calculated as $F_F = S_1^2/S_3^2$ with $k = r - 1$ and $k = (v - 1)(r - 1)$ degrees of freedom, and also $F_R = S_2^2/S_3^2$ with $k = v - 1$ degrees of freedom.

The calculation gives the values of $F_F = 2.2$ and $F_R = 2.4$. The table values for the 5% significance level are $F_F = 7.71$ and $F_R = 6.39$ respectively (Bolshev and Smirnov 1983). Thus the influence of these factors on the average SST at the test area is not significant, which means that airborne observations performed during a 3-hour interval can be used for comparison with satellite information obtained instantly.

The next flight at the test area was performed on 13 June from 9 hours 31 min till 11 hours (i.e., 20 hours after the beginning of the previous flight). Dispersion analysis of the results obtained, together with the data of the previous flight, showed considerable differences in the estimated values of the average temperature at the test area. Therefore, henceforth airborne observations were compared only with the satellite data obtained on the same day within 3 hours after the beginning of the aircraft survey.

Similar studies were done in April 1982 during the first USSR–GDR COSE in the Baltic Sea. At that period the research aircraft IL-14 was surveying the temperature at the test areas of a size commensurable with the ground resolution of the IR radiometer on board the METEOR satellite. On 18, 19, 24 and 26 April the survey was carried out at the test site located in the open sea to the south of the point 19° East, 56° North. On 23 April it was carried out at the square bordering the coast.

On 18 and 23 April several flights at the test sites were performed to estimate the influence of diurnal variability of SST on airborne measurements. The analysis of the obtained data showed that in the open sea the measurements could be considered quasi-synchronous during 4 hours, while in the coastal zone it was during 1.5 hours.

It seems advisable to estimate the AINS of airborne measurements within the period of durable COSE regularly taking into account the synoptic variability of oceanographic processes. While planning subsatellite experiments it is expedient to use the regional oceanographic database (see Section 3.8) in order to analyse the *in situ* measurements time series of hydrophysical parameters collected from different carriers and platforms and make preliminary estimations of AINS. These values of AINS should be updated in the course of COSE.

3.6 Calibration of satellite sensors using natural 'standard' test sites

3.6.1 Lake area as a natural 'standard' of temperature

Some natural test areas suggested for the selection of *in situ* sea-truth subsatellite data have been mentioned in Section 3.3. The measurements at these test areas were

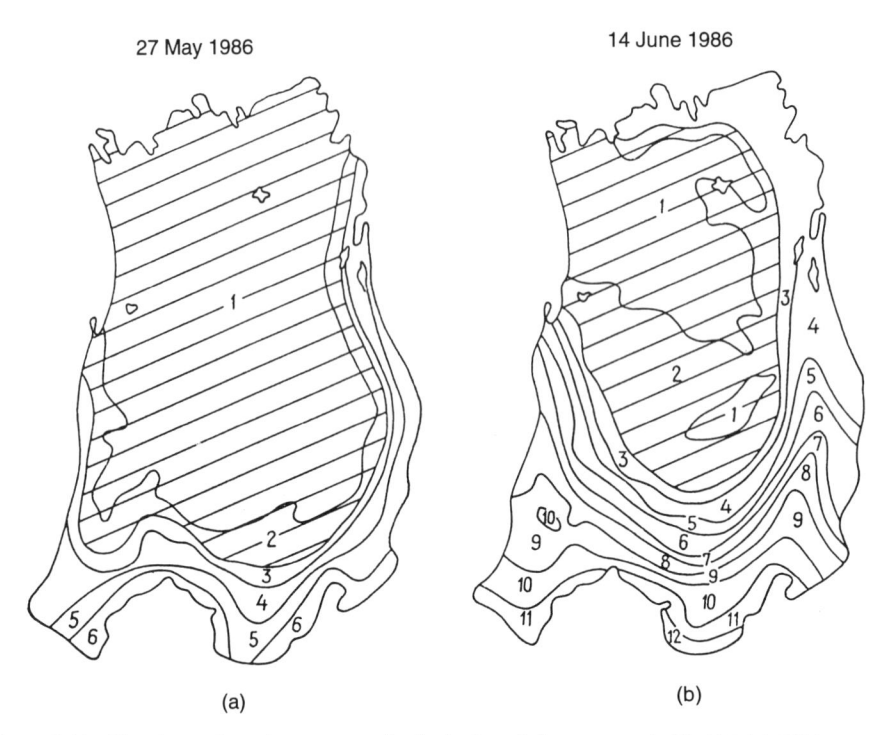

Figure 3.16 The thermobar phenomenon in the Ladoga Lake as recorded in NOAA SST imagery for (a) 27 May 1986 and (b) 14 June 1986. Figures show relative (non-calibrated) temperatures. The thermoinertial zone is shaded. (After Bychkova et al. 1989a.)

supposed to be carried out within test sites with homogeneous characteristics of oceanic parameters. During the satellite overpass these oceanographic parameters were to be measured by means of airborne or shipborne equipment or by buoys. The problem arising from non-synchronicity of measurements in the satellite–aircraft–ship system was discussed in Section 3.5. The suggested technique for the determination of the allowed interval of non-synchronicity (AINS) was tested in the course of Complex Oceanographic Subsatellite Experiments as described in Section 3.4.

In a limited number of cases and under specific conditions the complicated problem of *in situ* data collection can however be simplified. It seems worthwhile to discuss here one of these cases.

Based on the results of an analysis of satellite data collected in the course of experimental satellite monitoring of the Baltic Sea and its catchment area (in accordance with the concept of 'Baltic Europe' – see Chapter 5 for details), carried out in April–September 1986, Bychkova *et al.* (1989a,b) suggested calibrating satellite sensors of the IR band using a natural 'standard' – the phenomenon of the 'thermobar' in the lakes. The method was tested at the Ladoga Lake which is part of the Baltic Sea catchment area (see the maps in Chapter 5).

According to Tikhomirov (1982) the thermobar was discovered by Forel in the nineteenth century. The thermobar is a thermohydrodynamical phenomenon resulting from the basic physical fact of the existence of a peculiar temperature point of 4°C for pure water. During the so-called 'hydrological spring' (which lasts for about $2\frac{1}{2}$ months in the period of May–July) there is a narrow zone (or band) in the lake in which the water temperature is equal to the temperature of the maximum water density. Within this band the vertical circulation of water down to the bottom can be observed. In spring, due to the incident solar radiation, the temperature of the surface in the shallow coastal area increases faster than in deeper areas. With this heating the density of water increases up to a temperature of 4°C (the maximum density). Heavier water masses descend down and permanently mix with the underlying water masses. This creates the thermobar – the band with a temperature which corresponds to the temperature of maximum water density, exactly 4°C. During the spring the thermobar moves toward the central deep part of the lake. This process continues till the moment when all the water mass in the depth acquires the temperature of 4°C. During its lifetime the thermobar divides the whole lake into two parts separating the so-called 'thermoinertial zone' (with very low gradients of temperature) from the coastal 'thermoactive zone'.

Figure 3.16 shows the position of both zones for two dates in May–June 1986.

Estimates of the velocity of thermobar movement and the area of the 'thermoinertial zone' obtained from AVHRR NOAA-9 SST data are given in Figure 3.17 together with the air temperature data measured at the coast for May–July 1986.

In the 1980s we gained some experience in the study of thermobars in the Ladoga and Onega Lakes and also the upwelling in these areas (Bychkova et al. 1989a). This experience helped us to obtain a new vision of the dynamical processes in the coastal zone of the Baltic Sea. In the southeastern part of this sea with low-salinity waters we have managed to trace the thermobar phenomenon as recorded in satellite imagery of the IR band (Bychkova et al. 1987).

The large area of the thermoinertial zone (thousands of km^2), the low velocity of thermobar movement (about 1 km per day) and the fairly permanent temperature of the lake surface within this zone (see Figure 3.16) enabled us to suggest the use of the thermobar itself and the thermoinertial zone as natural 'standards' for calibration of satellite IR sensors.

3.6.2 Calibration of satellite sensors of the IR band using thermobar and thermoinertial zones in the lakes

The thermobar itself can be used to calibrate future sensors with high spatial resolution (about 100 m), while the thermoinertial zone can be used as a standard for current satellite sensors. The temporal variability of temperature in this zone is very low – about 1–2°C during the lifetime of thermobar. The interannual variability is also low – during the 1957–1962 period the mean value on 31 May was 2.3°C, with a value of 2.0°C after a severe winter and 2.7°C after a mild winter (Tikhomirov 1968).

Bearing in mind that the available procedures for the on-line correction of satellite SST data for the atmosphere provide data with 0.6–0.7°C accuracy, the use of the thermoinertial zone offers a unique and simple method for correction for the atmosphere and calibration of the satellite IR sensors during the period from May to the middle of July.

Bychkova et al. (1989a,b) suggested using the thermoinertial zone of the Ladoga Lake (and the Onega Lake located nearby) to make corrections for the atmosphere

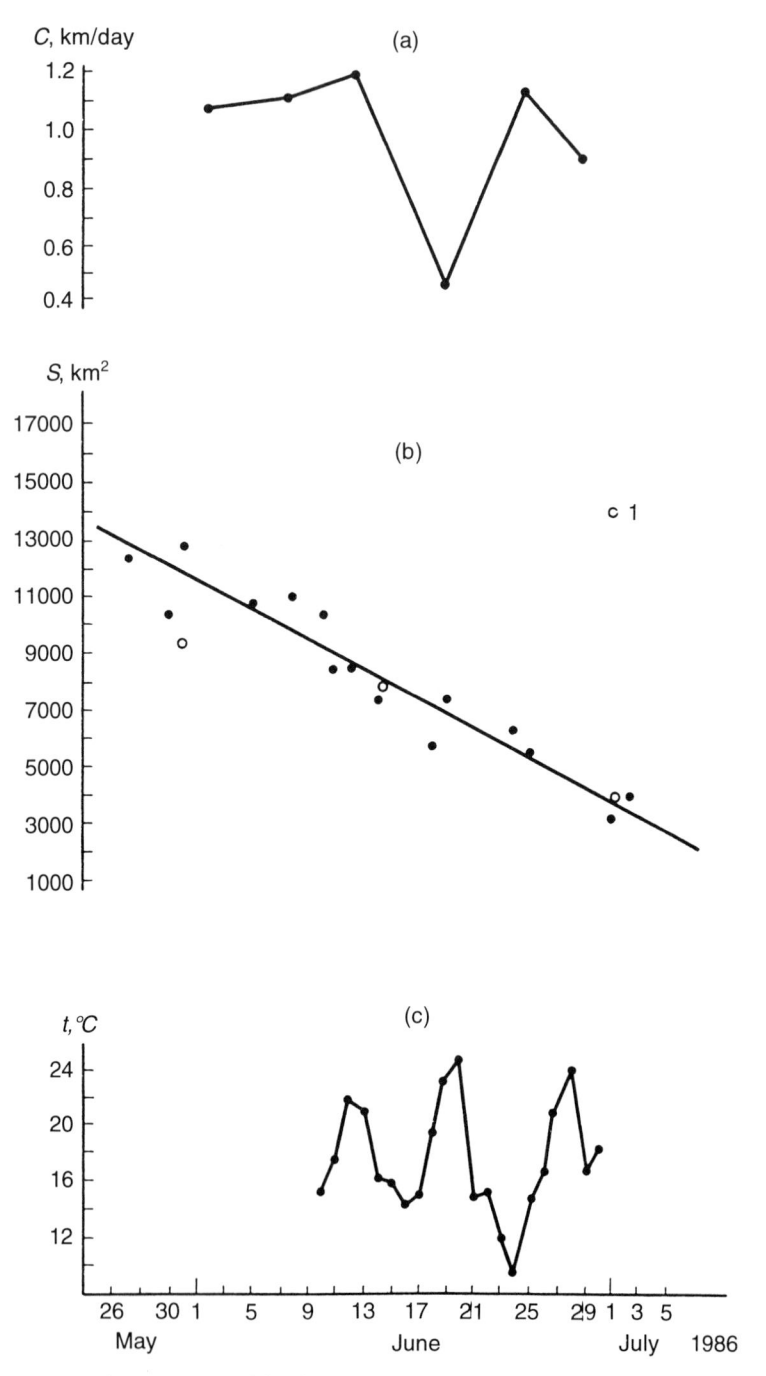

Figure 3.17 Some characteristics of the thermobar phenomenon. (a) The velocity of thermobar movement in km per day. (b) The area of the thermoinertial zone in km^2 in the Ladoga Lake in May–July 1986, as revealed from NOAA SST imagery. Empty circles (designated with 1) show averages for 1957–1962 (as recorded by the local coastal hydrometeorological station). (c) Air temperature in June 1986, data from coastal hydrometeorological station. (Bychkova et al. 1989a.)

of the NOAA SST data for the Baltic Sea. The Ladoga is located about a hundred kilometres from the Gulf of Finland (see maps in Chapter 5). The distance between the geometric centre of the Ladoga Lake and the Baltic Sea is about 1200 km which is a considerable distance. So while implementing this method one should estimate the difference in the optical paths between the satellite and two different objects. This estimation and the relevant calculation were carried out and their results presented by Bychkova *et al.* (1989b). The mathematical equations were published later by Bychkova (1994). The assumption of horizontal homogeneity of the atmosphere was used.

To illustrate the method some results of our estimations follow. It was shown that if the value of the atmospheric correction as obtained from viewing the Ladoga Lake in nadir is small (2°C), the corresponding correction for the Baltic Sea area is also small: 2.4–2.8°C in the temperature range 2–20°C. In this case while viewing the Baltic in nadir one can use the same value of correction for the whole sea area within the accuracy of the method. For the value of the atmospheric correction over Ladoga of 6°C the values for the Baltic are 7.5–8.3°C for the same interval of measured temperatures. In this case one should take into account the actual variability of the temperature over the sea.

This technique was tested during the period 1986–1990 while processing AVHRR imagery of the Baltic Sea in the Laboratory for Satellite Oceanography, State Oceanographic Institute, St. Petersburg. The method was compared to the method of correction for the atmosphere based on regular *in situ* measurements at the networked stations (regression technique). The results obtained showed that the method of natural standards yielded SST values with an accuracy of 0.7°C while the accuracy of the regression method appeared to be 1–2°C (Bychkova 1994).

Thus during the 'hydrological spring' the suggested method of calibration of the satellite sensors of the IR band using natural 'standards' at the test areas enables us to carry out the correction of current satellite data for the atmosphere in operational mode without collection of regular *in situ* data, without atmospheric sounding from balloons, without using additional satellite measurements.

3.7 The use of satellite imagery of IR band in cloudy situations

3.7.1 Some notes on cloud-free situations in the Baltic region

As is well known, remotely sensed data of the IR band make it possible to obtain an instant spatial pattern of the SST distribution on this or that water area. For regions with slight cloud cover recurrence, some interesting observations of dynamical processes in different water areas has been done with the help of SST charts; see for example the study of frontal zones in the Mediterranean Sea (Champagne *et al.* 1982a; Millot 1982), the migration of rings of the Gulf Stream (Spence and Legeckis 1981), etc. (For more examples see Chapter 4.)

At the same time there exist regions insufficiently studied either by satellite or *in situ* means. This fact can be explained by the specific characteristics of such regions – dynamical processes there are, as a rule, local, sporadic and have a very short lifetime. Furthermore, the interpretation of satellite imagery in these regions presents certain difficulties because the images are often taken in cloudy situations. One such region is the Baltic Sea and its catchment area. (The discussion that is to follow may be interesting and useful for users in other seas; our approach to the problem seems to be universal, though its implementation may need modification, taking into account the local features of the cloud cover and oceanographic fields.)

The use of satellite imagery of the visible and IR bands in the Baltic region is limited by unfavourable meteorological conditions, in particular by high cyclonic activity. In addition to increasing cloudiness, it causes significant inhomogeneity of water vapour distribution in the atmosphere column over the sea and may result in the appearance of false thermal and brightness gradients in satellite imagery. It is especially difficult to acquire satellite imagery during a period of west winds as the western transfer of air masses over the Baltic Sea generally brings about the development of heavy cloud systems. All this makes the interpretation and analysis of satellite images taken by radiometers in partly cloudy situations in the Baltic Sea a difficult problem.

The regularity of getting satellite imagery of visible and IR bands strongly depends on the cloud factor. What do we actually know about the availability of cloud-free scenes in this region? Bychkova et al. (1986a) composed Table 3.4 presenting statistical information on cloud-free scenes according to various sources. Data on cloud-free AVHRR NOAA scenes for the Northern Atlantic (including the Baltic region) for 1980 as recorded in twice daily acquisition in Lagnon (France) were taken from Champagne et al. (1982b). The same table gives statistical data on cloud-free scenes for the Baltic Sea itself – as revealed from AVHRR NOAA data received in Dundee (Scotland, UK) in 1981–1982 (Horstmann 1983) and in Leningrad (USSR) in 1983–1984. The satellite data acquisition centre in Lagnon received 20 images suitable for further oceanographic interpretation each month on average. For comparison, there were 37 images a month on average for the Mediterranean Sea for the same time period (Champagne et al. 1982b).

The then existing sets of observations seemed to show that even in summer the monthly averaged quantity of informative satellite images of visible and IR bands for the Baltic Sea is no more than 6–7 for each spectral channel. In the years of most unfavourable cloudy conditions about a dozen images could be obtained (see the table data for 1981). Similar data were presented by Gossmann (1982): only 25 IR images of Europe in cloud-free situations were obtained from the AEM-2 satellite during two years. (The data were received twice daily; it is characteristic that

Table 3.4 Availability of cloud-free satellite images of the IR band for the Baltic Sea for the period 1980–1984 (Bychkova et al. 1986a). (Dash (–) means no data on regular satellite data acquisition were available.)

Year	1	2	3	4	5	6	7	8	9	10	11	12	Total
1980*	17	16	16	22	16	20	28	27	30	13	10	24	239
1980	–	–	–	–	5	9	4	3	4	–	–	–	25
1981	–	2	–	1	5	1	3	1	–	–	–	–	13
1982	1	–	–	4	7	6	7	7	4	4	–	–	40
1983	–	–	–	–	–	5	8	4	–	–	–	–	17
1984	–	–	–	2	5	6	–	4	2	–	–	–	19

* For the North Atlantic (including the Baltic Sea).

cloud-free scenes both at night and in the daytime were recorded in only two or three cases.)

When we analysed the data presented in Table 3.4 our feeling was that these data could not be considered as reliable and comprehensive as there were too many uncertainties in their evaluation. (For example, the dash in this table indicates that there was no information about whether the satellite data were received on a regular and routine basis.) Moreover they were probably incomparable as the techniques of their evaluation could be different, and we knew from our experience that the technique used could affect the results considerably.

So we tried to tackle the problem of the statistics of cloud-free scenes for the Baltic region and, as soon as it was possible, we managed to study the problem using our own long-term time series of observations. In 1986 the staff members of the Laboratory for Satellite Oceanography operated the autonomous satellite data acquisition station located in the village of Lesnoye near Kaliningrad. About 500 images from AVHRR NOAA for the spring–autumn period were received. These data brought new knowledge in the context of feasibility of regional satellite monitoring of the Baltic Sea. The availability of cloud-free days and cloud-free images of the sea is shown in Figure 3.18.

Figure 3.18 presents some real data on the availability of 'useful' scenes for this area. Our approach was as follows. We divided the whole Baltic Sea into five regions: the Gulf of Bothnia, the Gulf of Finland, the Gulf of Riga, the Baltic Proper and the southeastern part of the sea. Practically every day AVHRR data in APT mode were received from two NOAA satellites; images could be obtained from up to 8–10 satellite passes daily if necessary; expert estimates were made of the 'usefulness' and 'cloudiness' of each scene. A certain image was considered 'cloud-free' if there were no clouds at least in one of five regions. In this sense the upper part of Figure 3.18 shows the number of cloud-free days in each month, while the lower part presents the number of cloud-free satellite scenes. For example, in June 1986 only 20 days out of a total of 30 were cloud-free, and only 47 passes out of a total 103 recorded were cloud-free at least for one of the five regions.

We think that our approach is practical and local user-oriented. The results of our study show that the real situation with cloudiness on a regional and local scale is not so bad as it might seem. The results also teach us that one must not neglect even a single satellite pass – the image should be recorded even if it is very cloudy at the site where the satellite data acquisition station is located, as another region may be cloud-free. The results presented also support the idea of regional autonomous satellite data acquisition stations as a flexible tool providing access to data transmitted in APT mode. The over-centralised system of satellite data acquisition will never become so flexible and local user-oriented.

The year 1986 also saw the appearance of another updated estimation of the number of cloud-free images available for the Baltic Sea. Horstmann (1986) wrote:

> Cloud conditions over the Baltic Sea allow an annual average of 40 images to point out sea surface pattern of water color, that can be used to conclude on the horizontal distribution of suspended matter and chlorophyll ...
>
> SST images can be obtained with frequency of an annual average of over 120 scenes of the Baltic, respectively parts of it ...

This ends a brief review of the problem of cloud-free situations for the Baltic Sea. It was meant to show that the estimations based on various datasets may sometimes

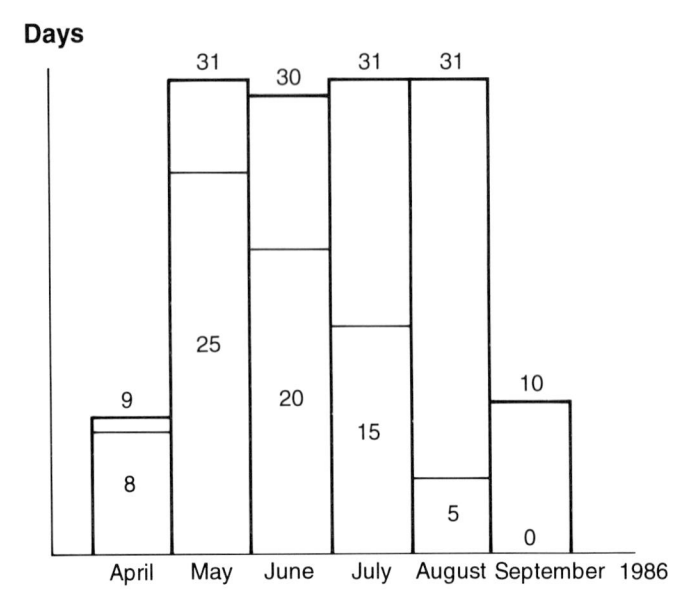

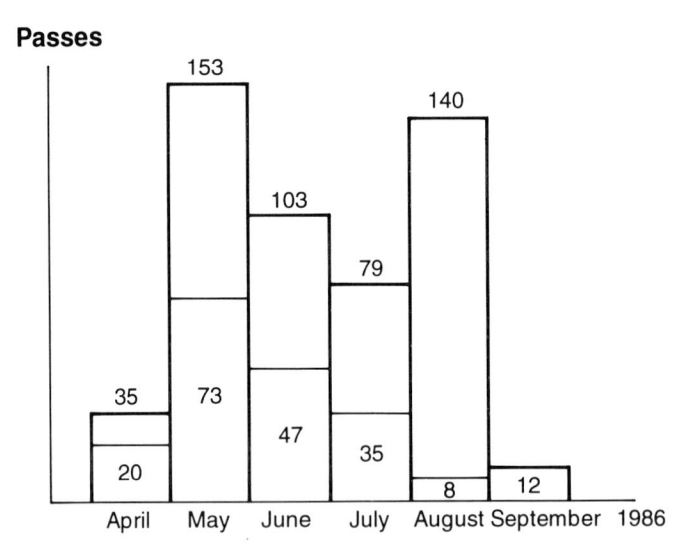

Figure 3.18 Availability of cloud-free satellite images of the Baltic Sea for April–September 1986 (Victorov et al. 1993a).

differ considerably; one should pay attention to the details of the techniques of those estimations.

It would be a mistake to believe that the remotely sensed data of the IR and visible bands, sporadic as they are, or seem to be, cannot be used for the study of water areas with high-frequency appearance of cloudy situations. For such areas it is necessary to use special satellite data processing techniques which make it possible to precisely identify the areas of heavy cloud and, which is most important, to reconstruct the real values of SST which were affected by (contaminated with) partial cloudiness. The spatial coherence technique developed by Coakley and Bretherton (1982) for the open ocean was modified by Bychkova et al. (1986a) to fit the problems of the Baltic Sea study and in accordance with the regional peculiarities of this sea.

3.7.2 Spatial coherence technique

As is shown by the experience of remotely sensed data processing, the class of radiative temperature SST in most cases can be described by multidimensional normal distribution functions. Hence it is necessary to divide such a class into subclasses with a single mode and nearly normal SST distribution functions. For example, for a cloud-free scene it could be the subclass of upwelling waters and the subclass of the waters in the central part of the sea, or the subclasses of warm waters and cold waters corresponding to the system of cyclonic and anticyclonic eddies. In the case of cloudy scenes it is necessary to distinguish between the class corresponding to cloud-free situations and the class corresponding to heavy cloud cover.

The pattern recognition approach is not suitable for the solution of the problem because the radiative temperature of the upper boundary of clouds is a very variable feature dependent not only on the type of cloud, but also on its height, on the temperature of the underlying surface located directly under the clouds, on the season of the year, etc. It makes creation of a reliable classification of cloudiness impossible. We believe that the most promising approach here is cluster analysis, ensuring that uncontrollable classification is especially effective in this particular case, as the number of 'teaching' selections (obtained from cloud analysis at the meteorological stations) is limited.

While analysing the data of radiation measurements it is expedient to operate not with the values of radiative temperatures themselves but with dispersions of temperatures at local squares, since different natural objects can have the same temperature, for example, cold water and low patchy cloud, or fog. The use of dispersions helps to reveal the natural structure of the data in the course of the cluster analysis. The analysis of different scenes shows that the dispersion for the regions with unbroken homogeneous cloud is of the same order of magnitude as for water surfaces with clear skies. Partial cloud over the sea surface causes a sharp rise in the temperature dispersion because the values corresponding to different distribution functions get within the same square. We come across a similar situation when the square under study includes both the water area and the land. The advantages of the spatial coherence technique will be illustrated by means of a case study of upwelling at the southern coast of the Gulf of Finland following Bychkova et al. (1986a).

3.7.3 Use of spatial coherence technique in the study of coastal upwelling

The upwelling along the Estonian coast of the Gulf of Finland in cloud-free situations was registered in satellite IR images in July 1980 and 1982 (see Horstmann

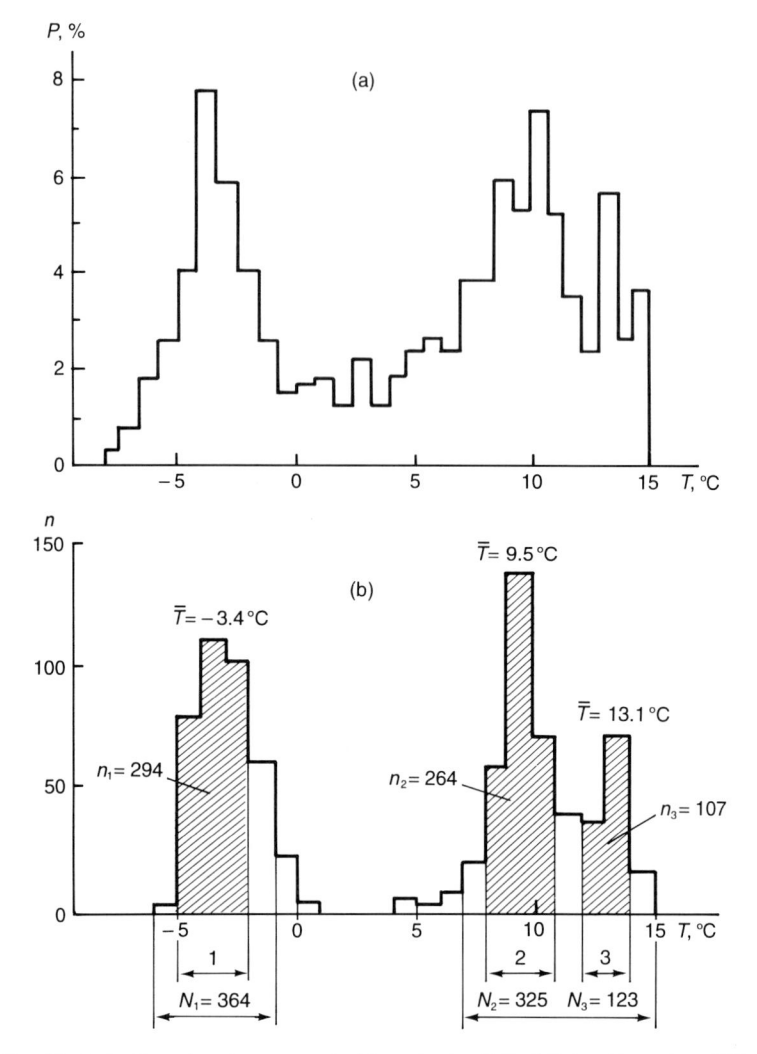

Figure 3.19 Clusterisation of various natural objects in the Gulf of Finland based on SST values and dispersion data. NOAA satellite. 30 May 1984. (a) Initial histogram of radiative temperature of underlying surface; (b) histogram of averaged values of radiative temperature in local squares of the scene with dispersion less than the critical rate. (Bychkova et al. 1986a.)

1983). Routine oceanographic measurements repeatedly registered the upwelling events in this area (Tamsalu 1979; Pappel 1983). In May 1984 using AVHRR NOAA data received at our autonomous satellite data acquisition station in Leningrad we managed to observe the development of the upwelling along the whole southern coast of the Gulf of Finland in partial cloud situations.

Application of the spatial coherence technique enabled us to single out three types of natural objects:

unbroken cloud	($-5°C < T < -2°C$),
open sea waters	($8°C < T < 11°C$),
warm coastal waters	($12°C < T < 14°C$).

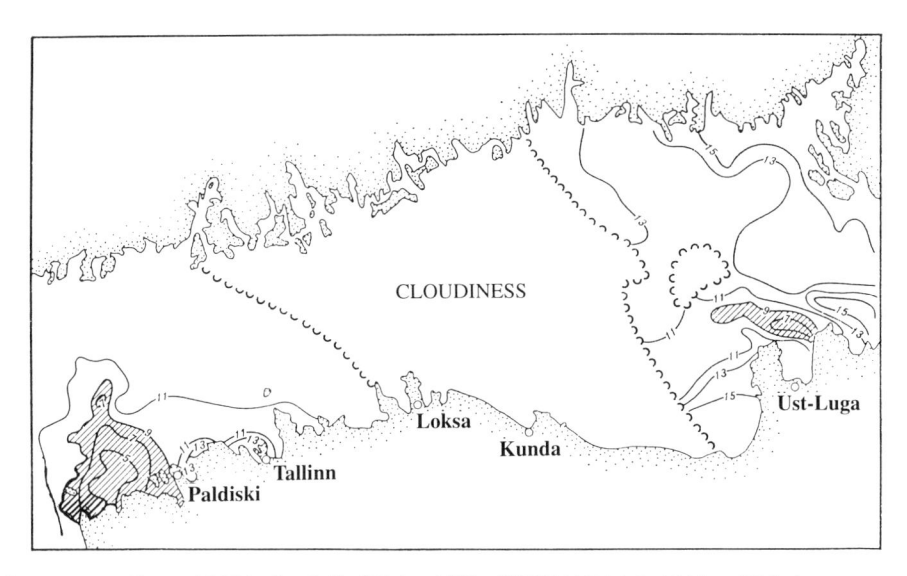

Figure 3.20 Chart of SST in the Gulf of Finland (°C). AVHRR NOAA-8. 30 May 1984. Autonomous data acquisition station, Leningrad. (Bychkova et al. 1986a.)

These three clusters are shown in Figure 3.19(b), with the initial histogram of SST shown in Figure 3.19(a).

The average temperature and the root mean square deviation were calculated for each square of 12 km × 12 km size, each of them having 15 values of SST. The critical value of the deviation revealed from the scattering diagram, amounted to 2.0 K. The clusters singled out by means of the spatial coherence technique are shaded in Figure 3.19(b). The total number (n) of SST values in each subband is indicated along with the total number (N) of SST values in each band symmetrical with regard to the histogram peaks. The decisive rule for settling out the cluster was: $n = 0.8 N$.

One of the peculiarities of the upwelling in the Gulf of Finland is the absence of a continuous front of cold waters at the southern coast of the gulf. The study carried out at the State Oceanographic Institute, Leningrad, by means of airborne IR radiometers in May 1984 showed that the upwelling waters are located in several regions: (1) from cape Ristna along the northern coast of Khiyumaa island, (2) from Piisaspea up to Paldiski, (3) in the region of Loksa–Kunda, (4) in the Luga Bay to the west of Kolgompya, and (5) in the Neva Bay near cape Shepelevo (see the section in Chapter 5 on upwellings). These locations of cold waters can be explained by the complicated structure of the coast-line. The upwelling was registered near the coast oriented in parallel to the wind at the relatively linear sections of the coast-line about 40 km long in the deeper parts of the gulf and about 20–30 km long in shallow water areas.

The development of upwelling in the Gulf of Finland on May 1984 was as follows. Cold water intrusion was registered on 26 May with the western wind 5–10 m s^{-1} at the Loksa hydrometeorological station (on 25 May the water temperature was 11.3°C, and on 26 May, only 4.0°C). The aircraft survey on 26 May registered

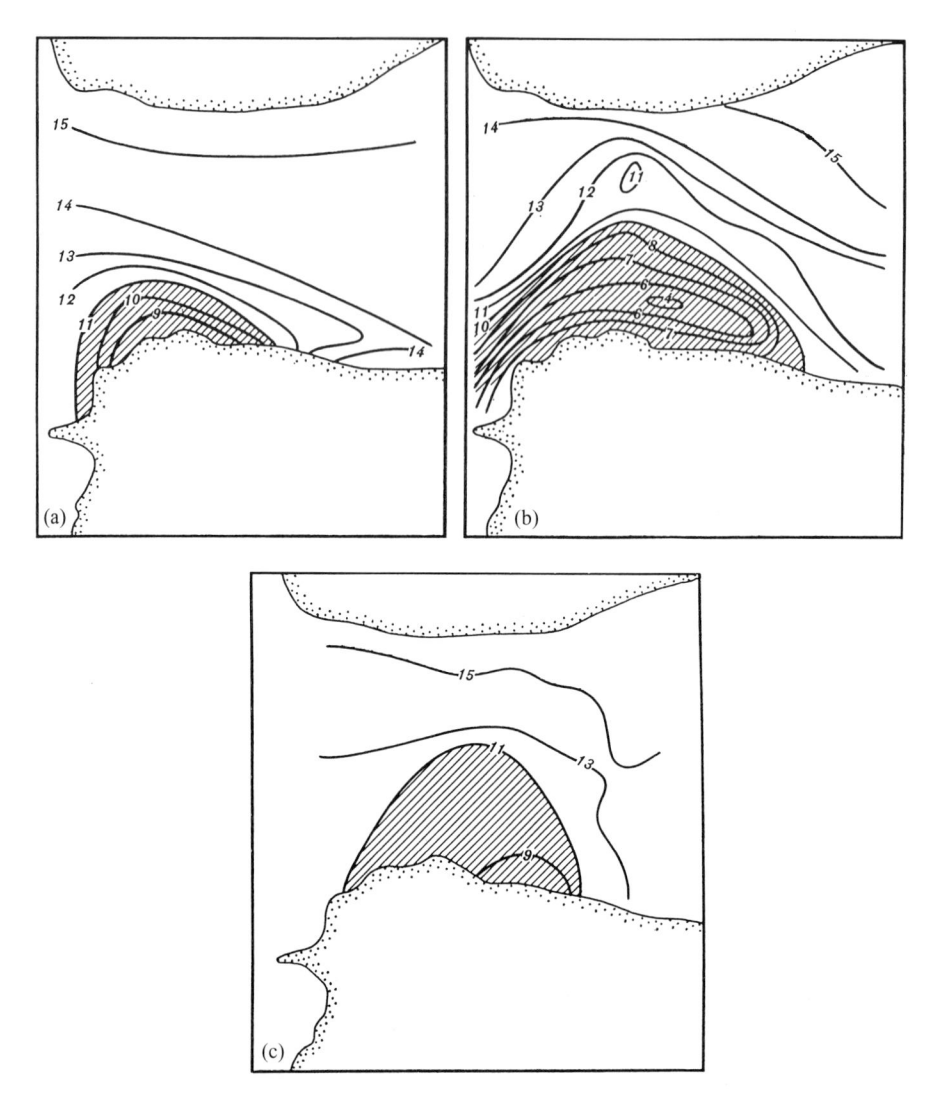

Figure 3.21 Upwelling event in the Neva Bay (eastern part of the Gulf of Finland) in May 1984. Figures show temperature in °C. (a) The beginning of the event as recorded during an aircraft survey on 26 May; (b) maximal phase of the upwelling development as recorded during an aircraft survey on 30 May; (c) maximal phase as recorded in the NOAA-8 image of the IR band.

the appearance of cold waters near cape Kolgompya where the temperature fell to 6.8°C (it was 12–15°C over most of the water area) and near cape Shepelevo (where the temperature was 8.1°C). Routine measurements at the coastal hydrometeorological stations and the research vessel *Orbita* enabled us to determine the phase of the upwelling maximal development on 28 May near Piisaspea (a minimum temperature of 5.6°C) and on 27 May near Loksa (4.5°C). The satellite SST chart was received on 30 May (see Figure 3.20).

The scene is heavily contaminated with clouds. Still this image with partial cloudiness could be analysed using the spatial coherence technique. Two zones of upwelling were identified.

Cold waters near Piisaspea–Paldiski showed gradients in the frontal zone of 0.5–1.0°C per km. Cold waters spread into the Gulf as far as 40 km. The location of the front as registered in the satellite image coincided with the synchronous airborne survey chart within 4 km accuracy.

The second zone of upwelling in the western part of the Neva Bay is shown in Figure 3.21. In the figure the process of the upwelling development is presented for the period 26–30 May. Both satellite and airborne data were used to reconstruct the process of upwelling in this region. It is interesting to note that visual observation from aircraft showed that the temperature front was also the border between two water masses with different colours. This proves that the upwelling water mass contained deep waters with higher concentration of nutrients.

The upwelling event in the Gulf of Finland in May 1984 as a whole lasted for seven days from 26 May to 1 June. The satellite data could be used in the study of this event only because the spatial coherence technique was implemented in order to make clear the SST chart with partial cloudiness. The spatial coherence technique was used in satellite data processing regarding the open ocean (see, for example, Coakley and Bretherton 1982; Darnell and Harris 1983). Bychkova *et al.* (1986a) showed that this technique can be used in more complicated cases of a semi-enclosed sea with a cut (chopped) coast-line generating fine features in the SST field. Cluster analysis based on the spatial coherence technique enables us to divide the SST data into several classes: land, water and cloudiness of various types. If necessary, one may divide a class into subclasses, e.g., class 'water' can be subdivided into subclasses of 'warm coastal waters', 'cold waters of the open sea', etc.

One should remember, however, that the spatial coherence technique is effective for homogeneous and extensive cloudiness with the number of cloud-free elements not less than 10% of the total number of pixels within water area. In the case of two- or three-layer cloudiness it is expedient to use the spatial coherence technique for two or three channels of the radiometer simultaneously (Coakley 1983). In general the spatial coherence technique increases the effectiveness of the use of remotely sensed data of the IR band in regions with high-frequency appearance of cloudiness situations owing to the lower number of rejected scenes and higher reliability of the results of satellite data analysis.

3.8 The data processing, database and GIS approach in regional satellite oceanography

3.8.1 On image processing

As was stated in the Introduction to this book, it was not the author's intention to present here the algorithms of satellite digital data processing or the relevant procedures of image processing and pattern recognition. A lot of good textbooks exist on various aspects of this problem area, and even a comprehensive review of various techniques involved would comprise another book. Hence the following 'notes' should be considered as rather an introduction to the problem, with some remarks on the oceanographic peculiarities in their regional context.

The whole scope of operations known as 'image processing in remote sensing' can be presented as a three-step analysis consisting of '(1) detection and delineation,

(2) classification, and (3) identification' (Elachi 1987). Let us briefly discuss each of these steps in the RSO context.

Detection One can easily see three major *procedures* in tackling remotely sensed data.

1. *Rearranging* the image(s), transforming the two-dimensional image(s) into another projection to adjust other materials (images, maps, etc.) available for this region. The 'density slicing' technique (a presentation of a digital image not in the whole range of numbers but in a selected range of digital numbers) can also be regarded as part of this procedure. The procedure can be combined with precise geocoding of the image using reference points with well-defined geographic coordinates. Various modes of data presentation (displaying) are also included in this procedure.

   If the same satellite payload is used to provide imagery for users interested in land sciences applications and for marine sciences applications (as was the case with RESURS satellites in the former USSR) the ground-based centralised facilities for satellite data acquisition and storage (with no digital processing) as usual were tuned to provide good imagery of land, which is much brighter than the sea. If many levels of grey scale are allotted for land, the sea appears black in almost all the images, with no details and no features on the sea surface. 'Density slicing' in its analogous form could have been applied to the satellite signal in order to select only the lowest part of it (corresponding to the sea) with subsequent expanding of this part of the signal for the whole dynamic range of the recorder.

2. *Authorised editing* of the image with an attempt to find some patterns and features which an analyst is looking for (preoccupied, predetermined vision, when an analyst is actually preprogrammed/trained to look for certain patterns). Another option is that analysts may find something that may trigger a cell in their knowledge base (either in their own memory, in their brain or in the artificial computerised memory) suggesting that they have found something that might be relevant to the information recorded in that cell, thus initiating the analytical process of finding associations, etc.

3. *Overlapping* (or *superimposition* of) the images, combining them with other data and subsequently manipulating multilayer imagery with the aim to prove the knowledge obtained in the analysis of a separate image or to improve the presentation of this knowledge or to obtain completely new knowledge (synergetic effect). (Some authors call this procedure 'image fusion'.) A good example of this procedure is the combination of satellite imagery with digital elevation maps resulting in a product giving an impression of a three-dimensional scene.

In most cases the combination of hardware and software is not able to bring more information on the image – in terms of *detection* – than can be obtained when thoroughly viewing it. But still it is wise to check whether the feature or detail or a patch you want to detect is actually recognisable, or whether the detail or a patch you want to ignore can be neglected. For that purpose you can use all the available mathematical algorithms and techniques and very sophisticated software. In the case of multilayer information (multichannel imagery or a set of images from various sensors of different bands) the ability to use comprehensive technical facilities with sophisticated software becomes vital.

The high dimensionality of imaging spectrometry data makes it necessary to seek new analytical methods which are *deterministic* in nature in comparison to *statistical* techniques commonly used in processing of images with few spectral band. (Elachi 1987)

With multidimensional imagery it is reasonable to operate in a transform domain using such deterministic transforms as Fourier, Walsh–Hadamard, Haar, or Chebyshev polynomial transforms.

Two basic operations are to be done in the course of image processing at the stage of *detection*: image enhancement and image smoothing. For both operations it is useful to generate one or more histograms for an image from the digital data. When a person looks at the image with some interest, actually a number of spots are being indicated and selected for some reason. Often a histogram is just a visual pattern (or a sign) showing how various elements of an image are distributed in space; this is especially the case for simple one- or two-dimensional histograms. A histogram of an image (or of its part) can be used as an indicative characteristic of an image.

Depending on your intentions you may try to get rid of some details (patches) in the image or/and you may want to highlight some other details (features, patches) in the same image. In the second case you may implement a 'contrast enhancement' technique. Depending on the histogram various types of transfer functions may be used to transform the initial image into the modified one with better contrast(s). The 'edge enhancement' technique may be applied to increase contrast in the boundary zone between two neighbouring areas in the image with different values of 'brightness'. Thus the boundary becomes sharper and more recognisable. A clear explanation of some image transform techniques is given by Cracknell and Hayes (1991) with examples illustrating the results of the implementation of different techniques.

Classification After some units in the image (details, features, patches) have been detected and delineated, the next step is to classify them using some sets of criteria. This is a common procedure in all branches of science based on involvement of remotely sensed data. Hence in oceanography and related sciences one may use routine classification techniques with correction for higher temporal variability of oceanographic processes as compared to those on the mainland. Perhaps the simplest example of the classification of a satellite image is its segmentation into three classes: the land, the clouds and the sea. Quantitative analysis of the spatial distribution of suspended sediments and algae in the coastal waters of the sea will need more sophisticated techniques of classification.

Identification Here we may speak about the unique identification of the delineated structures from two viewpoints. On the one hand, one may try to identify the type of substance that formed a patch on the sea surface. This will require a good knowledge of spectral signatures of all the suspected substances and precise measurements of spectral characteristics of the patch (including correction for the atmosphere). This is a complicated task: in many cases it is much easier to produce a chart with various levels of concentration of a *known* substance. (This problem is similar to the problem of *classification*.)

On the other hand, one may *identify* certain types of oceanographic *phenomena* in the satellite imagery of different bands. To some extent the problem is simplified because there is a restricted number of oceanographic phenomena that manifest themselves in remotely sensed imagery of various bands. (Some examples will be presented in Chapters 4 and 5.) But if a combination or superimposition of various phenomena is recorded in the image or a process – that is, the dynamics of a certain phenomenon – is to be studied, these complicated tasks will require an adequate level of modelling and comprehensive satellite data assimilation procedures.

These problems relevant to *identification* actually need the use of databases and knowledge bases, either in the form of the analyst's experience or in the form of specially designed data and knowledge bases as part of Geographical Information Systems (GIS).

3.8.2 On databases and GIS

The GIS world is becoming a useful (though sometimes just a fashionable and sometimes a self-consistent) tool to manage data in many branches of science, industry and the economy on the international, national, regional and local levels. There are many definitions of GIS. Nowadays it is common knowledge that GIS is a tool for the management of large amounts of geo-referenced information. GIS is a combination of hardware and software meant to integrate datasets and knowledge on a certain region providing an interactive mode of operation. According to Marble (1980) GIS is a system comprising four necessary subsystems: (1) subsystem of data input; (2) subsystem of data storage and search; (3) subsystem of data manipulation and analysis; and (4) subsystem of data input.

For some basics of GIS see Marble (1980). Interesting applications of GIS are presented in Hogg and Gahegan (1986), Guptill (1989), Ball and Babbage (1990), *Landscape Ecology and Geographical Information Systems* (1993). The problem of time in GIS is discussed in Langran (1992). Oceanographic GIS are presented in brief in Bivins and Palmer (1989).

It seems worthwhile to note that in the late 1970s–early 1980s the term 'Geographical Information System' was not known to remote sensing and oceanographic communities in the former USSR but all the basic elements of GIS could be seen in some papers published at that period. In this context let us briefly review a paper by Victorov and Tishenko (1982) dealing with the design of the *in situ* data 'bank' meant to be used in an interactive system of satellite oceanographic information processing. The authors used the term 'bank' and not 'base' to suggest a structure in which two types of data are circulating.

- Type 1 data – the stored (a) *in situ* sea-truth data collected at the special test areas and test sites, and (b) *a priori* data characterising the region: maps, annual reports, various tables of oceanographic and meteorological records, etc.

- Type 2 data – the professional knowledge, experience and even intuition of experts helping them to reach heuristic decisions. This information as a whole can hardly be formalised, though some items could be formulated as a set of simplified 'decision rules' and could become part of the data bank. Reviewing this part of the paper by Victorov and Tishenko (1982) it is strange to note that some 12 years later Wei Ji *et al.* (1994), describing the so-called 'innovative decision support methodology for coastal and marine environmental resource management', wrote that in their system '... domain experts' knowledge is represented as explicit rules and encoded into the database to provide knowledge-based decision support functions' (Wei Ji *et al.* 1994).

Victorov and Tishenko (1982) insisted on the system of oceanographic satellite data processing being interactive, with an experienced oceanographer being part of the entire informational structure (Figure 3.22).

The oceanographer–operator in such an interactive system is not the same as the ordinary operator of a mainframe computer. This person was meant to be an experienced professional oceanographer specialising in some type of oceanographic phenomena or processes or, which is more realistic, having a good knowledge of a region. This expert could carry out the so-called 'supervised classification' (an image processing technique based on the process of 'teaching' the computer by an expert). This expert could also provide manual detection, classification and identification of objects and phenomena in the so-called 'sample frame' (representative segment of the whole frame) with the subsequent spreading of the results over the whole frame. What is 'representative', how to select the 'sample frame', and whether the results could be spread, are questions that could be answered only by the expert in this particular region.

According to Victorov and Tishenko (1982) the *data bank* itself consists of the stationary and the operational modules, both connected to each other, with the manager–oceanographer, and with the satellite data processing system. The informational content of both modules is discussed using SST data as an example.

The stationary module contains:

- data of the ground-based calibration of satellite sensors;
- a digital map of the coast-line;
- climatic values of corrections for atmosphere for various models of atmosphere, different seasons, and regions;
- threshold levels to discriminate cloudiness, a set of histograms characterising typical cloudiness situations for the region, data on the cloudiness trends;
- measured values of SST collected at the previous satellite overpasses;
- tables with mean SST values (monthly and seasonal averages), data on SST trends and gradients;
- tables (or set of equations) connecting radiative and thermodynamical temperatures of water in the region at various seasons;
- coordinates of test areas and test sites where the sea-truth *in situ* data are being collected during this or that time interval;
- digital 'blank maps' of the region;
- encoded users' requirements for the output informational products and the users' addresses.

The original paper which is being reviewed contains also a detailed scheme of data flow with 13 stages of data processing and automatic and manual control. Here it is interesting to note that Stage 8 provides control of the output informational product based on a general knowledge of spatial and temporal variability of oceanographic parameters in the region and on a knowledge of the basics of oceanography, which corresponds to the modern term 'knowledge base'.

Part of the raw and processed current satellite data is to be transferred from the 'operational' module to the 'stationary' module, thus continuously supplying the database with new knowledge. Hence SST data obtained at the nth pass will be

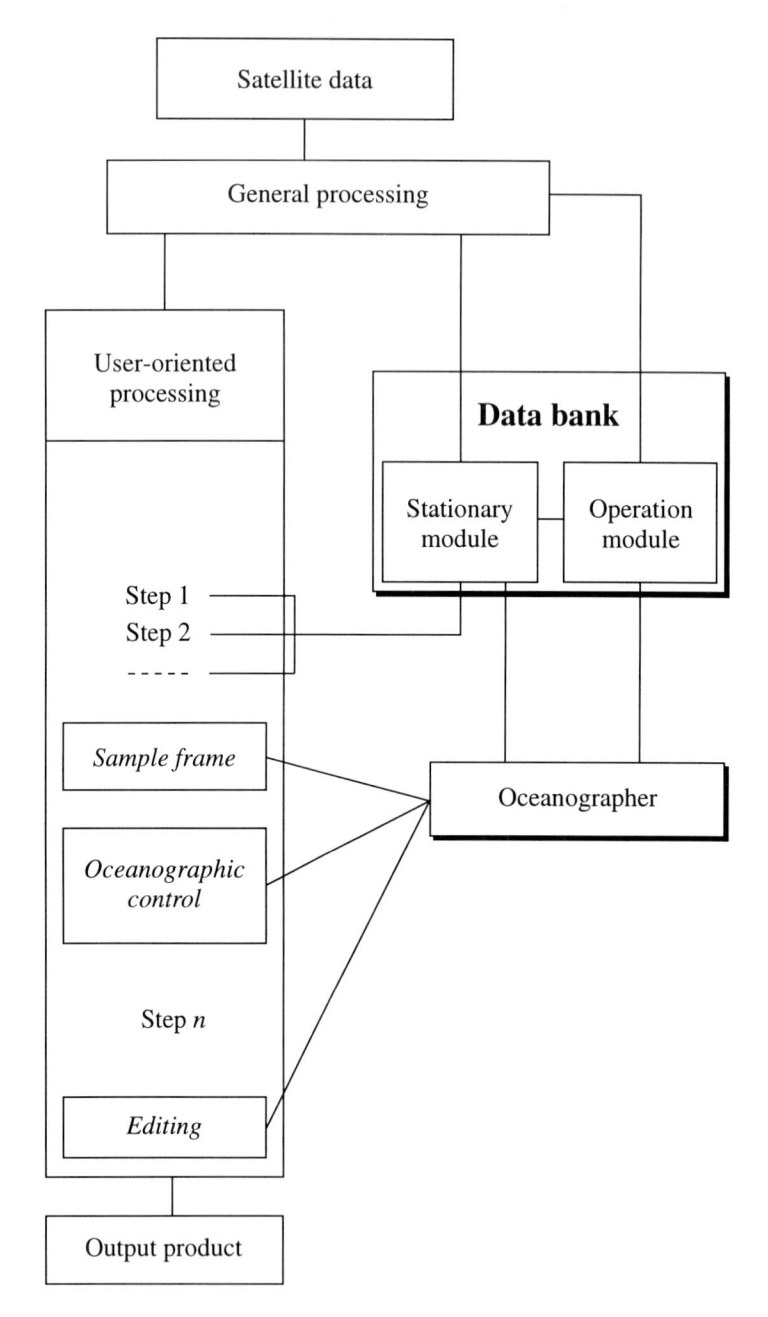

Figure 3.22 Data flow in an interactive system of oceanographic satellite data processing with a data bank (after Victorov and Tishenko 1982).

transferred from the 'operational' module into the 'stationary' module of the database before processing the new satellite data obtained at the $(n + 1)$th pass.

Of a total of 13 stages of data processing, nine are based on the oceanographer's personal involvement which once more supports the idea of participation of the oceanographic community in the design and running of an oceanographic satellite system presented in detail in Chapter 1.

This review shows that in the absence of a knowledge of what the terms 'GIS' or 'knowledge base' mean, just based on the systems analysis approach and 'common sense', one was able to design a system of satellite data processing and analysis in which different types of information bases (databases and knowledge bases) have been properly incorporated.

Nowadays the concept of GIS is commonly accepted, as well as the role of GIS in operational use of satellite oceanographic data. To be more exact, GIS are being considered as the structure, the framework providing proper integration of remotely sensed data with *in situ* data and conventional knowledge. Such GIS, comprising both remotely sensed data and all the other types of data, are sometimes called 'integrated GIS'.

GIS appears to be a suitable vehicle to develop remote sensing and to implement satellite information in many branches of science, industry and the management of natural resources, to move into the existing structures, to penetrate into them. As discussed in Chapter 1, the large amounts of remotely sensed data could not be tackled using conventional tools and techniques, and something should have appeared to help. This 'something' actually appeared, and its name was GIS.

In fact GIS help to use satellite information in regional oceanographic studies and in the management of marine resources in coastal zones (some examples will be given in Chapter 4).

3.8.3 On technological aspects of image processing and the running of integrated GIS

Using the proper combination of hardware and software is vital for activities in regional satellite oceanography. Nowadays it is just a problem of adequate finance to obtain sophisticated software packages meant specially for satellite data processing. I cannot use this book to advertise the products available on the market. In this respect it seems reasonable to present a quotation:

> At the top end of the market there are now a large number of proprietary turnkey image processing systems that are readily available, are competitively priced and well supported in terms of maintenance of the hardware and software on a worldwide basis. At the lower end of the market the recent developments in terms of microcomputers means that one can now put together a reasonable image processing facility with a microcomputer. Indeed microcomputer-based systems are approaching the situation in which they can be used as serious research machines in lieu of some of the very expensive proprietary systems. (Cracknell and Hayes 1991)

So it is the choice of regional/local users to choose their way of setting up and running their system. Both of the above-mentioned approaches have their own philosophical reasons. Interesting discussions on this topic appear from time to time in various journals (e.g., in *Earth Observation Magazine*).

There was an attempt in the former USSR to pave our own way in image processing technology. The well-known lagging behind of this country in computer techniques and informational technologies revealed itself in this issue too. The idea to create an up-to-date operational computer system for satellite data processing 'on-line' (Yu. Khodarev, the late B. Nepoklonov and A. Tishenko with colleagues,

State Research Centre for Studies of Natural Resources, Moscow – the then main contractor for the Earth Observation System) with some modules common to both the land and oceanographic information, was not a success, though in the course of these efforts the image processing computer system was designed in collaboration with German experts. Some tens of these 'DISK' minicomputer-based stations have been manufactured by VEB 'ROBOTRON' in the former German Democratic Republic, which made it possible to install this equipment in many research institutions engaged in remote sensing activities. The software package EURICA (the late S. Sazhin with Russian and German colleagues) was also provided for users. This was indeed an enormous step forward in terms of the development of remote sensing activities in various research institutions. This system was not meant for operational purposes. 'Unfortunately the present national level of operational satellite data handling is still far from modern requirements' (Victorov 1992a). Later this statement was once more backed by the document of the Russian Federal Agency for Hydrometeorology and Environmental Monitoring Board meeting held on 10 November 1993 (quoted in Section 2.4 of Chapter 2). With the access of the former Soviet researchers to personal computers the remote sensing and the oceanographic communities have naturally found themselves in the same technological environment as their colleagues elsewhere. So nowadays it is their choice (driven by the available finance) to use one of the two options mentioned by Cracknell and Hayes (1991). It seems to me that the tendency is to use worldwide known market products (both hardware and software), though there are still attempts to manufacture both using imported parts. As far as I know there were no successful attempts to create home-made integrated GIS packages.

It is not so easy to set up and to run an integrated GIS. Provided that you can afford to buy the software package, a number of problems are still to be solved: filling the integrated GIS structure with satellite images and relevant *in situ* data; design, creation and installation of your unique knowledge base and modelling module; testing and tuning the system; and training personnel. Nevertheless the impressive examples of regional integrated GIS in Chapter 4 give strong evidence that if many satellite images are to be involved in the joint analysis with three-dimensional *in situ* data, only modern sophisticated GIS are able to provide volumetric (three-dimensional) visualisation and quantitive analysis needed for the comprehensive interpretation of such complex data. Moreover up-to-date GIS provide the 'animation sequence' technique thus helping to display tremendous amounts of data in a most effective way (Irwin and Manley 1994).

CHAPTER FOUR

Regional satellite oceanography in action

In this chapter a brief review of worldwide activities relevant to regional satellite oceanography will be presented. Firstly, several characteristic examples of multidisciplinary or topic-oriented systems (of national sounding) will be given, then some integrated regional GIS and satellite databases will be described, mostly for coastal zone management applications. Finally, the most remarkable and characteristic examples will be presented showing the utilisation of satellite data for studies and monitoring on a regional scale of specific oceanographic phenomena and coastal zone features.

4.1 Regional systems

In this section the Norwegian operational ocean monitoring and forecasting system (HOV), the Russian space-based sea ice informational system and the satellite-borne data implementation within the framework of the US Coast Guard will be described. Dutch and Swedish operational systems producing informational products for regional marine applications are also presented here.

4.1.1 Norwegian operational ocean monitoring and forecasting system (HOV)

A brief discussion of the general outline and satellite segment of HOV is based on the presentation by Guddal and Strom (1993). This system is meant to harmonise the collection, processing and assimilation, in various models, of data relevant to

- marine meteorology (winds/pressure, heat fluxes, radiance, marine precipitation);
- physical oceanography (waves/swells, tides and surges, currents, ice parameters);
- water quality (discharges, transport and consumption of nutrients and pollutants);
- biological processes (including algae blooms and bacteria).

There are three different time horizons: monitoring (or 'nowcasting'); forecasting (1–10 days); and evaluation of climatological trends (months to a century). HOV covers three types of areas: *large area* (all ocean and sea ice areas north of 50° latitude, east of 30° W longitude and west of 50° E longitude, except for the Baltic Sea); *coastal regions* (200–500 km; along the coastline, including the island of Spitsbergen); *exposed areas* (typical extension 100–200 km; area is defined as one exposed to storms, or hosting very sensitive sea operations, or threatened by oil spills, etc.).

HOV is in a preliminary operational mode and acts as a national central coordinating body, a 'synthesizer and value adder' to data from a number of oceanographic data sources transforming the different data contributions relevant to marine meteorology, physical, chemical and biological oceanography into user-oriented informational products. 'In fact, HOV can be regarded as a national equivalent to GOOS, and indeed one of only few such systems in the world today' (Guddal and Strom 1993).

It is worthwhile to note that 'HOV is a rather unique example of a satellite data user; it has already a comprehensive provision of *in situ* ocean data, it has a suite of numerical models that are interconnected and, thereby, HOV can rationalize the needs for satellite data' (Guddal and Strom 1993).

I would add that HOV is a very captious customer. The satellite data will be assimilated in the system only if the new data:

- replace or supply the existing data sources (especially when the conventional data are inefficiently delivered, are irregular, or too expensive; examples: better details of the marginal ice zone, more exact wave spectra during storm events);
- give birth to new informational products and services that cannot be provided by HOV on the basis of conventional data (examples: SAR data, SeaWiFS data);
- improve the total data acquisition system efficiency (better temporal resolution, larger areas covered, etc.).

HOV has a well-defined set of output informational products relevant to each of four major ocean disciplines (and their combinations); there are strong requirements on quality of data. The efficiency of the system in terms of 'costs and benefits' is clearly assessed. The end-user community (including the offshore petroleum industry and the growing fish farming industry) is established, and their needs are well defined. All these help to

demand the following features of the satellite data acquisition:
- the data give the correct parameter information in accordance with the HOV service it is intended for;
- the data are timely delivered, i.e., in real time;
- the data give 'cost/benefit' effects in relation to alternative data acquisitions;
- the data are delivered in a regular and reliable system of communication. (Guddal and Strom 1993)

In the Annex to their presentation Guddal and Strom (1993) give a list of output informational products of the HOV system with the definition of the product, the state-of-the-art of HOV regarding this product, the needs for satellite data, assessment of the product efficiency and priority.

Although still in its pre-operational phase the Norwegian ocean monitoring and forecasting system HOV seems to be a good example of a solid framework for assimilation of current and future satellite oceanographic data.

As sea ice monitoring is an integral part of the HOV infrastructure, it was natural that, with the appearance of an experimental satellite ERS-1 with a SAR instrument, a special real-time sea ice monitoring pilot project was started with the purpose 'to show potential Norwegian users of ice maps how ERS-1 SAR data can be used to improve the monitoring of sea ice' (Sandven *et al.* 1993). It is worthwhile to recall that to date the ice maps of the Barents and Greenland Seas are based on AVHRR NOAA and DMSP SSM/I data and *in situ* observations and are issued once/twice a week by the Norwegian Meteorological Institute. Also the Navy–NOAA Joint Ice Center (USA) issues ice maps of this region (Sandven *et al.* 1993).

The demonstration project was based on the Tromso Satellite Station's (TSS) unique capability of acquisition, processing and dissemination of up to ten SAR scenes per orbit covering about 1000 km × 100 km area in total. In autumn 1989 the author had the pleasure of visiting the TSS located in a romantic site on a hill among pine trees. The high latitude position of TSS

> makes it possible to read out data from 10 out of 14 orbits per day from a typical polar orbiting remote sensing satellite. The station is covering the important areas of Southeast Greenland, Arctic, Barents Sea, North of Russia, the North Sea, the English Channel ... TSS is today the fastest ground station in the world for acquisition, processing and distribution of ERS-1 SAR data. TSS is also receiving data from NOAA AVHRR, ERS-1 ATSR and JERS-1. Future missions on radar and ocean colour are also planned to be read out at TSS, to be able to provide a multi-mission service to marine applications. (Landmark and Hamnes 1993)

The demonstrational project was carried out in January–August 1992, with SAR raw data received and processed to a 'low-resolution' (100 m) informational product at TSS. The latter transmitted this product via conventional ground-based data networks to the Nansen Environment and Remote Sensing Centre (NERSC) in Bergen where quality control, analysis, interpretation and fusion with other data were performed. The distribution of ice charts to ships was made via the INMARSAT telefax service (Sandven *et al.* 1993). The test areas included the Greenland Sea, the areas near Svalbard, Franz Josef Land, Novaya Zemlya Island, the Kara Sea, and the Barents Sea.

As a result of the project the general conclusion was made that ERS-1 SAR imagery could be very useful for 'special monitoring of smaller regions to assist icebreakers, fishing vessels, tourist vessels, in offshore drilling, pollution control, rescue operations and other marine activities in the Arctic' (Sandven *et al.* 1993). This rather obvious conclusion was complemented with recommendations which, in the author's opinion, were more important and useful. They dealt with the improvement of the ordering procedure for ERS-1 SAR data (a month in advance term was considered as 'difficult' for a customer), the problems of optimal data fusion from various satellites and sensors to provide mosaics for larger-scale ice mapping, the development of data transmission techniques to ships (providing transmission of good quality *images*, not only schematic ice charts), and other practical issues (Sandven *et al.* 1993).

4.1.2 Russian space-based ice information system

The system was presented by Nikitin (1991) and, in brief form, was described by Victorov *et al.* (1993a). This system consists of three main parts:

Figure 4.1 Western part of the Arctic. Satellite image of 12 June 1981. METEOR-2, visible band. (Bushuev 1984).

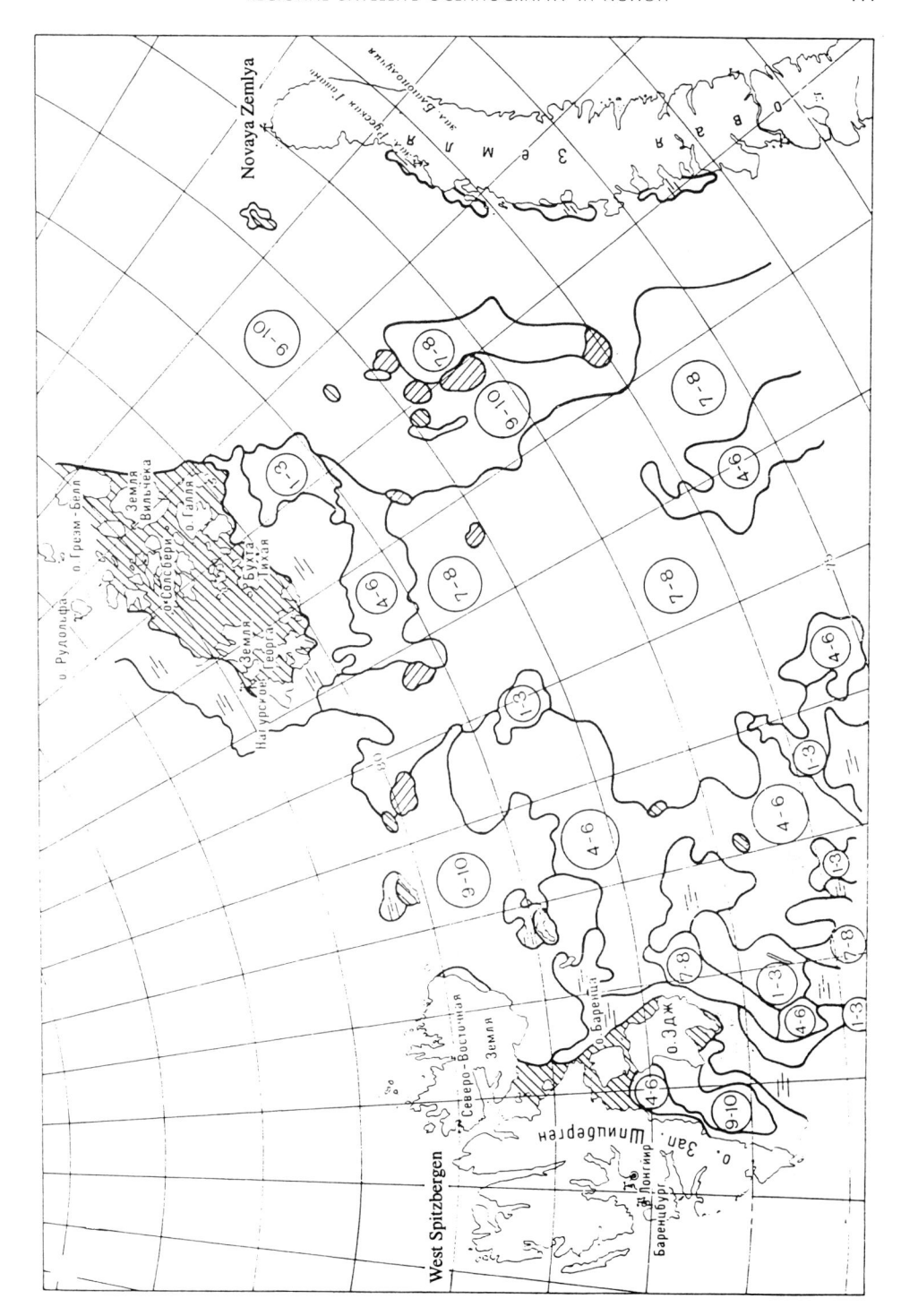

Figure 4.2 Western part of the Arctic. Ice chart based on the satellite image of 12 June 1981 (Bushuev 1984).

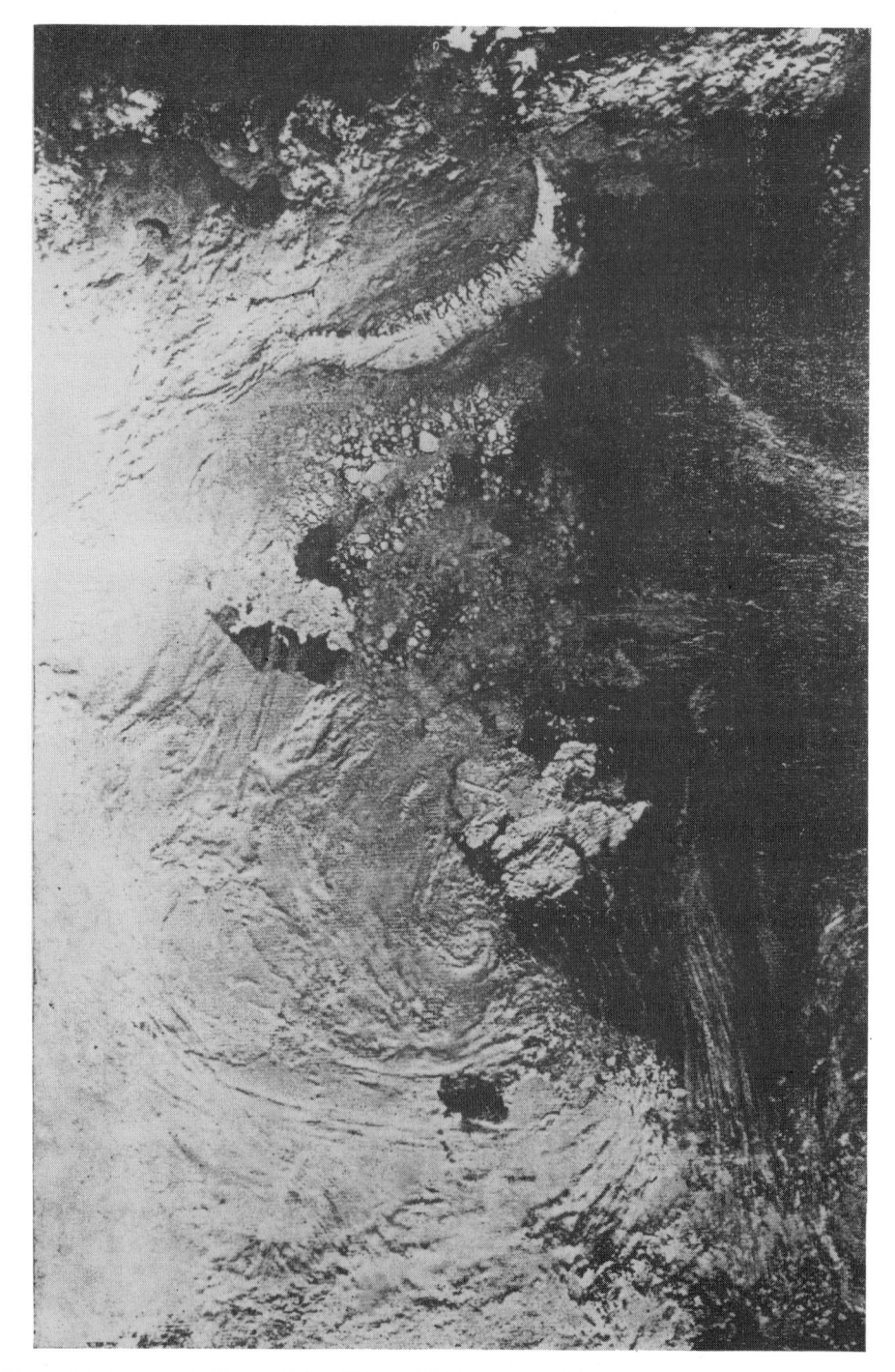

Figure 4.3 Same as in Figure 4.1 for 17 June 1981 (Bushuev 1984).

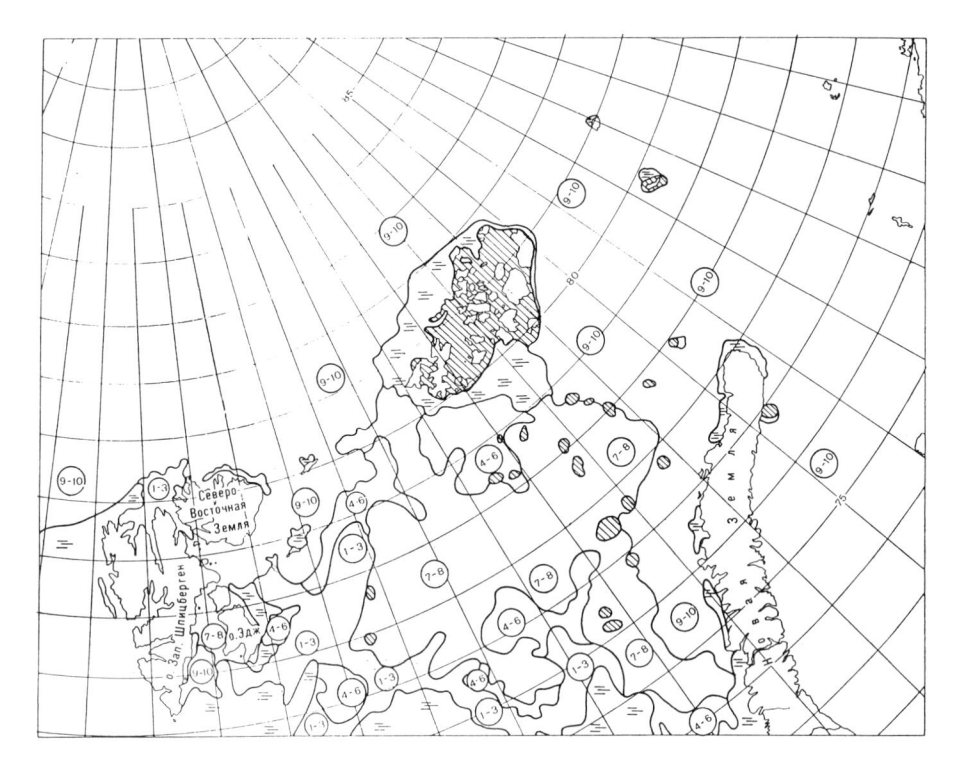

Figure 4.4 Same as in Figure 4.2 for 17 June 1981 (Bushuev 1984).

1. satellites as the source of raw data;
2. a ground-based analytical centre;
3. communication links (either ground-based or via satellites), including end-user equipment in the Arctic.

A brief description of the technology used in the system follows. Raw information from various types of satellites is used as the input products (radar imagery, visible and IR imagery, from satellites of the OKEAN, KOSMOS and NOAA series). Data processing and analysis are performed using the interactive image processing technology consisting of the following steps:

1. general analysis of scene, quality assessment, selection of the most interesting zones;
2. optimisation of visual presentation: modification of histograms, palettes, etc.;
3. segmentation of tones: smoothing, threshold transformations, construction of isolines;
4. identification of special structures: spatial differentiation and lineament analysis;

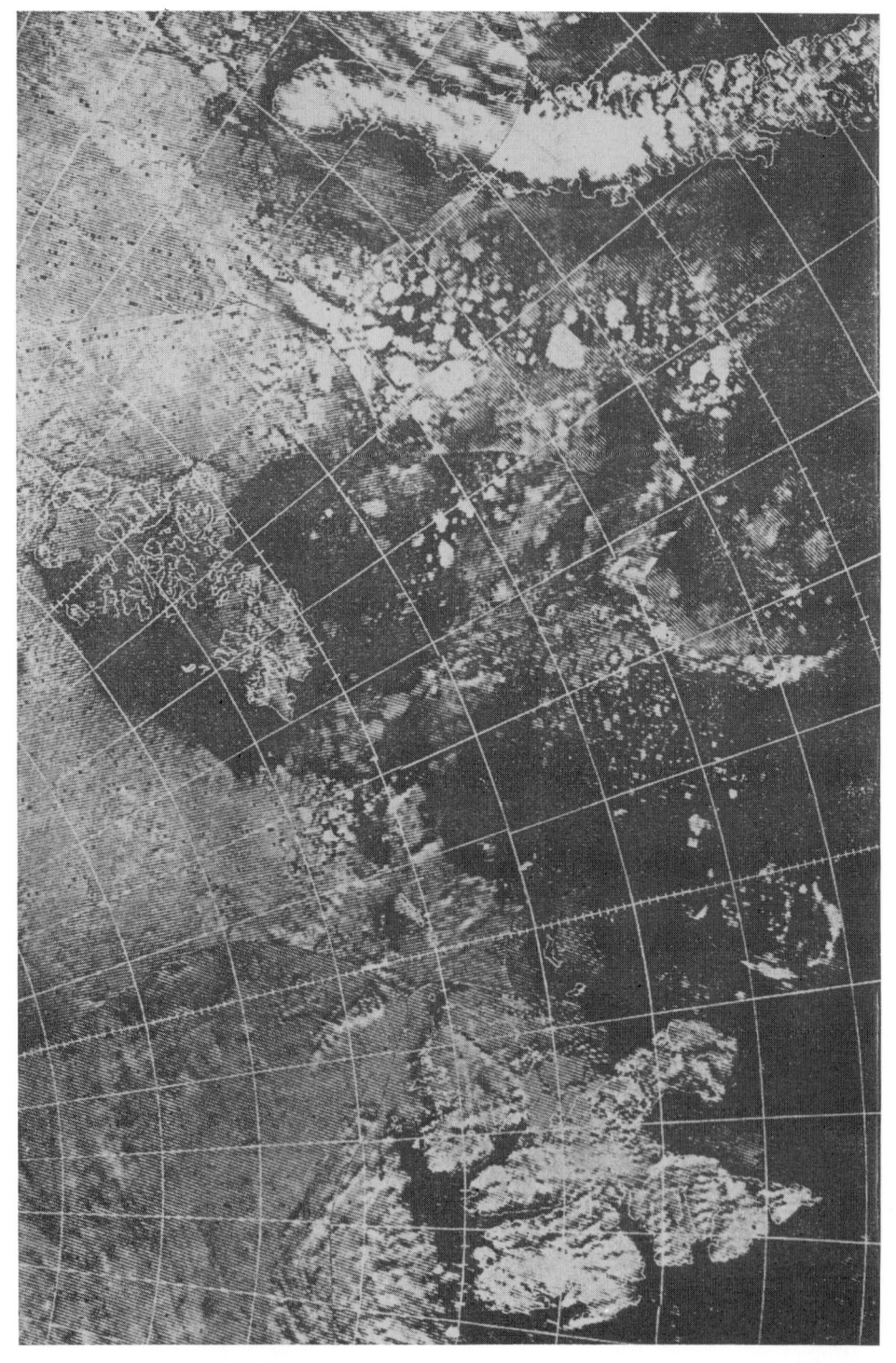

Figure 4.5 Western part of the Arctic. Ice chart (photographic mosaic) based on satellite images of 17–18 June 1981 (Bushuev 1984).

5. integration into chart format: transformation into a chart projection, superimposition of a coordinate system and a coast-line;
6. thematic analysis and editing based on all the available operational data and the information from the database;
7. production of annotations – short items of text attached to the output imagery, explaining their features.

The output products of this system are sets of colour display frames and comments. They are transmitted over conventional satellite TV channels (EKRAN, Moscow) or electronic communication channels (GORIZONT, INMARSAT) and thus may be made available in nearly all parts of the world ocean including the icebreakers operating in the Arctic region.

Figures 4.1–4.6 show a set of informational material related to the ice situation in the western part of the Arctic region during a week in June 1981, consisting of raw satellite images of the visible band, ice charts, a composite ice photo-mosaic and the ice drift chart for the period covered.

As the ice component of the regional marine environment was well centred in Sections 4.1.1 and 4.1.2, it seems reasonable to proceed with another example of activities related to this topic.

4.1.3 The US Coast Guard's recent experience in using satellite data on icebreakers

In June 1993 the US Coast Guard installed new satellite data acquisition systems on the polar icebreakers *Polar Star* and *Polar Sea* – the most powerful non-nuclear icebreakers in the world, to provide real-time large area coverage ice images (Jendro and Bernstein 1994).

The system TeraScan (SeaSpace Corporation, San Diego, CA, USA) uses a 1 m diameter steerable reflector to provide receiving signals at 1.7 and 2.2 GHz. A SUN Microsystems SPARCStation workstation computer provides all the operations with the antenna, image processing, storage and interpretation. A GPS (Global Positioning System) module is also incorporated. Output informational products are available in digital formats and also as greyshade and colour hard copies. Sophisticated software enables users to apply all the modern image processing and GIS techniques; animation of image sequences function is also available (Jendro and Bernstein 1994).

Similar systems are currently used on other ships including *Polar Stern* (Germany), *Kaiyo Maru* (Japan), two US university research vessels and a NOAA ship. The peculiarity of the two systems installed on the US Coast Guard icebreakers is that they are the only vessels capable of receiving DMSP satellite data. The satellite data-flow to these icebreakers consists of

- AVHRR visual and IR data (spatial resolution 1 km) from NOAA satellites;
- Tiros Operational Vertical Sounder (TOVS) meteorological data from NOAA satellites;
- Argos buoy data relayed via NOAA satellites;
- Operational Line Scanner (OLS) visual and IR data (spatial resolution 500 m) from DMSP satellites;

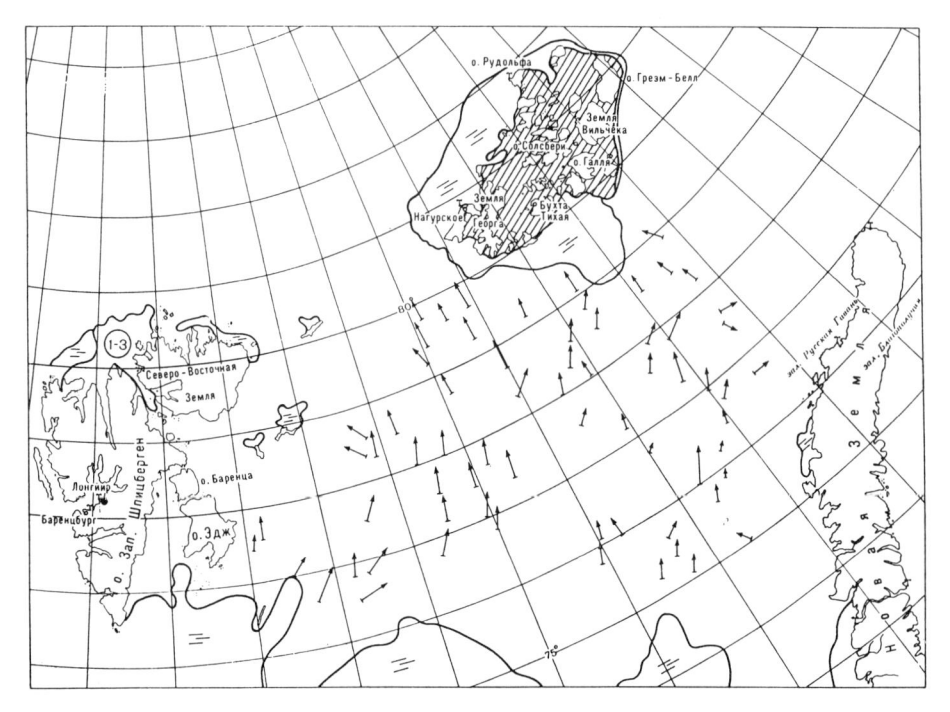

Figure 4.6 Western part of the Arctic. Chart of drift vectors for the period 12–17 June 1981 as revealed from satellite imagery from METEOR-2 (Bushuev 1984).

- Special Sensor Microwave/Imagery (SSM/I) imagery (resolution of 12–40 km) from DMSP satellites, which are processed into total ice concentration and multi-year or first-year ice percentages using the Calvalieri *et al.* (1990) algorithm. The operations with DMSP data required a 'decryption device, located in the ship's secure communications room' (Jendro and Bernstein 1994).

Perhaps the most interesting scientific result of the summer 1993 campaign (when the data from three NOAA satellites and two DMSP satellites provided new ice images about every hour during the daylight period) was that 'sufficient cloud-free images overcame concern that regional climatic cloudiness would obscure the ice from visual satellite sensors. It appeared that the high frequency of satellite passes allowed images through occasional cloud breaks' (Jendro and Bernstein 1994).

This result reminded the author of the discussion about the availability of actually cloud-free time intervals and the general cloudiness situation over the Baltic Sea (Section 3.7.1) in the summer period. Indeed, the more frequent the satellite images, the less dangerous are the clouds for satellite regional monitoring activities.

Frequent satellite imagery acquisition plus the 'extremely effective' procedure of differentiating sea ice from the common optically thin cloud cover (based on a technique described by Lee *et al.* (1993) for AVHRR channel 1 and channel 2 data) enabled users to improve the efficiency of polar operations in summer 1993; useful remarks on the technical aspects of the new equipment performance have been made, and relevant recommendations have been worked out (Jendro and Bernstein 1994) aimed at further development of ice mapping in the Arctic region.

4.1.4 Dutch operational system for marine applications

A brief description of the operational system set up in the Royal Netherlands Meteorological Institute (KNMI) in De Bilt is based on the presentation by Roozekrans (1993). The system has been operational since January 1990; nowadays it uses AVHRR/NOAA images. The KNMI HRPT-mode receiving station is the only operational one in The Netherlands; it is equipped with an HRPT station (VCS Engineering, Bochum) and a VAX-II/GPX workstation with colour printer and an optical disk archiving device.

AVHRR data from all overhead orbits are being initially processed in automated mode using the APOLLO algorithms of the British Meteorological Office. The final manual operations include data quality inspection, exact navigation, production of weekly and monthly composite images and production of quantitative imagery using *in situ* data. Satellite image-based informational products are produced for the North Sea area between 50° and 60° North and for the Ijsselmeer.

The following informational output products are currently produced on a routine basis:

- Sea surface temperature charts (using channels 4 and 5) with an accuracy of 0.5°C.
- 'Red' reflectance of the water column (REF), based on channels 1 and 2 with relevant models for atmospheric correction. REF images show qualitative turbidity patterns near the sea surface caused by mineral particles (silt, sand) and *Coccolithophore algae* in the upper water layer.
- Total Suspended Matter (TSM) concentrations, based on quantified REF images with the involvement of suitable (in terms of dynamic range of TSM concentrations and the synchronicity of data collection) *in situ* data for regression analysis. The accuracy is assumed not to exceed 50% of the true concentration values.
- Normalised Difference Vegetation Index (NDVI) based on channels 1 and 2. The value is normally negative for water surfaces, and positive for the floating layer of blue algae. This index is used to monitor floating algae in fresh-water bodies, like the Ijsselmeer.
- Ice-cover maps based on the difference between channels 4 and 5.

The output informational products are distributed as paper hard copies or on floppy disks by mail, reaching the user on the same or next day of a NOAA pass. Of the above listed products the cloud-screened, atmospherically and geometrically corrected SST, REF and NDVI images are regarded as standard products. Their weekly and monthly composites and also largely cloud-free single-orbit images of the North Sea are archived.

The high operational potential of AVHRR-derived information was shown in three application areas:

- 'the real time monitoring of dynamical processes at the sea- or lake-surfaces (floating layers of algae, ice-cover, navigation of frontal systems);

- a retrospective data-source in studies of specific processes at the Earth-surface (i.e., the outflow of the river Rhine into the North Sea);
- assimilation in numerical meteorological and oceanographic models and information systems (input and verification)' (Roozekrans 1993).

4.1.5 Swedish operational system for marine applications

Operational marine applications services are carried out by the Swedish Meteorological and Hydrological Institute (SMHI) in Norrkoping. No general description of the system used in SMHI is available, hence the following brief presentation is based on the original papers by the SMHI leading remote sensing experts published in 1993–1995 and referred to below.

NOAA AVHRR and Meteosat data are received at SMHI on a routine basis. Some 10–12 AVHRR scenes in HRPT mode are received daily. PROSAT – the operational image processing system – consists of the image processor TERAGON 4000 connected to a dedicated VAX/VMS computer. The raw images are automatically geometrically transformed to different projections and scales (currently 15 options are available). For marine applications two Mercator products based on AVHRR channels 1, 4 and 5 with maximum geometrical and radiometrical resolutions are generated in the form of red–green–blue (RGB) pictures (Thompson et al. 1993; Moberg and Hakansson 1993; Hakansson et al. 1995). The main application areas are ice mapping, SST mapping and blue–green algae monitoring.

In the *ice mapping* area SMHI uses AVHRR images to produce ice charts 'identifying leads, total ice cover and to some extent the quality of the ice' (Hakansson et al. 1995). Ice informational products are generated within one hour after the satellite data registration. The technology of ice mapping is highly automated but still needs the involvement of experienced operators. They have graduated from universities and have been trained for some years in the course of their practical work at SMHI. Nowadays the ice service at SMHI is an impressive example of a well-balanced operational informational system with proper fusion of satellite and *in situ* data. Further development of the system is associated with the use of the SAR data. Results of the pre-operational experimental use of SAR images in 1994 are reviewed in the recent article by Hakansson et al. (1995).

There was nothing unexpected during the experiment period from 18 January to 31 March 1994, as the experiment had been well prepared. Still two features will be interesting for the reader. The first feature is related to SAR images interpretation. The Swedish national station provided by the European Space Agency for delivery of full-resolution ERS-1 SAR images was installed at SMHI in 1992. But joint Swedish–Finnish field experiments in validation and calibration campaigns of airborne and satellite-borne (ERS-1) SAR images have been carried out since 1987. A special Baltic Experiment for ERS-1 (BEERS-92) was performed in January–March 1992 in the Gulf of Bothnia with the field programme carried out from an icebreaker (*in situ* ice and snow parameters, regional roughness variations) and from helicopter (photography of sea ice cover with a Hasselblad camera; a video camerarecorder was also used) (Thompson et al. 1993). The experience gained and the results obtained enabled users to create an analytical/empirical quantitative model which relates the ERS-1 SAR backscattering coefficient (for incidental angles 20–26°) and the surface roughness which is applicable during cold weather conditions in the Baltic Sea (Hakansson et al. 1995). Graphical presentation of this model is given by Hakansson et al. (1994). The details of the model can be found in Carlstrom et al. (1994).

This issue again highlights one of the cornerstones of regional satellite oceanography – the importance of sea-truth experiments which provide knowledge of regional peculiarities of various parameters relevant to the analysis and interpretation of satellite data.

Another interesting feature is the temporal characteristics of various phases of SAR data delivery, processing and the transmission of ice informational products to the end-users. During the experiment with the ERS-1 SAR data of 1994 (by that time the internal technological procedures at SMHI had been tested thoroughly), the processing in the Ice Centre itself took about two hours. There were two variants of the SAR data delivery. Images obtained from Tromso (100 m resolution, total volume 2.52 Mbytes per scene) were available in SMHI two hours after satellite acquisition time, and it took another 1.5 hours for various processing steps before the images reached the Ice Centre itself. The so-called 'fast delivery' data with 25 m resolution (63 Mbyte image transmitted via satellite communication link from Kiruna, Sweden and Fucino, Italy) were available at the local computer typically 19 hours after satellite acquisition. The average transmission time between the Ice Centre in Norrkoping and the icebreakers in the Gulf of Bothnia, using the Iceplott system and the mobile telephone net, was 40 minutes per frame (Hakansson et al. 1995). These figures show that indeed, in some operational problems of regional satellite oceanography, the efficiency of satellite data use may strongly depend on the available communicational links and facilities.

Speaking of sea surface temperature based informational products it is worth mentioning that for quantitative analysis of fronts an automated procedure was worked out by Kahru et al. (1995) based on a modified algorithm developed by Cayula (1988) and Cayula and Cornillon (1992).

Another remarkable routine satellite data-based product at SMHI is the charts of algae blooms in the Baltic Sea. Special filtering and correction for varying sun illumination angle is undertaken (Moberg and Hakansson 1993).

The above material on the topic of ice cover enabled the author to summarise the state-of-the-art of this issue. Figure 4.7 shows the generalised scheme of data flow in a satellite-based operational system of ice-cover monitoring. There are two ways of providing operational help to ships and icebreakers working in seas covered with ice. The application of each option depends on the general availability of satellite imagery on board the vessel and the quality of this imagery, as well as on the level of the crew expertise in satellite data use.

The first option is based mainly on the services of an Ice Centre which receives satellite images from different types of satellites, analyses them together with relevant *in situ* data and produces ice charts (and processed satellite images) to be transmitted to the ships. The latter may provide *in situ* data useful for the Centre. Data flows in this option are shown with thin lines in Figure 4.7. This option was demonstrated in Sections 4.1.1, 4.1.2 and 4.1.5.

The second option is based on the activities of the on-board personnel in producing ice charts using satellite imagery received directly on the ship (data flow shown as a bold line in Figure 4.7). This option was successfully tested, as presented in

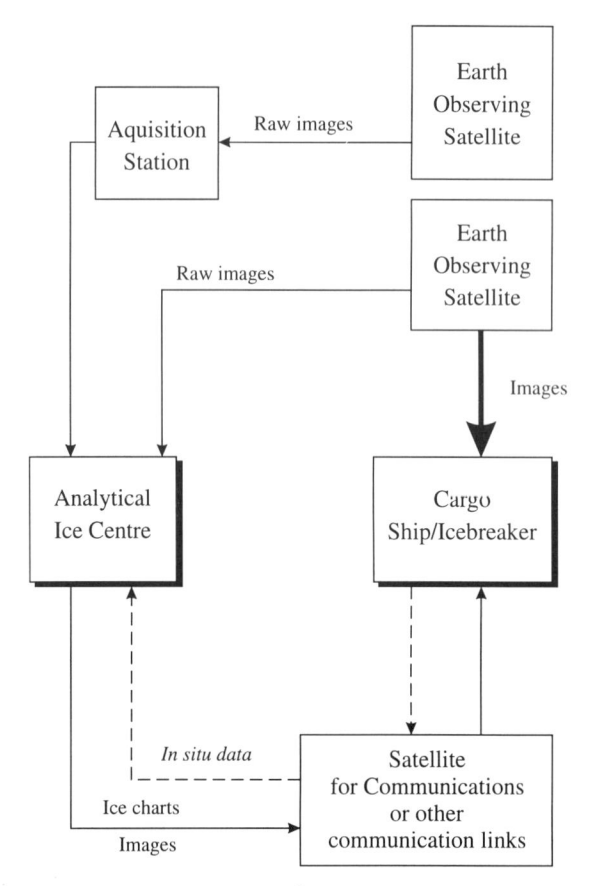

Figure 4.7 Data flow in operational satellite-based sea ice monitoring systems. This generalised scheme is applicable to Norwegian, Russian, US and Swedish systems described in Section 4.1.

Section 4.1.3. To ensure safety and efficiency of transport operations in the polar regions a combination of both options is preferable.

4.1.6 US NOAA's CoastWatch Program

NOAA's CoastWatch Program, with support from the NOAA Coastal Ocean Program, in conjunction with NOAA Line Offices, delivers satellite informational products and *in situ* data to federal, state and local marine scientists and coastal resource managers (NOAA CoastWatch, 1993). CoastWatch was established in 1988 (for the history of the Program, data flow and technical details of communication links see Pichel *et al.* 1991) with the general objective in the field of product access and distribution which was formulated as follows:

> Provide access to quality-controlled, near real-time, and retrospective satellite, aircraft, *in situ*, and analysis/forecast model data and derived products for the coastal and Great Lakes regions of the United States by supporting an operational distributed communications and data storage network. (Coastal Ocean Program CoastWatch 1994)

Informational products available from CoastWatch include digital satellite surface temperature imagery, digital satellite turbidity imagery and others; future informational products will include ocean chlorophyll and turbidity from the SeaWiFS Ocean Color Satellite. There are eight regions covered by the CoastWatch network: Southeast (focal point at Beaufort, NC), Northeast (Narragansett, RI), Great Lakes (Ann Arbor, MI), Gulf of Mexico (Stennis Space Center, MS), West Coast (La Jolla, CA), Caribbean (Miami, FL), Central Pacific (Honolulu, HI) and Alaska (Anchorage, AK). The data are accessible via Internet (high speed 50 kbaud), dial-up modem (9.6 kbaud), FTS-200 system (9.6 kbaud) and by mail (on diskettes).

The Gulf of Mexico Regional Node, which became fully operational at the end of 1992, provides CoastWatch data for the entire Gulf of Mexico. High-resolution (1.47 km) satellite imagery is available for any area of the Gulf usually four times a day within three hours of the satellite overflight at a Mercator map projection (Gulf of Mexico Regional Node 1994). Section 4.2.1 presents the regional activities relevant to oceanography and coastal zone management in the Gulf of Mexico.

CoastWatch is complemented by NOAA's Coastal Assessment Framework (CAF) which provides 'a consistently derived, watershed-based digital spatial framework for managers and analysts to organise and present information on the Nation's coastal, near-ocean, and Great Lakes' resource' relevant to estuarine capability (NOAA's Coastal Assessment Framework 1994).

4.2 Use of regional integrated GIS and satellite databases for marine applications and coastal zone management

4.2.1 Gulf of Mexico

Perhaps the most impressive public demonstration of powerful *regional* satellite database for marine applications and coastal zone management was the development of an 800 m bio-optical database from the level-1 Coastal Zone Color Scanner (CZCS) data meant to present the spatial and temporal variability of the Gulf of Mexico basin (Oriol *et al.* 1994). This work was based on *regional* optical algorithms developed specially for ocean colour satellite data in the Gulf of Mexico waters (Gould *et al.* 1994). Together with GIS-based spatial analysis support for mapping the Gulf of Mexico estuaries and their contaminants (Johnston *et al.* 1994), these studies give an excellent example to follow in complex research and management of coastal zones on a regional scale. Besides, the methodologies developed for this region will be used in other regions and with other sources of satellite data, which makes these efforts still more valuable. Some features of these studies will be presented here.

Optical relationships in Case II waters of the northern part of the Gulf of Mexico have been studied in detail by Gould *et al.* (1994) with the purpose of applying this knowledge to satellite monitoring of coastal processes through the improvement and further development of remote sensing algorithms. For these purposes a set of *in situ* optical measurements was carried out in April 1993 at 28 stations representing three different water masses associated with river discharge areas. These regions were off the Atchafalaya Basin, the Mississippi River Delta and the northwest Florida Shelf. The shipborne measurements were performed at stations located at 15 nautical mile intervals along seven transects 40–50 nautical miles long running offshore. It is common knowledge that *in situ* sea-truth optical measurements are often made at a wavelength that does not correspond to those of satellite sensors. So during this cruise spectral measurements of upwelling light were made at 2 nm intervals, and the scientists were able to develop a conversion technique to estimate the subsurface upwelled radiance at one of the CZCS channels (520 nm) from *in situ* measured upwelled radiance at 531 nm (Gould *et al.* 1994). This technique may be applied to other wavelengths. Biological and optical data collected by three institutions during 13 other cruises at 189 stations in this region have also been compiled into a database.

Regional optical algorithms linking apparent and inherent optical properties with upwelled subsurface water radiances have been developed for ocean colour satellite sensors. They are valid for the coastal waters of the Gulf of Mexico, and extend the range of diffuse attenuation coefficient K vs. radiance relationships from a K value of 0.4 m^{-1} (inverse metres) for 490 nm (such as presented by Austin and Petzold 1981) to 1.3 m^{-1}. It is clear that

> similar regional algorithms may be required in other coastal areas to optimize estimates there, if factors that affect the optical characteristics of the water (such as particle size, number, and composition) vary significantly from those in the northern Gulf of Mexico. (Gould *et al.* 1994)

These results were incorporated as a module into another regional project dealing with a satellite database for the Gulf of Mexico (Oriol *et al.* 1994). General scientific considerations show that to characterise the local marine environment in this area affected by tides, with the winds and tidal currents responsible for 'resuspension and flocculation of particles and redistribution of nutrients', where 'growth and decay of coastal chlorophyll can double within hours' (Oriol *et al.* 1994), a synoptic scale representation of those features is required. The spatial scale of data should cover 'bays, estuaries and river discharge plumes' (Oriol *et al.* 1994). But, whatever the dreams are, the eight-year time series of satellite data of bio-optical parameters could be created only from the CZSC 800 m resolution images. The Gulf of Mexico was selected as a test area for the development of comprehensive GIS multi-parameter databases of bio-optical and related characteristics. The future SeaWiFS images will be also stored in similar databases. Moreover, using the technology developed for the Gulf of Mexico project, the Naval Research Laboratory at Stennis Space Center, Mississippi

> plans to develop additional databases from CZCS and SeaWiFS imagery for areas of greater tactical interest to the Navy such as the Yellow Sea or Arabian Sea. Future plans include the development of a database of sea surface temperatures from the Advanced Very High Resolution Radiometer (AVHRR) and of the Solar Irradiance Field that is coincident with the CZCS database. The automated processor will drive the development of both these databases and provide the Navy with a GIS style package that will give Naval oceanography a new perspective on the world ocean and our environment. (Oriol *et al.* 1994)

The Gulf of Mexico is a water body representative of three major water types: namely the Mississippi Delta waters are highly absorbing; the West Florida Shelf waters are highly scattering; and the Keys represent clear shallow waters. About 850 CZCS images from the Nimbus-7 satellite were available for this area (31.0° N–15.0°

N, 105.0° W–80.0° W) from November 1978 to June 1986. (For example, 40 images of a local area of the Mobile Bay entrance were available in 1981.) To handle this volume of data, automated processing software has been developed. The raw level-1 data (digital counts) could be processed into level-3 products (radiometrically and geometrically corrected) within 10 minutes. The automated processing package runs on a Silicon Graphics Crimson workstation (80 mps, 128 megabytes memory), while data display and database manipulation is performed under the Navy Satellite Image Processing System (NSIPS) (Gremillion 1993; Fetterer *et al.* 1993) which operates under PV-Wave (Oriol *et al.* 1994). The software in the automated processor can be easily modified to perform SeaWiFS processing in real time. A brief description of data flow is presented following a more detailed presentation by Oriol *et al.* (1994).

At the first step, five images of the Gulf at CZCS channels are created, which takes one minute to complete. At the second step, eight level-2 files are created: four radiance files are the result of subtracting the Rayleigh scatter from each level-1 file and aerosol radiance from the 443, 520, 550 nm level-1 files; images of the Rayleigh radiance and the aerosol radiance at 670 nm; and two bio-optical products generated from the radiance files, namely the pigment concentration and the diffuse attenuation coefficient at 490 nm. The latter is calculated using the regional K-branching algorithm developed by Gould *et al.* (1994). One satellite scene is processed at this step for two minutes.

Image registration (the third step) of eight files from step 2 and the 750 nm file (from step 1) is possible in 20 map projections. A Mercator projection was selected for a latitudinal range from 31° N to 13° N, and a longitudinal range from 98° W to 80° W. The degree-per-pixel resolution of 0.0074 was used to maintain the original CZCS resolution of 800 m. Nine informational products of the Gulf of Mexico are processed each to a grid size of 2430 samples by 1810 lines (40 megabytes). This step takes five minutes. On the fourth step, a $K(490)$ 'quick-look' product is generated for each scene in Mercator projection with a grid, coast-line, and label (containing date, orbit number, latitude and longitude range, file name and the geographical area of coverage). This procedure takes one minute to complete. 'A directory add function' is applied later, followed by an 'image subsectioning' procedure which provides the additional information necessary to select and analyse parts of the entire products. Finally the automated shifting function is applied to fit the standard projection of a coast-line (the routine operation which is usually performed manually in many other systems), which takes another three minutes.

About 90 hours of computing time were spent to create this co-registered nine-parameter database which 'provides the mechanism to understand the local coastal processes on a regional scale' (Oriol *et al.* 1994).

There are plans to use this unique regional database to develop procedures capable of predicting the optical properties of waters in seven coastal areas along the northern Gulf of Mexico (Oriol *et al.* 1994). It is clear that such a comprehensive database will help to gain many interesting scientific findings in the region, including the inshore waters and estuaries.

A conceptual and managerial framework of the US National Biological Survey's National Wetlands Research Center (NWRC) activities in the Gulf of Mexico was presented by Johnston *et al.* (1994). It was emphasised that, while more than 70 billion US dollars are spent annually on various environmental regulatory programmes, the means to assess the effectiveness of these programmes in protecting the environment and natural resources at national and regional scales and over the long-term do not exist. (Johnston et al. 1994)

To meet this challenge the US Environmental Protection Agency (USEPA) initiated a nationwide programme meant to monitor and assess the state of the nation's ecological resources, the coastal areas and estuaries being one of the components. The current development of a nationally based monitoring system over most of the coastal areas in the United States is being done in the framework of the EMAP-E (Environmental Monitoring and Assessment Program – Estuarian component). The EMAP-E activities currently cover the US East Coast, the Gulf of Mexico, and the southern California coast.

For the northern Gulf of Mexico a set of mapping protocols for the identification and delineation of submerged aquatic vegetation beds (about 500 1 : 24 000 US Geological Survey quadrangles) will be developed and implemented. The EMAP-E activities will include production of a Landsat Thematic Mapper (TM) satellite image backdrop which should display and highlight sample locations and data for the coast-line (Johnston et al. 1994). The whole work will be performed in the GIS context and yield valuable practical knowledge which could be of tremendous use for the world scientific community.

One more issue is of great importance when speaking about coastal zone mapping, monitoring and management with remotely sensed data being one of the data sources, namely the *coastal land cover classification system*. It seems to me that sooner or later the international scientific community will meet the problem of standardisation of mapping protocols, map legends and the unification of terms used in this growing and expanding area of remote sensing applications in the coastal zone. In this context it is worth presenting here the classification approach designed for use with satellite data (Klemas et al. 1994).

The classification system was developed by Klemas et al. (1994) in the framework of the activities covered by the NOAA Coastal Ocean Program, namely the Coastwatch Change Analysis Project (C-CAP) which was aimed at monitoring the US coastal wetlands, submerged vegetated habitat and adjacent upland cover and change in the coastal regions every 2–5 years (with annual monitoring of areas with significant change). Remotely sensed data are to be used to establish a coastal land cover database and monitor the changes on a regular basis. The C-CAP is currently developing a comprehensive, nationally standardised information system for land cover and habitat change in the US coastal regions (Dobson et al. 1993) based on the GIS approach, with both *in situ* and remotely sensed data used as the sources. Data from the Landsat Thematic Mapper, SPOT and other satellite sensors, as well as aerial photographs, will be incorporated. The output informational products will include digital images, hard-copy maps and tabular summaries. Land cover change is planned to be detected on a pixel-by-pixel comparison basis (Klemas et al. 1994).

The system utilises previous efforts in this area (Anderson et al. 1976; Klemas et al. 1987) and the experience gained during pilot studies in the Chesapeake Bay Watershed (Dobson and Bright 1991) and coastal North Carolina (Ferguson and Wood 1990). There are three level-1 superclasses: uplands, wetlands and water, and submerged land in the C-CAP land cover classification system. These superclasses are divided into classes (level-2) and subclasses (level-3). Detailed definitions of all classes and subclasses are given by Dobson *et al.* (1993). A table helping to understand the structure of a classification system is presented by Klemas *et al.* (1994), with special indication of those classes which 'can generally be detected by satellite remote sensors, particularly when supported by programmatically required surface level observations and ancillary data sources' (Klemas *et al.* 1994).

Another example of multi-year satellite observations of the Gulf of Mexico was presented by Stumpf *et al.* (1994), who studied the transport of fine sediments in the Alabama–Mississippi coastal area. The database included over 200 AVHRR scenes from 1990 to 1993; water samples and CTD (conductivity, temperature, depth and transmission) surveys on monthly cruises from 1990 to 1993; and salinity, temperature and transmissivity measurements from moorings from 1990 to 1992. Shipborne and satellite data were used to produce charts of suspended sediment concentration (SSC) and the diffuse attenuation coefficient for **p**hotosynthetically **a**ctive **r**adiation (K-par). It was shown that mineral material dominates the suspended load during winter and spring, while organic components become relatively large in summer. Temporal variations in the optical properties of suspended material were determined using comparisons of SSC to K-par values. Moreover, considerable variations in optical characteristics in the suspended material during late summer and autumn made estimates of SSC and K-par from satellites 'more problematic' (Stumpf *et al.* 1994) than in other seasons. This is a remarkable note, meaning that the analytical properties of the satellite-based regional environmental monitoring systems may be season-dependent.

The material presented in this section show that the Gulf of Mexico acts as a test area where new GIS and remote sensing application methodologies are being developed, and new concepts of coastal zone management are being implemented. A large number of scientific institutions are located on the shores of the Gulf. Thus the Gulf of Mexico can be regarded as one of the areas of worldwide importance, where activities in regional satellite oceanography are concentrated, and where new horizons and trends in its development can be studied.

4.2.2 Satellite-derived components of the environmental database for the US West Coast

Another impressive example of an up-to-date regional environmental GIS for marine and coastal zone applications is the one developed by Crout *et al.* (1994) for the West Coast of the USA. It is remarkable that there are two levels of spatial resolution in this database. A low-resolution (20 km) database is meant to provide an overview of the area which was defined as 10–45° N and 146–101° W. With an image size of 512 × 512 elements, and a spatial grid spacing of about 10 km, 'mesoscale and regional features of temperature and optics are readily observed' (Crout *et al.* 1994). A higher-resolution (1 km) database for coastal and nearshore regions, providing observation of patterns of upwelling structures (tongues of cold water), covers the area 42–47.12° N and 128–122.88° W.

The lifetimes of the phenomena under investigation were taken into consideration when choosing the temporal resolution of the database. For the low-resolution database, monthly averaged images for each year are used, thus providing a study of the variability of temperature and optics during the year and on a year-to-year basis. All the individual images are stored in the higher-resolution database. The prototype of this database was described in detail, and the sources of satellite and *in situ* data were indicated, by Crout *et al.* (1994).

4.2.3 Remote sensing component of the ASEAN–Australian Coastal Living Resources Programme

This brief description of the programme is based on the original presentation by Mohamed (1994). Since 1984 a network of scientists from five ASEAN (Association of South East Asian Nations) countries, Indonesia, Malaysia, Philippines, Singapore and Thailand, and Australia are running the regional programme dealing with three major marine and coastal ecosystems – mangrove forests, coral reefs and the soft bottom communities. Over the whole region standardised sampling methods and data management procedures are being used, and a large amount of sea-truth data have been collected.

The remote sensing component of this programme has two main objectives:

- 'To develop a network of remote sensing specialists in the region competent in the application of remote sensing of the marine and coastal areas;
- To explore and develop the applications of remote sensing to environmental problems and management of the coastal zone and marine areas' (Mohamed 1994).

Satellite images of visible bands from LANDSAT MSS, TM and SPOT, and AVHRR are being used in order to overcome the well-known obstacles of dangerous underwater observations, the limited amount of manpower capable of performing measurements and observations in this large-scale project, and time- and cost-consumption restrictions. Current activities in the implementation of satellite data in *coral reef* studies focus on the following issues:

- shallow water mapping (actually the mapping of shallow water features – coral reefs and sandy and rocky substrates in clear waters down to 30 m depth);
- classification of coral reefs (using the depth–species zonation relationship);
- water quality determination (estimations of concentrations of suspended matter and chlorophyll); sedimentation rate studies are closely connected with this issue;
- water circulation studies, including the search for zones with low current speed where the coral reefs are likely to grow.

Current activities in the use of satellite images in *mangrove* studies include the mapping, determination and delineation of mangrove areas along with attempts to classify mangrove species and determine the state of mangrove in wild and managed mangrove forests.

During the first stage of the Coastal Living Resources project baseline information on the coral reefs of 10 island groups from the Malacca Straights to the Sulu Sea was obtained. In the second five-year phase a more detailed study of the ecology of coral reefs in connection with coral reef fish is being performed. Redang Island archipelago was chosen as a test area, with Pulau Redang Island Marine Park used as a test site.

Encouraging results were obtained and used as a practical instrument in coastal zone management activities, including the assessment of the impact of intensive development of the Pulau Redang Island Resort infrastructure on the fragile marine ecosystem. The scientific team faced the problem of timely acquisition and availability of cloud-free satellite imagery during the northeast monsoon months (November to March) (Mohamed 1994).

4.2.4 A nearshore information system along the coast of Brittany, France

This system, meant for mapping eutrophication sensitive zones (Urvois et al. 1994), is an example of rather common GIS with a modest satellite data component; some of them were presented at the EARSeL Workshop on Remote Sensing and GIS for Coastal Zone Management (October 1994, Delft, The Netherlands).

A regional GIS is meant to support studies and monitoring activities in the coastal zone of Brittany, including the mapping of eutrophication sensitive zones affected by proliferation of both green macrophyte seaweeds (Ulva) and phytoplankton. The annual mass production of Ulva represents 100 000–200 000 tonnes which accumulate on the shore during spring and summer (Urvois et al. 1994). The range of the study area (2600 km of coast-line) dictated the scale of 1 : 100 000 to be chosen as an intermediate resolution between various datasets, which included base maps (1 : 100 000 scale and Lambert 2 projection), coast-line data (high tide level), the stream network, and the watersheds. The digitisation of the coast-line, the islands, the stream network and the 750 local watersheds was carried out to a precision of 150 m. Historical nearshore (down to 10 m underwater) data on sediments were mapped on a 1 : 100 000 scale with eight granulometric classes at a subregion (from the Mont Saint-Michel bay to the River Loire estuary). A network of 200 polygons was used to map phytoplankton, and two special aerial true-colour photographic surveys were performed in 1988 and 1991 along the entire coastline of Brittany to collect data on Ulva proliferation. In all, the map preparation and digitisation work took 340 hours using Microstation software run on PCs (Urvois et al. 1994).

The listed vector data were complemented by satellite imagery and a hydrodynamic two-dimensional model providing 'resident time' values (an evaluation of the transit time of a water parcel within a grid cell of the model). Residence time of water masses in the coastal zone appeared to be a good parameter indicating the sensitivity of water areas to eutrophication. Zones with residence time exceeding two days were regarded as 'potentially sensitive to eutrophication' (Urvois et al. 1994). The only satellite-borne data used in the regional GIS were the AVHRR/NOAA sea surface temperature time series. It is interesting to note that 'the initial images taken at a resolution of 1000 m by 1000 m were resampled using the krigging method at a resolution of 250 m by 250 m' (Urvois et al. 1994). A set of 19 preselected cloud-free images of the IR band were used as representative for the period from the end of winter (February–March) to the end of July over six consecutive years from 1986 to 1991. The SST patterns were visually compared with the results of modelling. The published images of the SST field, the residual time chart (for a local test area) and the chart of main algae proliferation sites shows some similarity. The total thematic content of the regional GIS covering an area of 75 000 km^2 at a 150 m resolution for vector data and a 250 m resolution for image data was presented by Urvois *et al.* (1994). The rather limited use of satellite data in this regional environmental GIS seems to be characteristic for the present stage of the implementation of remotely sensed data in many common cases of marine-biology-oriented studies.

4.2.5 Satellite component of fisheries forecast activities in the northwestern waters of India

AVHRR/NOAA digital data (channels 4 and 5) acquired at Ahmedabad are currently used to support activities in long-term fisheries forecast by means of the studies of interannual and interseasonal variability of thermal patterns 'and its effect on the distribution and aggregation of fish in the northwestern waters of India in the Arabian Sea, precisely off Saurashtra region' (Kumari *et al.* 1994).

Raw data processing includes radiometric normalisation, geometric correction and SST retrieval. Data from low-elevation passes are being radiometrically corrected using a radiance normalisation approach suggested by Narayana (1992). Geometric correction is based on satellite orbital parameters and on a ground control points technique. SST values were obtained using McClain's multichannel SST approach (McClain 1985); the relevant algorithm was tuned using *regional* climatological data. SST images were generated in steps of 1°C. The total range of SST values measured in the seas around India is 21–31°C. A standard colour scale was used to provide further analytical work with multi-year datasets.

Ten satellite images for 1991–1993 were analysed by Kumari *et al.* (1994) resulting in better understanding of such oceanographic features as thermal fronts, warm or cold core rings, eddies, meanderings, etc. 'With the availability of near real time NOAA data, the oceanographic features and various phenomena which influence the fish stocks, their distribution, migration, schooling, success or failure in spawning, early survival, etc., can be monitored' (Kumari *et al.* 1994). At the same time the scientists faced the problem of inadequate knowledge of the basics of fishery forecasting: unknown causes of many phenomena, uncertainties of various estimates and relationships, etc.

4.2.6 Some trends, technological novelties and common problems of the use of integrated GIS and satellite databases in regional marine and coastal applications

In the course of developing regional GIS and satellite databases one may face some problems (and find some useful approaches), which will be briefly presented in this section. With the development of remote sensing and GIS technologies *observation*, *geolocation* and *communication* must be considered as a whole. Proper integration of these three items is becoming critical in terms of costs and benefits.

Lotz-Iwen (1994) pointed out that nowadays the utilisation of satellite data in global and regional applications in various fields was not adequate for the potential of inherent information for the following reasons:

- 'the missing knowledge of potential users about the existence and applicability of remotely sensed data;

- the user-unfriendly concepts and realisation of most data archives and interfaces' (Lotz-Iwen 1994).

It was announced that 'in the context of future international programs on Earth observation from space (EOS of NASA, ENVISAT of ESA)', the German Remote Sensing Centre was about to set up an Intelligent Satellite Data Information System (ISIS) which would provide various improvements as compared to the existing informational databases and archives. The ISIS thesaurus contains about 4000 descriptors representing scientific terms. For detailed information on ISIS see Lotz-Iwen (1994).

Another piece of information which may be of interest (and could become useful) for the oceanographic remote sensing community, is the image compression technique developed, in response to user's needs, by the NOAA Ocean Products Center (Nault 1994). The standard NESDIS-preprocessed Coastal Ocean Program image products are 512 × 512 pixels (ranging from 1.47 km per pixel to 5.87 km per pixel). A full-resolution frame with coast-line needs 558 080 bytes (Nault 1994). The new NOAA QuickLook software produces viewable, displayable, ultracompressed images typically between 2 and 2.75% of the original image size. The QuickLook display software was designed to run on a standard IBM-compatible PC within a Windows environment. It allows up to 10 images to be viewed simultaneously at a preliminary stage of examination before the user decides to order the full-resolution image (Nault 1994).

The next note relates the marine and coastal zone GIS issue with that of GPS (Global Positioning System) technology. Nowadays GPS is an established industry and a part of routine technology in geology, the construction industry, forestry, etc. Actually one cannot use an environmental GIS to their full power, nor can we use satellite data-based mapping techniques, without the involvement of up-to-date GPS technology. *Earth Observation Magazine* published in Aurora (Colorado, USA) has a monthly column 'GPS Consumer Series' meant to explore and discuss various issues associated with GPS data collection, including those relevant to satellite-borne remote sensing applications.

Mick *et al.* (1994) pointed out that often in the course of the construction of an environmental GIS with their multilayer structure, the control-point intersections of several data layers do not overlap with the desired accuracy. Nowadays this is a situation which many common users have not yet met, but they inevitably will face this problem in the course of further development of their regional integrated GIS using data from various sources: satellites, aircrafts, maps and registers. It simply means that the accuracy of georeferenced satellite data and the accuracy of *in situ* data within the regional GIS should be harmonised. In the USA 'until very recently, accurate geolocation knowledge was either quite expensive to acquire ... or was unavailable as a practical matter (e.g., plus/minus 40 feet point accuracy was the best available national map USGS Quadrant quality). Satellite remote sensing data ... is rarely registered to the earth at even coarse accuracies' (Mick *et al.* 1994). The GPS itself and industry innovations in GPS equipment, allowing the collection of data from various moving platforms, have 'revolutionized the means to acquire geolocation information ... within the past five years' (Mick *et al.* 1994).

As a part of US nationwide activities in creating High Accuracy Reference Networks (HARN), the Stennis Space Center, Mississippi is involved in the design and implementation of a HARN in Mississippi state, including the regional Gulf of

Mexico Coast reference network. This activity is part of current efforts aimed at studying land–water interaction along coastal regions. Due to the highly dynamical nature of the coastal environment, geolocation of various features and geo-referencing of satellite high-resolution imagery is of vital importance. In many cases in wetland areas even small fluctuations in water levels may result in large changes in areas of dried/flooded land. Some details of the Stennis HARN infrastructure and its applications in coastal Louisiana, where six separate local GPS subnetworks have been set up, were presented by Mick *et al.* (1994). In addition to monitoring wetlands, deltas and estuaries, other GPS/HARN marine applications include mapping of rivers for navigation and navigational hazards; surveying and mapping of wharfs and harbours; and surveying of levees, dikes, dams, and drainage systems, etc. Further development of the HARN system in the USA, and the growth of digital portable GPS equipment will allow 'sub-pixel accuracy reference' in satellite imagery 'to be built quickly and at low cost' (Mick *et al.* 1994).

Time, the fourth dimension in integrated GIS, in the context of remote sensing data applications to coastal zone monitoring, was discussed by Mulder (1994).

As more and more remotely sensed images are being used as routine informational products in various data systems and application services, the problem of image transferring arises. One of the solutions currently used in regional satellite oceanography applications is the use of commercial mobile telephony. One of the examples was already given in Section 4.1.5 dealing with the transmission of imagery and charts from ground-based ice centre to icebreakers. Recent analysis of mobile communications technologies available to the marine user community indicated that the least expensive is cellular telephony (Whitehouse and Landers 1994). Several organisations in Canada and the USA tested this option and determined its usefulness. The results demonstrated the technical and commercial feasibility of cellular telephony in transferring AVHRR NOAA and ERS-1 SAR imagery to the coastal zone.

By incorporating data compression techniques, AVHRR SST images were transmitted at a maximum rate of 21 kilobits per second. Processed SAR imagery could be transmitted at significantly lower rates, with transmission range varied from site to site due to coastal topography, state of the atmosphere and the quality of ship-borne facilities. Reliable transferring can be done at distances up to 160 km from shore, sometimes at greater distances. At ranges greater than 240 km alternative technologies must be used (Whitehouse and Landers 1994).

Some new trends in coast-line management in the Netherlands (about 350 km long; 254 km of dunes; 34 km of sea dikes; 38 km of beach flats; and 27 km of boulevards and beach walls) based on the GIS approach with the involvement of remotely sensed data were presented by Van Heuvel and Hillen (1994). Since the middle of the 1960s annual coast-line measurements are performed by means of remote sensing (on land) and sounding (offshore) techniques, namely at 200–250 m intervals a coastal profile is measured, extending from 200 m landward to 800 m seaward. Thus a unique database of annual monitoring is being created.

In 1990 the Dutch parliament decided on a new national policy for coastal protection: *Preservation of the Coastline of 1990*, which marked a new era in coastal defence policy in the Netherlands (Van Heuvel and Hillen 1994). The concept of the 'basal coastline' has been developed by Hillen and De Haan (1993), which is the 'coastline-to-be-preserved'. Annual monitoring activities should bring knowledge on the difference between the actual coastline and the 1990 'basal coastline' in order to plan the relevant actions. A total of 105 charts (4 km × 4 km) are required to cover the entire coast-line on a scale of 1 : 25 000. They are produced using a GIS based on a professional production programme, with satellite images of 30 m resolution (or a scanned topographic chart) acting as background. Annual updating of 105 charts takes about three weeks including checks (Van Heuvel and Hillen 1994).

The recently developed GIS application SHOMAN (SHOreline MANagement tool) includes about 450 oblique photographs of the Dutch sandy coast in digital form (on CD-ROM) helping to assess the situation in the framework of a decision-support system. This rather new aspect of the GIS environment is actually 'a first step in the direction of a Multi-Media Decision Support System for the Dutch Coast, which can include sound, voices and short video shots' (Van Heuvel and Hillen 1994). It is expected that new developments in remote sensing and GIS technology will bring major improvements in coast-line and coastal zone management by the year 2000.

4.3 Topic-oriented use of satellite data in regional marine and coastal zone environmental and development activities

In this section a set of concrete original papers will be reviewed, each dealing with certain oceanographic phenomena (or a cluster of parameters characterising natural or man-made impact), with an indication of the accuracy of certain measurements, where appropriate. Traditional water dynamics parameters and phenomena will be presented, along with less traditional satellite data-based issues related to shore, coast and bottom topography. General notes on marine pollution monitoring and the practical implementation of the 'environmental sensitivity index' concept will precede the discussions of oil slick detection, turbidity/sediment transport and river plumes issues. As the sea-ice topic has been discussed in Section 4.1, ice parameters will not be included in this section. However, non-traditional activity will be discussed here: ship traffic monitoring (in the national coastal water/exclusive economic zone) based on SAR data.

4.3.1 Shore, coast and bottom topography

Zuidam (1993) published a paper containing a review of remote sensing applications in coastal zone studies, with an emphasis on the geophysical/geomorphological aspects of (1) sea currents, water quality and sediment transport; (2) near-shore bottom topography; (3) shore topography; and (4) coast topography. This is a descriptive review of various techniques, with no examples of original concrete studies. The reader can use it as a short introductory text on remote sensing (both airborne and satellite-borne) technique applications in marine and coastal zone studies. Perhaps the most interesting feature of this review are the tables with the author's expert rating of the usefulness of various satellite sensors in respect of determination and analysis of various parameters and features of coastal zone phenomena. Zuidam (1993) used three qualitative grades: very useful, useful, limited useful, regarding LANDSAT TM and MSS, SPOT, ERS-1 SAR, JERS-1, NOAA/ AVHRR and Nimbus/CZCS. The evaluation diagrams of usefulness include the following observing characteristics:

- water classes (case I or case II);

- suspended and/or floating material/water quality (chlorophyll, sediments, yellow substances, pollutants);
- currents (wind waves/currents, tides, density, river and ground water outflow);
- nearshore bottom topography (through penetrating blue–violet light and through the analyses of waves, colour/tonal variation, sea-surface height and slope);
- beach/shore topography (foreshore or beach face by low tide, crest of berm, lower berm, beach scarp, upper berm, bluff or escarpment);
- coast/mainland topography, namely
   - cliff coasts (bluff, notches, abrasion platforms, springs);
   - coastal dunes (initial dunes/fresh sand accumulations, vegetated dunes, non-partly vegetated dunes, blow outs, environmental destruction, depressions/lakes);
   - barrier islands and related landforms (cheneers, spits, tombolos; sand/silt flats (non-partly vegetated); tidal flats);
   - marine terraces and related landforms (cliffs, terrace flats, springs) (Zuidam 1993).

Practical experience of coastal mapping in the Canadian Arctic using integrated satellite data was obtained by Tittley *et al.* (1994) who presented an example of coastal sensitivity mapping using LANDSAT TM, SPOT PLA and ERS-1 SAR datasets. They consider classification of the coastal zone in terms of its geological and geomorphological characteristics, an important constituent part of any assessment of possible development impacts, and especially in the sensitive Arctic regions. Their methodological study was carried out at a small (25 km^2) section of the North Head region on Richards Island, in the Mackenzie River delta, centred around 69° 41′ N and 134° 15′ W. The aim of this work was to study the potential of integrated satellite imagery (in the visible band and SAR) in detecting and mapping the previously defined, geologically significant coastal types.

This work has a clear motivation. The Canadian Beaufort Sea has potential oil and gas reserves, and future development needs good knowledge of the complex characteristics of the coastal zone. 'Rapid coastal erosion, thaw settlement, sediment transport, and other processes along the coastline can have a significant impact on the stability of pipelines and shore-based facilities' (Tittley *et al.* 1994). The use of techniques other than satellite-borne is very expensive in this region.

Six coastal features were identified as being 'potentially mappable' (Tittley *et al.* 1994): deep water in lakes, embayments and the Beaufort Sea; tundra; and four accretional types, namely, beaches/spits, flats, lagoons, and dunes. Accretional features could be well detected by both radar and optical sensors. It is important to note that the success of the detection of these features was based on the synergetic effect of using both types of data, as SAR provided 'additional information relating to bedforms and grain size distribution' (Tittley *et al.* 1994) through the mechanism of surface roughness sensitivity.

As for erosional coastal types (they are the dominant coastal landforms in the Beaufort Sea), their detection 'was generally not successful' (Tittley *et al.* 1994). Further investigations are being planned using the application of a detailed digital terrain model and more sophisticated pattern recognition techniques (for example, neural network algorithms) in an attempt to detect erosional features based on integrated satellite datasets (Tittley *et al.* 1994).

The results of the previous study could probably be better understood using another paper dealing with coastal evolution monitoring from a more general point of view. De Lisle et al. (1994) examined the combination of spatial scales and temporal intervals of satellite data that are appropriate to the study of coastal erosion/sedimentation. Coastal evolution monitoring is to a larger extent the monitoring of the advances and retreats of the shoreline. It is a challenge to substitute (or to complement) the existing time- and manpower-consuming precise field measurements with remote sensing techniques. The practical question immediately arises of what would be the accuracy of satellite-based measurements? It is obvious that the answer will be based on a pixel-size approach, and this was clearly demonstrated by De Lisle et al. (1994). They used imagery from the optical sensors LANDSAT TM and MSS, SPOT PLA and MLA and radar images from ERS-1 SAR, Radarsat (simulations) and airborne SAR to monitor the coast-line evolution of the 70 km long island Iles-de-la-Madeleine located in the centre of the Gulf of St. Lawrence (Canada). The island was chosen as a test site because the coastal processes here were studied in detail and on a quantitative level: 'the northern portion of the island is characterised by continued erosion (7.5 m/yr) on the north-western shore and continued sedimentation (up to 10 m/yr) on the south-eastern shore' (Drapeau 1980).

Profiles across the littoral zone were selected to compare the responses from different individual satellite sensors and their ability to discriminate land-water and give information on other morphological characteristics of the coastal zone. As expected, LANDSAT TM channel 5 (1550 nm) data showed 'the broadest signal contrast between land and sea' (De Lisle et al. 1994). Channel 2 (520 nm), with its penetration into the water, was also considered to be of high interest as it can be used 'to delineate shallow water bedforms such as sand bars' (De Lisle et al. 1994). These data recorded in 1982, 1986 and 1988 were analysed with a bit-map technique, and it was shown that erosion during an 18-month period could be detected in this type of satellite image in this region.

SAR images appeared not to be as efficient in detecting the land–sea interface as the LANDSAT and SPOT systems. Some features, inherent to SAR, such as the incidence angle of swath and noise level, radar shadowing by coastal topographic features, 'make coastline detection difficult'. 'Wet sand can also be interpreted as calm water as its dielectric constant is similar' (De Lisle et al. 1994). 'If waves are shorter than radar detectability, the water surface becomes noisy due to velocity smearing' (Rufenach et al. 1991). Radar could also be affected by atmospheric effects, making land and water differentiation difficult (Thomson et al. 1992).

The above considerations, and the concrete results obtained by De Lisle et al. (1994) show that regional experience should be gained at the initial phase of satellite data-based studies of coastal processes providing the assessment of actual detectability of various coastal zone geomorphological features, and the estimation of the accuracy of measurements of linear and two-dimensional structures in a particular site/region, based on satellite imagery.

A remarkable set of studies related to the use of satellite observations for bathymetric surveys was carried out by the experts of Delft Hydraulics, Emmeloord (The Netherlands). Wensink et al. (1993) and Hesselmans and Wensink (1994) presented the results of the use of passive optical satellite imagery in the bottom topography problem area. The use of ERS-1 SAR to support bathymetric surveys was reported by Hesselmans et al. (1994b). Theoretical background and examples of the use of optical and SAR images to assess bathymetric information in coastal areas were also summarised by Hesselmans et al. (1994a).

Many kinds of activities in inshore and coastal waters, as well as offshore operations (including shipping, navigation, fishery, dredging, pipeline laying, harbours and platforms construction) require precise information on bottom topography. This information should be updated regularly, which is very time- and money-consuming. Usually the bathymetry survey is done by means of shipborne echo sounding techniques. In the Netherlands, with its traditionally high level of expertise in activities in its coastal zone, coast protection and dam construction, some research projects were carried out aimed at development of cost-effective operational techniques combining traditional and remote sensing methods.

The results obtained were based on theoretical background supported by advanced numerical modelling expertise. Bathymetry based on passive optical remote sensing uses the relationship between depth and light intensity as described by the two-flow model of Spitzer and Dirks (1987). This model relates the height of a water column with the amount of reflected radiation, reflectance of the bottom, attenuation coefficient of the water, and two empirical coefficients depending on the water composition (Hesselmans et al. 1994a). The unknowns of this model can be determined using *in situ* calibration measurements. Some assumptions are made: on the horizontal homogeneity of the atmosphere over the studied area, and on the constant composition of water and bottom surface. Wensink et al. (1993) consider remotely sensed bathymetric information giving a two-dimensional overview (though not always very accurate) as complementary to the traditional precise echo-sounding information at a point or along the ship transects. In their turn the remotely sensed bathymetric charts can be used to optimise the echo-sounding operations. Two examples of successful implementation of satellite optical bathymetry were presented by Wensink et al. (1993).

The bottom topography information was needed within an area of 10 km × 40 km to provide a coal transport study for the thermal power station near Kayamkulam (India). The panchromatic SPOT image of 1 February 1988 (spatial resolution of 10 m) was used to produce a bathymetric chart, with some survey data used to calibrate the depth algorithm. When later a new bathymetric chart was constructed, it was compared with the satellite image-based chart. Comparison

> showed differences of about 1 dm, being the measuring error. The SPOT imagery and processing cost $12 000, whereas the limited survey cost $60 000. If the whole ... system would have been surveyed by traditional means the survey costs would have doubled and it would have taken much longer. (Wensink et al. 1993)

Depth information over an area of 200 km × 200 km was needed for a site near Mirfa (United Arab Emirates) on the shores of the Arabian Gulf. For an area of 7 km × 10 km near the power and desalination plant the information was needed with 1 dm accuracy. A bathymetric chart was compiled using traditional echo-sounding survey data of 1992 (spatial resolution of 200 m) to calibrate the depth model, and the LANDSAT TM image of 15 July 1989. A site visit also brought some useful information on the high transparency of water, the constant composition of the bottom, and the stretches of algae along the shores. The difference between the satellite image-based bathymetric chart and the soundings

> turned out to be of the order of 1 or 2 dm ... Note that remote sensing data are used for an area of 15 000 sq. km, whereas the survey data cover an area of just 80 sq. km ...

The depth estimate based on the remote sensing imagery cost $30 000, whereas a survey of the same area would have cost $350 000. (Wensink et al. 1993)

Bathymetric charts based on optical remote sensing technique have been used in several projects conducted by Delft Hydraulics, with an accuracy approaching that of traditional methods (under suitable conditions), and can be constructed 'on an almost routine basis'. 'It has been shown that bathymetry based on optical remote sensing is both technically and commercially feasible. This technique is used on a commercial basis in bathymetric surveys world wide' (Wensink et al. 1993).

Another interesting example of the use of a similar technique in a sand inventory in the coastal waters of the Caribbean island Aruba (10° 30′ N; 70° W) up to a depth of 30 m was presented by Hesselmans and Wensink (1994). The objective of the study was to find suitable areas for mining sand and to assess the size of these sand resources, its quality and the consequences of sand mining for the marine environment.

Earlier Hesselmans (1990) demonstrated that in clear coastal waters the bottom features could be observed up to a depth of 40 m, with the relevant assessment of the composition of the top layer of sea bottom. The aim of the reviewed study was to investigate the technical and commercial feasibility of the technique for sand inventory studies. The LANDSAT TM image of 8 September 1988 was used (28.5 m × 28.5 m pixel size), along with an Admiralty Chart based on soundings performed in 1970–1972, and the *in situ* data on the occurrence of sand in the top soil layer of the sea bed in 927 sample sites in the depth range between 10 m and 30 m. As a result a chart of sand concentrations in the top layer of the sea bottom based on the satellite image (and calibrated by means of *in situ* data) was constructed. A comparison of this chart and the field data chart 'shows a remarkable correspondence' (Hesselmans and Wensink 1994). Some differences in the shapes of three sandy areas can be explained easily, if one takes into account that *in situ* data were collected in the area between 10 m and 30 m depth, while the satellite data cover shallower waters as well. In general, remotely sensed data were shown to be more suitable to provide estimates over larger areas, but not to predict sand occurrences at specific points.

In the course of further development of satellite-borne bottom bathymetry in Delft Hydraulics a technique based on SAR imagery has been implemented. Satellite bathymetry based on SAR remote sensing is related to shallow waters, for which the imaging mechanism can be presented as a sequence of three physical processes (Wensink et al. 1993): interaction between (tidal) current and bottom topography produces modulations in the surface current velocity; these modulations cause variations in the wave spectrum; and the variations, in turn, cause modulations in the radar backscatter. For each of these processes the relevant numerical models were developed (a more detailed mathematical formulation was presented by Calkoen et al. 1993). From these models, a model train was constructed which predicted the radar backscatter from bottom topography, depth-averaged current and wind. This model train can be inverted to estimate the bottom topography in shallow waters from radar imagery with the help of a data assimilation scheme. The Bathymetry Assessment System developed in Delft Hydraulics was briefly described by Hesselmans et al. (1994b) who also gave two examples of its implementation in the province Zeeland (a delta area in the south of the Netherlands) and in the Zeebrugge area.

In the Zeeland project, where the optical remote sensing bathymetry technique could not be used due to the high turbidity of the North Sea, an ERS-1 SAR image of 29 April 1992, taken in favourable hydrometeorological conditions (large tidal currents, and wind speed between 2 and 5 Beaufort), was used. The estimated bottom topography (calibrated by means of some sounding data) appeared to be 'in good agreement with the measured bottom topography' (Hesselmans et al. 1994b).

The project carried out in the Zeebrugge area included the execution of detailed multibeam bathymetric surveys to investigate sand wave activity and excavation effectiveness for the Zeepipe Development project, which included the laying of pipelines from a Norwegian gas field, one of the pipes coming to the town of Zeebrugge. 'The bottom topography of the area in question is characterised by numerous sand waves and bars which together with the strong regional tidal currents provide ideal conditions for the application of SAR imagery for bathymetric purposes' (Hesselmans et al. 1994b). SAR imagery of 1 September 1993 and depth information from Admiralty Charts were used to provide depth estimates of an area of 50 km × 50 km. The Admiralty Chart data were used as boundary conditions by the Bathymetry Assessment System (BAS). It was concluded that the available SAR data and the BAS made it possible to construct and to update large-scale depth maps based on remotely sensed data and existing Admiralty Chart data. 'Absolute accuracy achieved by combination of ERS-1 SAR and echo-sounding was of the order of 30 cm which is comparable to most industry-standard echo-sounders' (Hesselmans et al. 1994b). Thus the feasibility of optimisation of survey operations using SAR imagery was shown.

Wang and Koopmans (1994) demonstrated the usefulness of SAR imagery in a mapping method following the 'water line procedure' as applied to the monitoring of topographic changes in the bottom configuration.

4.3.2 Some notes on marine pollution monitoring and practical use of the environmental sensitivity index concept

A general approach to the problem of the use of remotely sensed data for marine pollution estimation was considered in detail by Victorov (1980b). The range of priorities of water pollutants is not strictly determined, and it can be substantially different for various regions of the world ocean. At the first international workshop on marine pollution monitoring held in Nairobi in 1974 a general classification of the most important pollutants was worked out on the basis of expert estimations. Petroleum hydrocarbons were considered to be of top priority among marine pollutants. Monitoring of this type of pollutant was meant to become an essential part of both regional and global programmes. Marine pollutants include chlorine-organic pesticides, phenols, detergents and heavy metals, among which mercury and lead were mentioned as most serious.

Except for oil slicks, these pollutants cannot be detected from currently used satellites. Only 'optically active' substances can be determined in sea water using remotely sensed data. The problem of detecting areas with an anomalous distribution of chlorophyll in the upper sea layer is similar in methodology to the general problem of detecting polluted zones. Chlorophyll should probably be classified as a specific kind of pollutant. Marine pollutants should also include the so-called suspensions – solid particles of different substances which get into the seas during dredging and bottom-deepening operations and also with river discharges (Victorov 1980b).

> Nowadays there are no operational specialised spacecraft systems meant for studying ocean pollution ... Though the sensors of the existing satellites comprising the operational and experimental spacecraft systems of various purposes enable to obtain some information about certain types of ocean pollution. (Victorov 1980b)

This was written in 1980, and 10 years later I had to say that 'the situation did not change dramatically' (Victorov 1990a). Now (1995) I am glad to see some progress in this problem area. With the ERS-1 SAR in orbit many scientific teams got the practical opportunity to participate in oil slick detection exercises based on the effect of changes in the high-frequency part of the wave spectrum (change in sea surface roughness or 'smoothing effect'). This process yielded a lot of publications which may give the impression that the problem of operational detection of oil slicks is about to be solved. Unfortunately, this is not the case, as the articles reviewed in Section 4.3.5 will show. During the past 15 years the geography of the use of satellite imagery of the visible band for monitoring sea areas with anomalous concentrations of suspended substances expanded considerably. This area of practical use of satellite imagery in regional monitoring activities is indeed becoming an operational, routine procedure in many seas, so that in Sections 4.3.3 and 4.3.4 only a few examples of the many dozens available will be presented in brief.

Now – in the context of complex monitoring of the marine and coastal environments – let us consider the practical implementation of the Environmental Sensitivity Index (ESI) concept. ESI was suggested by Gundlach and Mayer (1978) as a quantitative term to measure the value of the environment and a measure which could help to assess quantitatively the damage to it from man-made impact.

> While this concept is rather widely used today, its application varies according to operators and sites. Remote sensing is still seldom used in this respect (Jensen *et al.* 1990), although its value is generally acknowledged, especially in regions of the world with poor mapping material. This value has been recently enhanced by the availability of satellite radar images which allow access to areas with severe cloud cover. (Populus *et al.* 1994)

I selected the paper by Populus *et al.* (1994) to start this ESI-related discussion. Although this study was carried out using pre-GIS technology, and addressed the environmental sensitivity to oil pollution only, it raised and suggested reasonable solutions to some common problems related to satellite environmental monitoring on a regional scale.

The study area is the River Loire estuary, for which seven scenes from SPOT over a period of five summer months in 1990, one scene from LANDSAT TM of 28 August 1985, and one ERS-1 SAR image of 19 April 1992 were collected, complemented by airborne images of 27 and 18 August 1991. Relevant geological, biological and socioeconomic data have been collected in map and report formats. Though in some references the ESI was considered to be mostly a geomorphological concept, Populus *et al.* (1994), committed to maintain that concept, still believed that

> a complete ESI should be a combination of the following parameters (Dutrieux and Denis 1992):

- the shoreline index based on geomorphology and vegetation, which relates to the physical impact of pollution and is usually ranked from 1 to 10;
- the sediment type, an indicator of the restoration capacity;
- the wildlife, closely related to the quality of both vegetation and sediment;
- the water quality, regarding its living component (chlorophyll production);
- the human activities, economical as well as recreational. (Populus et al. 1994)

It was too difficult to present the whole set of data on a single map, and three maps were produced. The first map, resulting from satellite imagery, is mostly geomorphological and presents four sensitivity classes. The second map, resulting from *in situ* observations, deals with living resources, while the third map presents socio-economic features and regional infrastructure (roads, slips for access to the sea, harbours and marinas, administrative boundaries and demographic data including the number of residents and visitors, which is a very important factor in popular recreational areas). All the maps are 1 : 50 000 scale.

Now we come to the essential point – whether the 'three maps' (plus a few additional tables) match the requirements of a decision-maker (end-user)? The remarkable thing is that Populus et al. (1994) think there is a need to ascribe different weights to the various parameters involved. Their specific idea is to use the shoreline as the spatial 'support' for other parameters relevant to the ESI concept. (For details see Populus et al. 1994.)

To conclude, we must confess that the practical use of the ESI concept is actually far from being a clearly defined procedure. It requires deeper research and more experience of practical work with this concept. Even in the GIS context, actual science (or intuition) will be required when manipulating various data layers in the GIS structure, trying to give proper (?) weights to different layers. The solution will be strongly based on *regional* knowledge with maximal involvement of *regional* peculiarities, possibly including those not present in a GIS in a formalised file.

4.3.3 Turbidity determination and sediments transport studies

Water quality monitoring in the Golden Horn of Istanbul (Turkey) using LANDSAT-5 TM images as presented by Coskun and Ormeci (1994) is an example of the implementation of remote sensing techniques to the assessment of the environmental situation in a rapidly growing metropolitan area surrounded by water. The Golden Horn is the 8 km long arm of the Bosphorus in European Turkey forming the harbour of Istanbul. Water quality in the Golden Horn is affected by waste discharges from many industrial plants and municipalities. The polluting substances in this region listed by Coskun and Ormeci (1994) are total suspended solids (precipitating solids and small pieces of soil leading to a greyish water colour), particular organic substances suspended in water, humid organic material (which includes the products of biological decomposition of plants, living matter, and also industrial paints and pigments), and polyaromatic hydrocarbons (oil grease and heavy petroleum particles deposited into the water from the land by surface water and discharged from boats and ships). Coskun and Ormeci (1994) pointed out that 'these pollutants of a body of water are some of the most important parameters from an aesthetic point of view'.

Image processing of two radiometrically and geometrically corrected images (of 24 October 1986 and 9 September 1992) was performed using the image processing software Resource and the ERDAS System. Regression analysis was used to study the relationship between the water quality parameters and the reflectance values. It is remarkable to note that the results of the regression analysis were not very good when the non-synchronicity between the satellite data and *in situ* measurements was about a week; the results were considered as very good when the sea-truth data were taken on the same day as the LANDSAT overpass. The TM observed reflectance showed a strong relationship with total suspended solids, humic materials and polyaromatic hydrocarbons.

Charts of turbidity distribution were made, and some information on the surface flow carrying the polluted water mass was obtained (Coskun and Ormeci 1994).

An investigation of remote sensing technology for semi-closed coasts has been carried out by Maktav and Kapdasli (1994). They pointed out that semi-closed bays are more sensitive from the ecological point of view, because all the dynamical processes here are weaker than at the open coast, and recovery processes after man-made impact are not so efficient. At the same time semi-closed bays were very attractive for people to live at and to develop industry, trade and navigation leading to a high pollution load. Maktav and Kapdasli (1994) studied the eastern coast of the Fethiye Bay (Inner Bay) including the Fethiye Harbour, located at Southern Anatolia (Turkey), part of a protected area since 1988.

LANDSAT-5 TM data of 8 August 1984 and 27 July 1991 were used to study the suspended sediment patterns and investigate 'the environmental changes' (Maktav and Kapdasli 1994) in the area.

> Comparing both images of the dates 1984 and 1991, an increase of turbidity at the northern part of the inner bay can be recognised. It is a surprising result that at the more closed southern part, a significant change has not been seen. (Maktav and Kapdasli 1994)

One cannot comment on this result, as Maktav and Kapdasli (1994) did not present the data characterising comprehensively the hydrometeorological conditions shortly before and at the moment when the satellite images were taken. A lot of additional data are actually required to analyse the environmental changes using satellite imagery. The attempt of using only two images with a seven-year interval between them for the assessment of the state of the marine environment by means of comparison of turbidity patterns (with their high variability) does not seem to be a good example to follow. This important issue will be discussed in more detail in Section 5.5 in which multi-year satellite and airborne imagery – as applied to turbidity studies in the Neva Bay (Baltic Sea, Russia) – will be described.

Conceding their presentation, Maktav and Kapdasli (1994) wrote that though remote sensing is capable of providing monitoring of semi-closed bays on a synoptic scale, actually 'however, cloud cover and delays in data acquisition seriously diminish its usefulness for monitoring on anything less than a seasonal basis', which sounds too pessimistic.

Another paper, dealing with the use of satellite imagery of the visible band in studies of some aspects of estuary dynamics by means of turbidity pattern analysis, is based on the 15-image series from LANDSAT TM and SPOT HRV sensors (Zujar *et al.* 1994). Turbidity patterns were used as 'natural tracers' to obtain views of the circulation patterns under different environmental conditions in the Tinto-

Odiel estuary located in the Atlantic face of the Iberian Peninsula (Huelva, SW Spain). Parts of this area were declared natural protected areas by the regional government which launched the monitoring programme of the quality and dynamics of the marine waters and coastal zone in Andalusia using remote sensing techniques (Zujar et al. 1994). Within the framework of this programme, satellite imagery is regarded as (a) a complementary source in the study of coastal water dynamics and assessment of water quality, (b) an essential source of data for the regional environmental information system of Andalusia, and (c) cartographic documents meant to test the *in situ* data collected during field campaigns and the results derived by numerical modelling activities.

Fifteen satellite images were selected in an attempt to cover 'as many different hydrodynamic situations as possible' and Zujar et al. (1994) believe that those '15 satellite images correspond to 15 synoptic situations of the turbidity patterns, associated to different hydrodynamic conditions in the estuary (tidal phase and coefficient, wind ...)'. The behaviour of the estuary over the tidal cycle was traced and analysed in detail for various water levels from 0.41 cm to 2.59 cm (pressure corrected water height).

To improve the existing knowledge of estuarine dynamics and the dispersion and renewal processes of estuarine waters, the obtained turbidity patterns were compared with the established 'flow schemes' based on measurements taken from 20 October to 4 December 1981. The results were summarised 'in four flood and ebb situations, depending on the coastal waters eastward or westward [of the] dominant flow direction' (Zujar et al. 1994). The analysis of satellite imagery initiated useful discussion on the structure of currents in the Bay and its variability.

Along with the obtaining of valuable concrete oceanographic/environmental results of regional sounding, Zujar et al. (1994) managed to draw some more general methodological conclusions. Based on their practical experience, they pointed out that the use of turbidity patterns as natural tracers is more useful in ebbing situations than during the flood phase. This method is more useful in shallow well-mixed waters, while in stratified or deep estuaries the dynamic interpretation of turbid patterns 'would be much complex' (Zujar et al. 1994).

A comprehensive approach to the problem of monitoring and forecasting of sediment transport in coastal waters seems to have been demonstrated in the general outline of the regional project COAST (Coastal Earth Observation Application for Sediment Transport) presented by Peck et al. (1994a,b). It is a typical regional project covering an area of about 100 km along, and at least 20 km offshore at 'two areas of contrasting water quality and geomorphology, namely Christchurch Bay, West Solent (southern coast of England) and the Humber Estuary, Holderness coastline (eastern coast of England)' (Peck et al. 1994a). Maps of these sites are presented by Peck et al. (1994b). The two areas are called test areas, as the techniques and technologies developed within the framework of this project could well be used in other areas. The overall objectives are:

- 'to develop a robust and practical methodology for extracting coastal zone management data from Earth Observation (EO) data;
- to initiate an operational Water Quality Information Service providing water quality information pertinent to coastal zone management' (Peck et al. 1994a).

The Water Quality Information Service (commencing in 1996) has already started marketing activities and mailed a detailed questionnaire to potential users.

Figure 4.8 Fragment of the Caspian Sea. METEOR satellite image of 20 May 1984 (Sukhacheva and Victorov 1993).

The planned EO data are the SeaWiFS satellite imagery (for characteristics of this sensor see Section 2.3.1) and the CASI airborne data. CASI (Compact Airborne Spectrographic Imager) is an imaging spectrometer (up to 288 channels of 3 nm bandwidth in the 490–914 nm range) providing a spatial resolution of order 10 m from about 3 km height within a 5 km swathwidth (Peck *et al.* 1994a).

The planned COAST output informational products include regional and local maps of coastal surface sediment concentrations and relevant statistical data; water quality parameters (including algae blooms) and relevant statistical information; coastal suspended sediment vertical profiles. Another set of output products includes predictions (forecasts, nowcasts and hindcasts, of sediment transport, rates of accretion/erosion and changes in coastal morphology). The estimated forecast period is of the order of a few weeks (Peck *et al.* 1994a).

The operational fusion of SeaWiFS imagery with data from airborne imaging spectrometers and the relevant *in situ* data within tailor-made GIS is a remarkable feature of the COAST project providing its possible synergetic potential in marine and coastal zone research (Victorov *et al.* 1995).

4.3.4 River plumes studies

The general concept of regular observations/monitoring of patterns of water constituents dispersal from coastal sources and particularly from river plumes as formulated by Barale and Schlittenhardt (1994) is:

142   REGIONAL SATELLITE OCEANOGRAPHY

it is often impossible for the CZCS to distinguish the signature of biogenic pigments from that of the total load of dissolved and suspended material present in the water. However, the observation of such features provides important clues on coastal frontal dynamics and potential correlations with nutrient enrichment, sediment transport, pollution sources.

Long before this statement was made, patterns of river plumes were studied in European seas using radiometric sensors much coarser than CZCS. It is worth mentioning studies carried out by Horstmann (1983, 1986, 1988) using AVHRR/NOAA, LANDSAT and CZCS data regarding the Baltic Sea basin Rivers Odra, Wistula, Nemunas, and Venta; Victorov *et al.* (1984) using MSU-M and MSU-S/METEOR imagery regarding discharges and plumes from the Rivers Danube (at the western coast of the Black Sea) and Chorokh (at the southeastern coast), and the River Kura plume in the Caspian Sea. The latter has been studied since 1980 by means of multichannel aerial photography from AN-30 aircraft at 500–7000 m height (to take account of the atmosphere); later these studies were complemented by satellite imagery analysis (Victorov 1983; Sukhacheva and Victorov 1993).

The hydrological front in the northern Caspian Sea separates less saline waters influenced by the Volga River discharge, from the water mass of the middle Caspian Sea. In Figures 4.8 and 4.9 this front can be seen as the border between turbid shallow waters (light tone) and clear waters (dark tone). The River Volga delta is on the top left of both images. The hydrological front in the northern Caspian Sea is recorded in satellite imagery, with SST and suspended matter (turbidity) acting as tracers. The location of the front is governed by river discharge, winds, currents, etc. As a rule, the front is registered between 10 m and 20 m isobaths. The front is

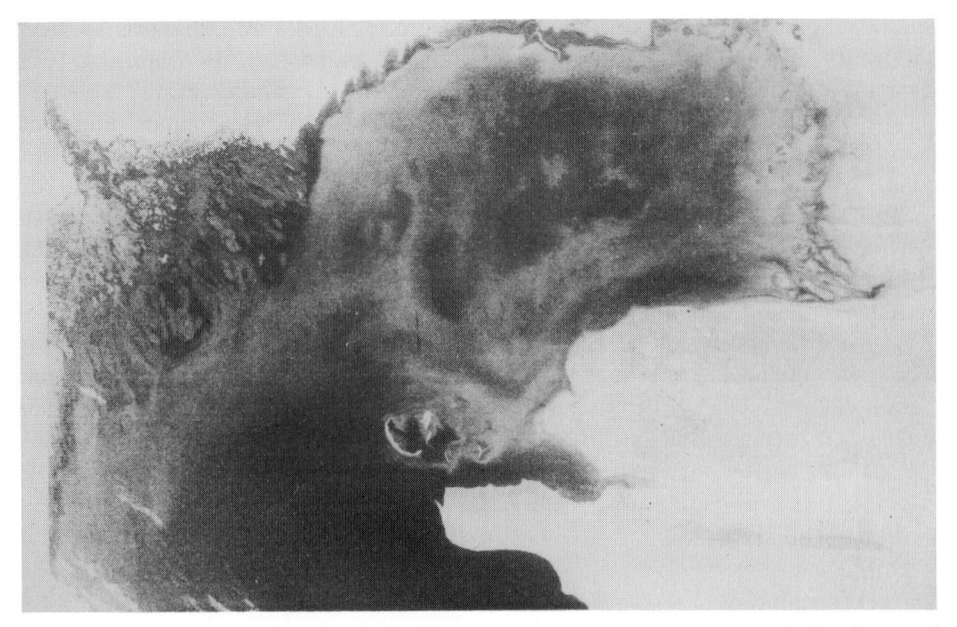

Figure 4.9 Fragment of the Caspian Sea. METEOR satellite image of 31 May 1986 (Sukhacheva and Victorov 1993).

shaped along them. For example, on 20 May 1984 the hydrological front followed the 10 m isobath (Figure 4.8) (Sukhacheva and Victorov 1993). In the top right part of Figure 4.9 the delta of another river, the Ural, can be seen.

Figure 4.10 shows the remnants of the Aral Sea. Two rivers, the Amu-Darja from the south and the Sir-Darja from the east, have been discharging into the sea for centuries. But nowadays they do not reach the sea. Though the decrease of their discharge due to inadequate water management (the water has been taken for irrigation purposes in tremendous volumes without proper control) was discussed in the official *Large Soviet Encyclopedia* (2nd edition) as early as 1950, nothing was done to prevent the disaster. In recent years the total area of the sea is decreasing rapidly, and hundreds of fishing boats are now found in the desert. The Aral Sea became a matter of international concern. Some elements of the satellite monitoring of the Aral Sea were carried out by Sukhacheva and Victorov (1990). Let us hope that, in decades to come, scientists will once more observe the river plumes in the Aral Sea. Nowadays, the image in Figure 4.10 shows not only turbid waters and bottom features but also (as compared with the imagery of 1984) the changes in coast-line configuration at the southern and eastern parts, and the increase of the total area of the island of Vozrozhdeniya (due to the decrease of the sea level). The former island of Lazarev is no longer an island, the Ajibai Bay disappeared, and the Small Sea bay actually became a separate water body (Sukhacheva and Victorov 1990). In the analysis of the Aral Sea satellite imagery one should take into consideration the changing coast-line, the new emerging features of the bottom topography, as well as the specific aerosol content (salt dust from the former sea bottom).

Now we come back to the living seas with their river plumes. Examples of coastal interactions due to fluvial runoff in the northern Adriatic Sea and in the Ligurian–Provenzal basin have been recently shown by Barale and Schlittenhardt (1994) using CZSC imagery. They also presented the plume features originated by the Rivers Po, Ebro and Rhone. Small plumes of 'high pigment waters' extending from the Bosphorus seem to trace the Black Sea outflow into the Sea of Marmara (Barale and Schlittenhardt 1994). These studies were carried out within the framework of the OCEAN project set up in 1990 by the Institute for Remote Sensing Applications, Joint Research Centre of the European Commission (Ispra, Italy) and the European Space Agency to take advantage of the existing CZCS time series. A set of 15 000 images was prepared in several formats for users (Barale and Schlittenhardt 1994).

Hakansson (1989) reported on studies of the Glomma River plumes in the northern part of the Skagerrak area based on the LANDSAT TM image of 13 May 1988. The total suspended matter concentration varied from 10 to 35 mg l^{-1}; the area of river plume was about 200 km^2; and strong synoptic variability of the plume features was shown using AVHRR data for May–June 1988.

Armstrong (1994) reported on the distribution of ocean colour in the northeastern Caribbean Sea (from about 14–20° N, 60–70° W) as recorded in the CZCS image of 1 September 1981. The dominant feature in this image corresponds to seasonal South American river intrusion in the east Caribbean, with the highest values associated with the Orinoco River discharge. As the high ocean colour signal associated with this plume resulted from a combination of chlorophyll, phaeopigments and dissolved organic matter, no quantitative analysis of the image was made (Armstrong 1994). The presence of Amazon/Orinoco river water in the Caribbean was first reported in 1964 based on the analysis of *in situ* salinity data; satellite-based data thus supported the results of the previous studies.

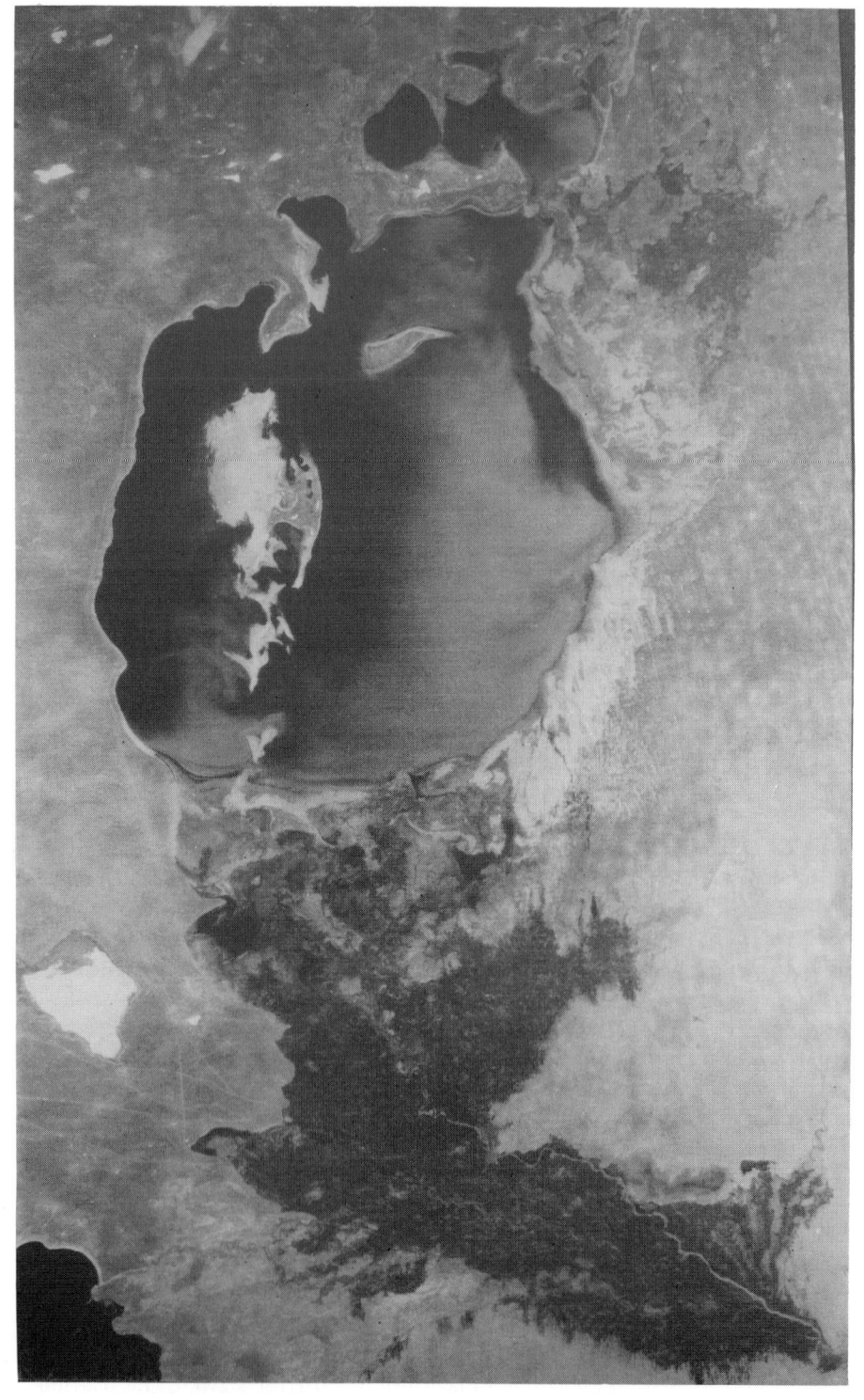

Figure 4.10 Aral Sea. Satellite image of 20 August 1989 from the OKEAN satellite, MSU-S, 0.5–0.7 m km.

The Caribbean sea 'which was thought to be devoid of significant biological activity, shows complex circulation patterns and phytoplankton patchiness that appear to be mediated, for the most part, by seasonal South American river water intrusions' (Armstrong 1994). The relative contribution of the Amazon River and the Orinoco River to seasonal and interannual ocean colour variability in the eastern Caribbean 'will probably be resolved using the SeaWiFS' (Armstrong 1994).

In recent years much attention has been paid to numerical modelling of currents and sediment transport in coastal waters on a regional and local scale. With the development of computer technology the grids used in these modelling exercises are becoming fine enough and comparable to the spatial resolution of satellite sensors. Still the efforts to combine the benefits of fine-grid models with those of high-resolution satellite imagery seem to be insufficient. I tried to discuss this situation with the experts in numerical modelling but could not get a clear answer. In many cases (too often, in my opinion) the results of numerical modelling cannot be proved/checked by *in situ* measurements. The modelling community usually says that there are no adequate measurements with proper spatial and temporal resolution to validate their models. (But why should they exist? If there were such datasets, nobody would need any models!) My message is: there is a gap between numerical hydrodynamical (dispersion) models and remotely sensed data, and the modelling community seems to hesitate (or not be able) to assimilate the remotely sensed data in their models.

Victorov *et al.* (1991) invited national experts in hydrodynamical modelling 'to use satellite data to establish or improve interior and boundary conditions and some parameters of numerical hydrodynamical models'. Blumberg *et al.* (1993) wrote:

> there is a need to make better use of available observations. Models require data to establish interior and boundary conditions, to update boundary fields, to validate the model physics and to verify the simulations ... One needs to blend the results from both circulation and water quality models with the available data to provide the best estimates of how water and materials are transported throughout a coastal system. The data assimilation, that is, the process of this blending, is undoubtably the most powerful tool presently available for extracting information and insight from the sparse coastal ocean data sets and the imperfect model results.

Having said this, and having demonstrated the comprehensive model work on transport processes in the coastal ocean, Blumberg *et al.* (1993) still used the only SST image just to illustrate the modelled upwelling event. I am sure the blending of sophisticated models with high-resolution satellite imagery should be given more attention. The remote sensing community and the modelling community should find a common language on a regional level.

4.3.5 Oil slick detection

In Section 4.3.3 presenting turbidity and sediment transport studies, and in Section 4.3.4 dealing with river plume studies, the emphasis was put on the optical (visible and near IR) bands of the electromagnetic spectrum. In this section, as well as in

Section 4.3.7, we will mainly deal with satellite radar imagery. Along with wave mapping in the coastal/inshore areas, and the mapping of coast-line features (see Section 4.3.1), satellite radar imagery is expected to become an important source of operational information for detection of oil slicks and in the application area which Werle (1993) called 'monitoring for anomalous marine activities' with a potential market related to 'coastal and inshore area surveillance for: illegal activities (e.g., transport of drugs, marine dumping); contravention of fisheries or defense regulations'. Characterising the SAR information products as 'new, with virtually no existing competition', Werle (1993) writes, that 'however, given the innovative nature of these products, market acceptance may be tentative at first and considerable emphasis must be placed on verification studies to build user confidence'. Some examples of current activities in this problem area will be presented below.

After ERS-1 SAR was launched, many demonstrational/pilot projects have been carried out.

> Research and development efforts have been carried out primarily in North America and in Europe ... When compared to other, more advanced SAR applications development efforts, e.g., in the disciplines of agriculture and forestry, investigations of coastal regions were restricted to using only a relatively narrow range of radar system and imaging parameters. Therefore, many coastal zone studies were not in a position to conduct a more complete assessment of SAR capabilities in terms of identifying optimum radar frequency, polarization, and incidence angle for a particular application or target of interest. (Werle 1994)

The reader should bear this remark in mind when analysing the following reviewed papers.

An operational infrastructure based upon the West Freugh (UK) ground station that 'would be capable of providing an information service to a number of end user organizations in Europe, such as the European Environment Agency' was presented by Sloggett (1994). The hardware component of this infrastructure is the Oil Slicks Detection Workstation (OSDWS) developed by Earth Observation Sciences Ltd (UK) for the Defence Research Agency at Farnborough and the British National Space Centre. The software algorithm for oil slick detection was verified using (a) the results of modelling the signature of oil slick on the sea surface, and (b) the results of the analysis of known 'test' slicks in SAR imagery. There are plans to increase the system functionality by introducing a predicting module, and providing information on the path of the slicks (Sloggett 1994).

The practical experience gained in the Netherlands in operational applications of airborne side-looking imaging radar was used in conducting experiments meant to study the potential of ERS-1 SAR for oil slick detection in the North Sea (Bos *et al.* 1994). Nowadays the North Sea Directorate of Rijkswaterstaat is using aircraft equipped with radar, IR/UV sensors, a photo-camera and a night vision camera on a daily basis with over 1200 flight hours a year, with flights scheduled randomly in time and area (Bos *et al.* 1994). Satellite SAR is considered as a potential additional system for oil slick detection in the context of the following user requirements related to an 'early warning' subsystem: time delay less than 1 hour; coverage – traffic lanes and coastal zone; repeat cycle less than 12 hours; success rate exceeding 75%; slicks with an area of 0.01 km^2 and more should be detected . It should be noted that these requirements are not met either by airborne radar (Bos *et al.* 1994) or by satellite SAR of ERS-1 type.

Plate 1 (Figure 3.16c)   Thermobar phenomenon in the Ladoga Lake as recorded in NOAA SST satellite imaging for 20 May 1988. The thermoinertial zone is shown in blue and dark blue. Photograph from display. (After Bychkarva et al. 1989a).

Plate 2 (Figure 5.9)   SST pattern in the Baltic Sea as recorded in the AVHRR/NOAA-9 image of 27 May 1985. Channel 4, HRPT mode. Photo from display.

Plate 3 (Figure 5.11)   Development of the upwelling at the eatern coast of the Gulf of Riga as recorded in SST AVHRR data. Moscow time. Photo from display (temperature scale unit in Kelvin degrees is indicated in brackets): (a) 13 May 1985, NOAA-6, 09:20 (0.5) (b) 15 May 1985, NOAA-9, 15:35 (0.75) (c) 16 May 1985, NOAA-9, 15:00 (0.25) (d) 18 May 1985, NOAA-6, 10:45 (0.5) (e) 19 May 1985, NOAA-6, 10:00 (0.25) (f) 23 May 1985, NOAA-6, 10:20 (0.75) (Bychkova et al. 1988b).

Plate 4 (Figure 5.30)   SST pattern in the south-eastern part of the Baltic Sea as recorded in the AVHRR NOAA image of 19 May 1985 at 7 h UTC. APT mode. Photo from display.

Plate 5 (Figure 5.43)   Stripes of algae in the northern Baltic Proper. Photo taken by the author on 4 August 1994 from the ferry *Anna Karenna*.

Plate 6 (Figure 5.65)   Digital processing of satellite image of the Neva Bay. KOSMOS-1939, MSU-SK scanner, 12 July 1988. Photo from display.

Plate 7 (Figure 5.78)   The Neva Bay. Photo taken from Soviet satellite on 9 July 1981, 9:53 Moscow time, height 280 km, original scale 1 : 200 000. Colour film. The image shows large areas of suspended matter fields *before* the construction of the Flood Barrier.

Plate 8a (Figure 5.82)   The Neva Bay. Satellite photographic image of 11 August 1983. The image shows large areas of suspended matter fields at the *initial* stage of construction of the Flood Barrier.

Plate 8b (Figure 5.88)   The Neva Bay. Photograph taken from Soviet satellite on 5 July 1989. 10:36 Moscow time, height 278 km, original scale 1 : 200 000. Colour film. The image shows large areas of a suspended matter field *after* construction of the Flood Barrier.

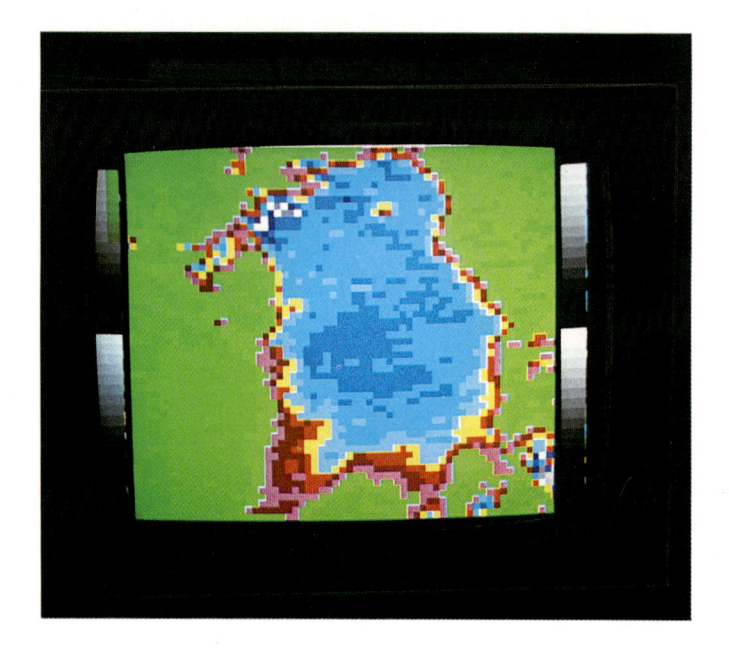

Plate 1

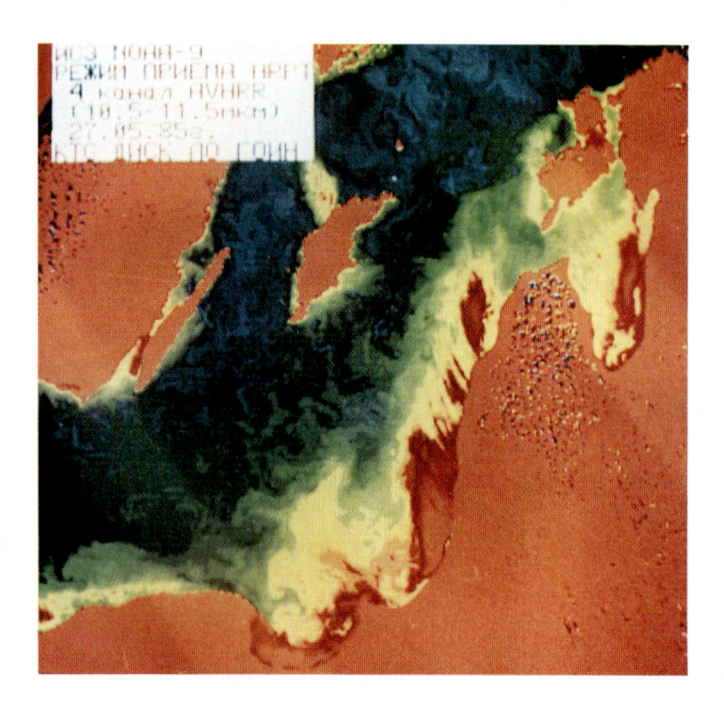

Plate 2

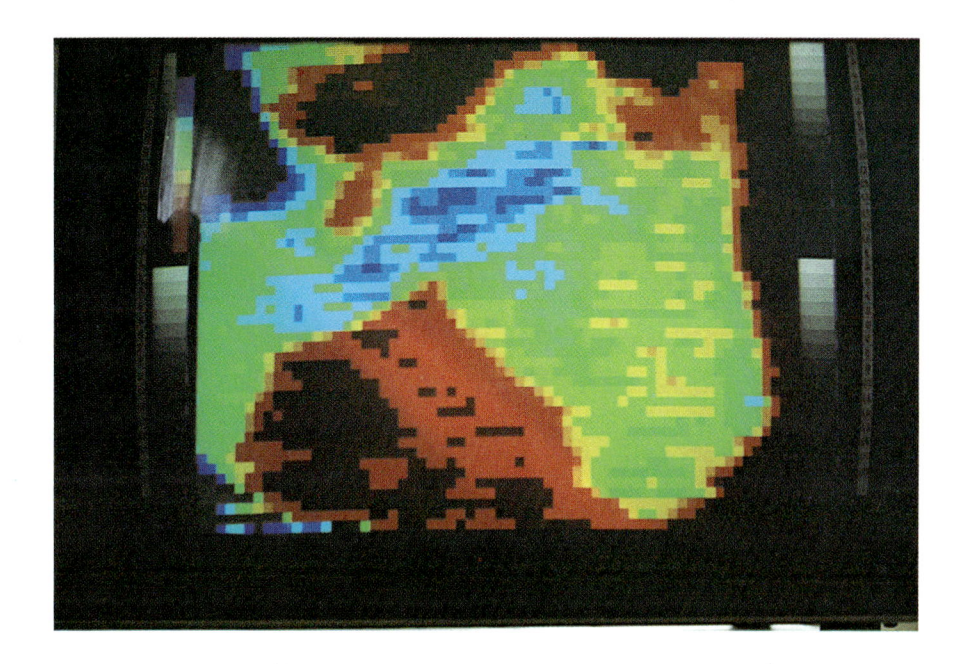

Plate 3a

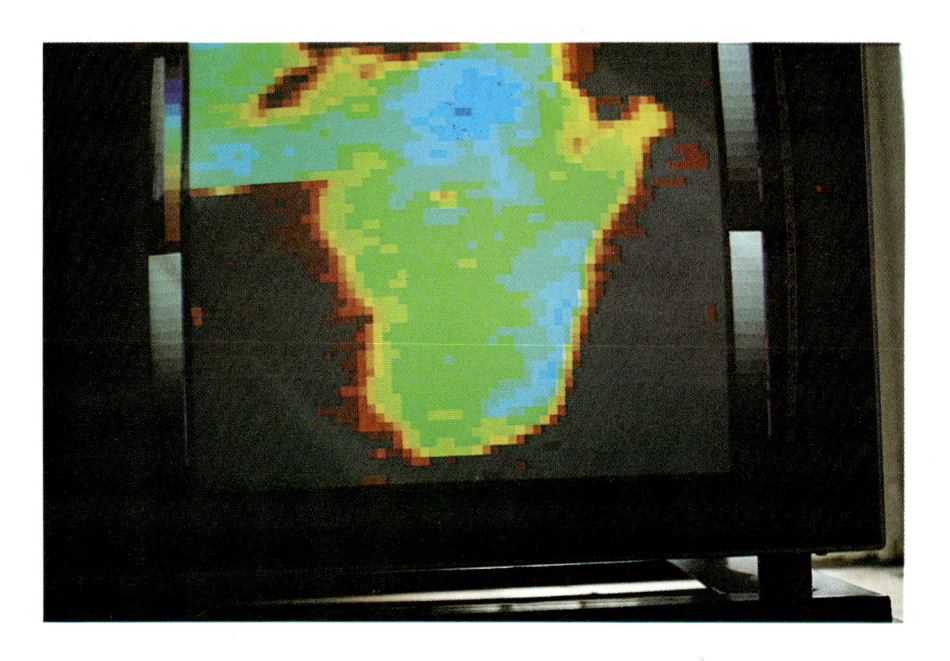

Plate 3b

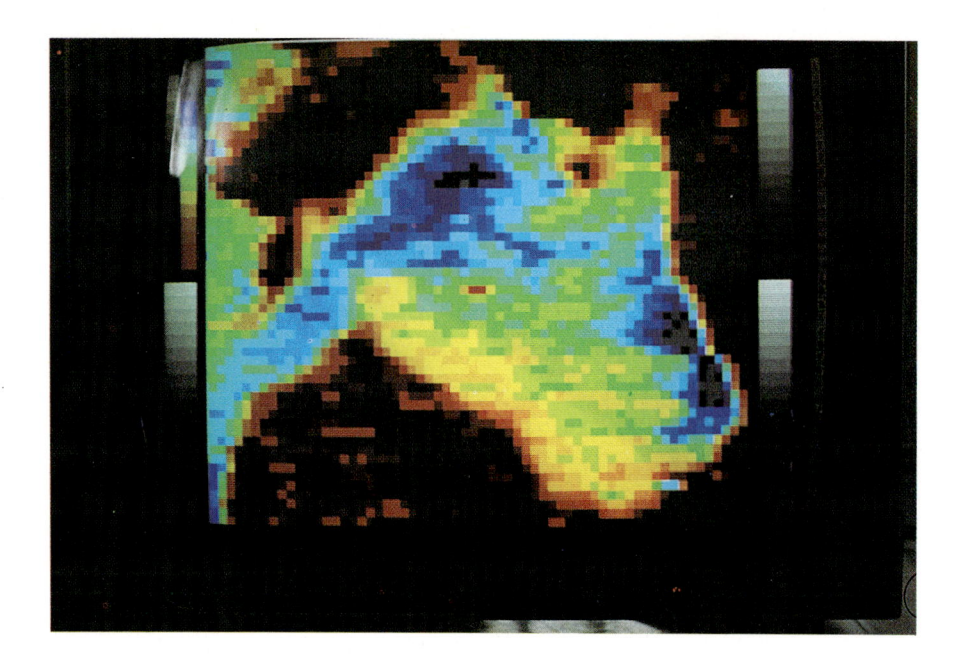

Plate 3c

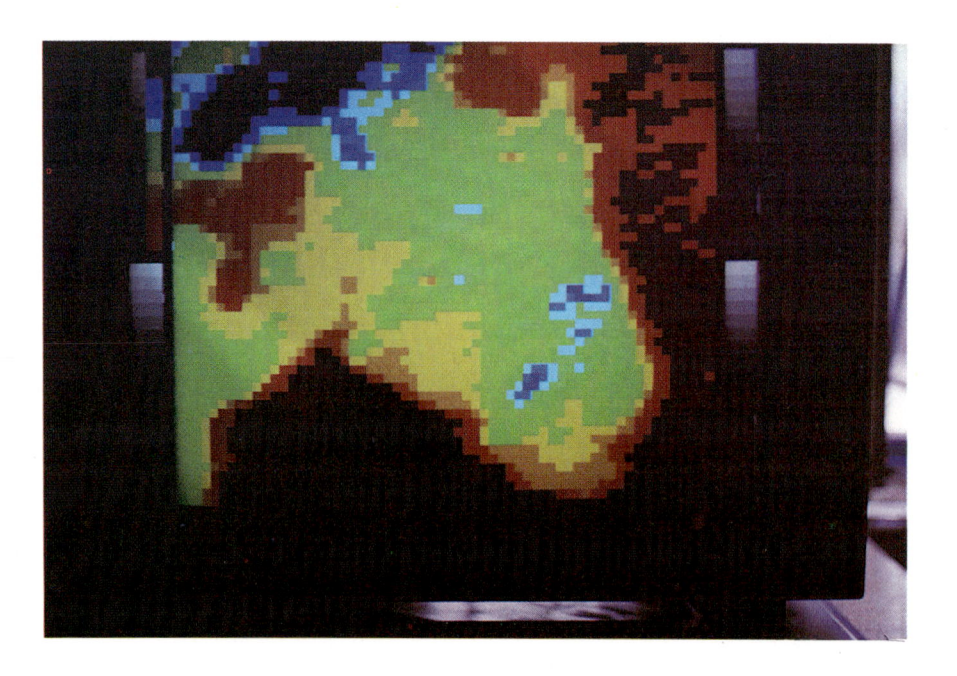

Plate 3d

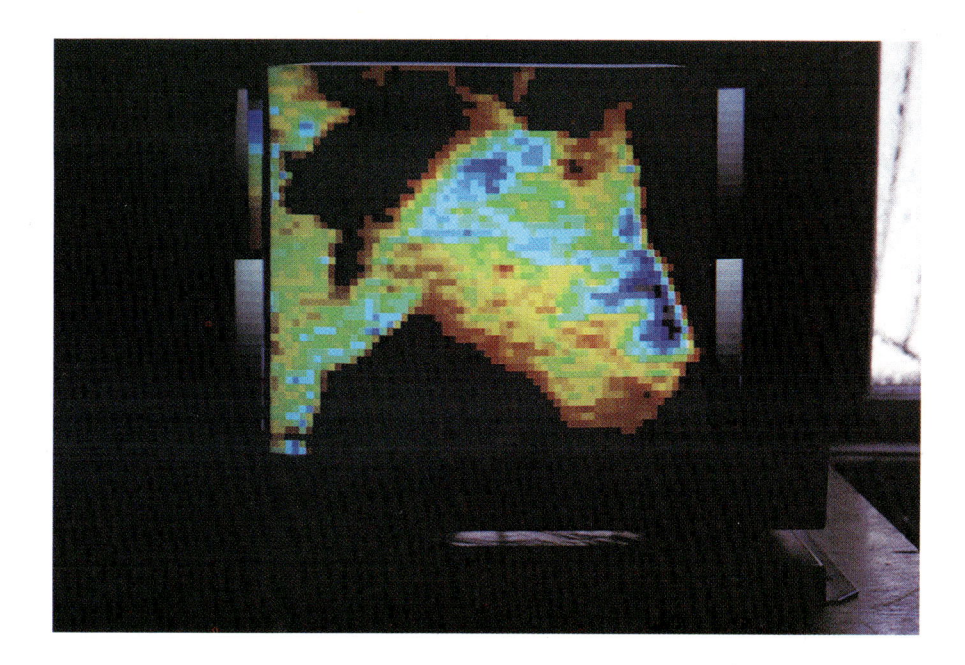

Plate 3e

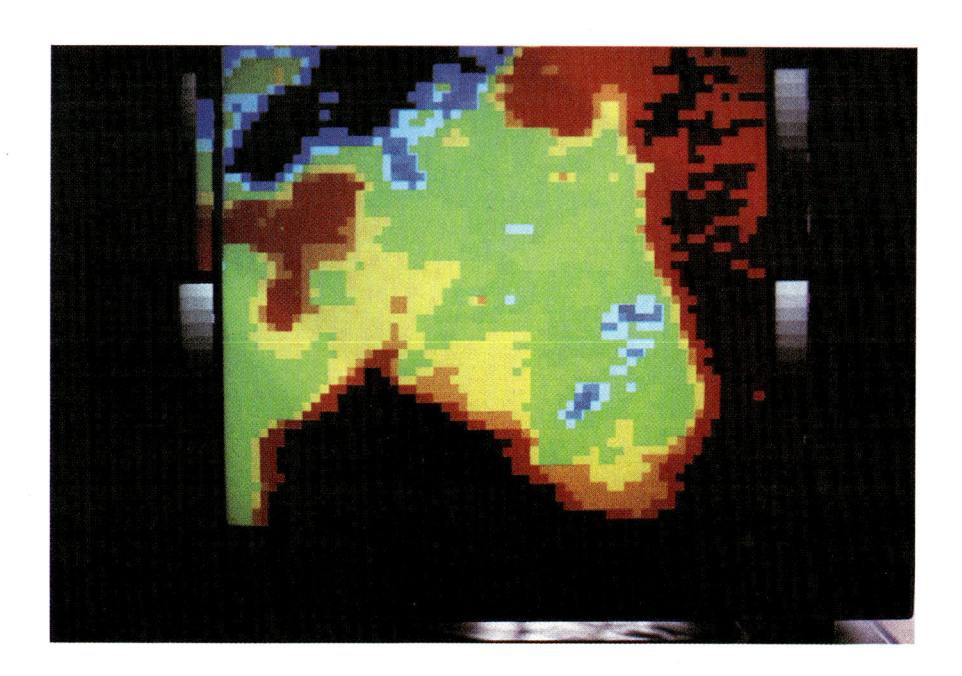

Plate 3f

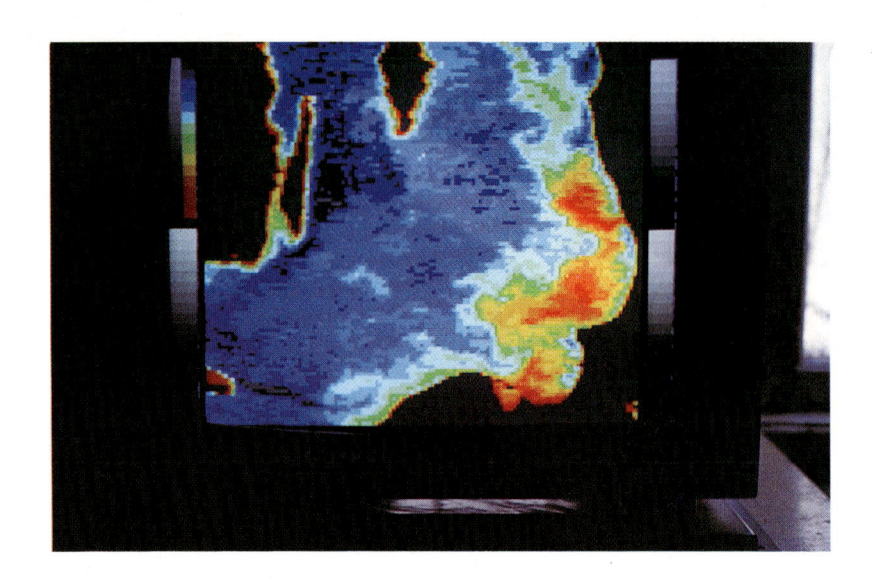

Plate 4

Plate 5

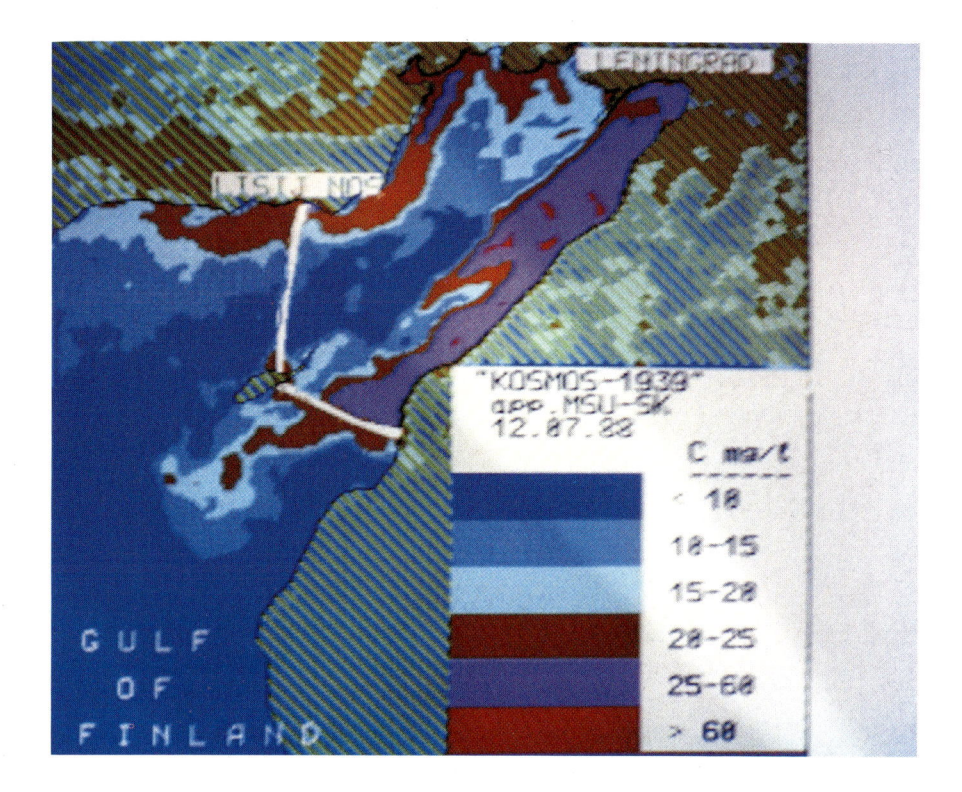

Plate 6

Plate 7

Plate 8a

Plate 8b

In the period from 1 June to 31 December 1993 a pilot experiment was carried out during which 171 ERS-1 SAR frames were obtained, and nine simultaneous flights were performed. A visual method for the interpretation of the SAR images was used, with a graphic workstation used only to perform routine image manipulations. Initially, the 171 frames were analysed by an operator 'with little practical experience in slick detection'. Later the procedure was repeated which resulted in 48 new slicks and 63 rejected events. A total of 192 slicks were detected (Bos *et al.* 1994). This result means that indeed the training of operators is a vital element in monitoring of oil slicks based on satellite SAR imagery. We will discuss this important issue later in this section.

The flights of aircraft PH-MNZ were carried out as three long parallel routes over the area covered by satellite SAR frames. Each flight lasted for four hours – the maximum allowed operational flight time of this aircraft. (The total flight time of the veteran IL-14 aircraft used in international Complex Oceanographic Subsatellite Experiments in the 1980s in the Baltic Sea (see Section 3.4) and in the national programme of oil slick monitoring in the Caspian Sea in 1975–1990 (see Section 3.3) was eight hours.)

It is important to mention that 'the nature of most slicks is not known because no ground truth data could be collected for all ERS-1 passes due to the restraint in available flight hours .... However at this moment we are confident that slicks, although of unknown origin (oil, other pollutants or algae), can be detected.' (Bos *et al.* 1994). I am afraid this approach means that one problem is probably being substituted by another one, much easier to tackle. In this respect it is worth discussing whether the flight routed over a regular grid was the best way to study the slicks and to help in the interpretation of satellite SAR imagery. It is possible that another technique, the so-called 'additional observations' approach (see Victorov 1980b) could be more fruitful. To introduce this approach briefly, let us assume a situation where the ground-based processing system and satellite sensors do not give unambiguous information on the nature of a certain anomaly on the sea surface. In this situation, additional and more detailed information provided by an independent system is required. It is reasonable to perform such an 'additional observation' by means of an airborne remote sensing system (Victorov 1980b). In the experiment under discussion, it seems that in some designated cases a well-equipped aircraft could have been used to perform 'additional observations' in order to distinguish between oil slicks and non-oil-slick events.

During nine simultaneous satellite/aircraft missions the airborne radar detected 36 'slicks' over 290 000 km^2 of sea surface during 29 hours in flight. In 34 frames of ERS-1 SAR imagery covering 270 000 km^2, 34 'slicks' were detected. Only 11 'slicks' were detected by both instruments, of which six were less than 1 km^2 in area (Bos *et al.* 1994). The practical experience gained during this subsatellite experiment led to the submission of proposals on the setting up of an operational system for oil slick detection based on satellite SAR data (Bos *et al.* 1994). In their opinion, commercially available image processing software 'with an application layer on top is in favour'; visual interpretation of satellite images 'is the best option for this moment', with a kind of knowledge base of oil slick imagery being used in the analysis, and 'the manpower needed for a maximum of 400 images a year will be about 50–100 hours for image processing and interpretation' (Bos *et al.* 1994).

The same infrastructure established in Norway for ship detection (see Section 4.3.7) was used 'to test and demonstrate near real time detection of oil slicks based on ERS-1 SAR images' (Wahl et al. 1993). It is known that the radar backscatter from the sea surface is plagued with speckle. 'The speckle distribution is known *a priori*' and thus – in principle – can be corrected for, using statistical detection techniques (Wahl et al. 1993). The philosophy behind the Norwegian activities in oil slick detection is that the oil spills from oil and gas platforms in the North Sea are usually reported quickly and treated properly according to routine procedures; what is required is the detection of illegal oil spills from ships. A total of 177 SAR images were analysed in the period from 1 July 1992 to 31 January 1993. The weather conditions were very unfavourable during this period, 'with winds much too high for detection of oil spills not forming stable emulsion' (Wahl et al. 1993). This statement could be illustrated by the 'Braer' accident in the Shetland Islands in January 1993, when 80000 tons of North Sea crude oil spilled from the ship; this was not detected either by airborne radar or in ERS-1 SAR images.

The experiment showed, however, that low-resolution (100 m) SAR images were suited well enough for detection of oil slicks, and in general the results were promising at low wind conditions. The presence of natural slicks 'will cause a high number of false alarms if a dark slick is the sole detection criterion. However, a well-trained human operator is in many cases able to discriminate between natural slicks and probable oil slicks', the latter being considered as an encouraging factor for the development of a fully automated detection system. Several Norwegian research institutes are reported to be working on the problem of automated discrimination between natural and true oil slicks (Wahl et al. 1993).

Now we come to the training aspects of the oil slick detection issue. McFarland and Hall (1994) discussed the scope and components of an aircraft observer training programme meant to aid the remote sensing effort in collecting truthful information during the operational phases of an oil spill response. The topic of this article is not directly connected with the topic of satellite data-based oil slick detection, but there is a remarkable point worth presenting in brief here. When the aircraft observer and the operator of a satellite image processor start their work, both of them are pre-programmed to find oil. 'Perception is often affected by what is expected' (McFarland and Hall 1994), which results in too many oil events being detected in the satellite imagery and to overestimation of oil spill features during airborne observations. Nowadays the problem of aircraft observer training is being discussed very seriously. It seems that the problem of training the satellite imagery interpreter is of equal importance.

4.3.6 Studies of water dynamics

Studies of boundary oceanic currents, and the associated fronts and eddies, historically were the first examples of the use of satellite imagery in oceanography. Satellite-borne sea surface temperature (SST) patterns for the first time showed an 'instant view' of the Gulf Stream current. The location of the main core of the Stream, the major eddies and meandering process – all these characteristics of the regional current system – were the subject of keen scientific analysis two decades ago. It is worth mentioning, among others, the studies of Gulf Stream rings carried out by Doblar and Cheney (1977) and Schmitz and Vastano (1977), and observations of the dynamical behaviour of the Stream as recorded in satellite images of the IR band (Maul and Baig 1977; Morgan and Bishop 1977; Niiler 1977). All those pioneering studies showed remarkable temporal and spatial variability of the Gulf Stream. An interesting feature of that phase of the Stream studies was the use of guided research ships meant to collect *in situ* sea-truth subsatellite data. The first experiments of that kind were carried out in spring 1973 and 1974, when the ship was guided to a ring near the North Carolina coast, followed by similar studies of 1976–1977 (Richardson 1980). Probably some hundreds of papers have been published during almost two decades, dealing with various aspects of the use of satellite images in the activities related to the Gulf Stream. This phase of scientific research and demonstration pilot projects developed into another, operational, phase. Nowadays the Gulf Stream is being monitored on a regular and routine basis, and Gulf Stream Analysis Charts are being issued; they are available in various formats. 'Major information on the location of the main core of the Stream and the major eddies are also broadcast daily over the marine radio network' (Needham 1983). The regional marine community, recreational and commercial fishermen, headboat and charterboat operators and sailors are the users of these charts. 'In 1982, the American Swordfish Association reported a saving of $2.25 million in fuel costs alone by utilizing these charts' (Needham 1983).

Satellite monitoring of the flow of the powerful west boundary current of the Pacific Ocean – Kuroshio – has also become operational. Satellite images are used to study mesoscale eddies, ocean fronts and other dynamical phenomena in the vicinity of Japan (Yamamoto 1994).

Satellite images of boundary currents were used to develop quantitative techniques of front detection and analysis (among the pioneering works see, for example, Legeckis 1977, 1978; Gerson and Gaborski 1977). The fronts and frontal features are detectable in SST imagery due to the considerable difference in temperature (more than, say, 0.5°C) as compared to the neighbouring waters.

Studies of regional (mesoscale) features of water dynamics, including estimates of current velocity, are based on the concept of 'tracers' (see Section 5.3.3.1 on visualisation of water motion in remotely sensed two-dimensional data). There are two major ways to determine the current velocity using time series of satellite images (or at least a sequence of two images). The first technique is based on the Lagrangian tracking of any 'labels' in the satellite-borne field (inhomogeneities of the SST field or the field of suspended matter can be used as 'labels'). The second technique is based on the numerical solution of the inverse problem describing the tracer's motion, provided that some assumptions are valid. The second method was used, for example, by Vukovich (1974) at the eastern coast of the USA; Bychkova *et al.* (1988b) reported on the utilisation of this technique to estimate the current velocity at the eastern coast of the Caspian Sea; recently Ilyin and Lemeshko (1994) reported on the use of this method at the southern coast of the Crimea peninsula in the Black Sea.

The SAR era opened new horizons for studies of regional water dynamics (for a recent review of radar remote sensing of coastal regions see Werle 1994). Fu and Holt (1982) used Seasat SAR imagery to show the detectability of mesoscale features associated with current system boundaries, fronts and eddies in the range of kilometres to hundreds of kilometres. There are lots of examples of the detection of dynamical phenomena at the sea surface in radar imagery of medium (1–3 km) and high resolution. (Sections 2.3.2.3 and 2.3.2.4 give some information on the less publicised radar imagery from the former USSR satellites.)

Now we turn to nearshore ocean wave patterns. Werle (1994) emphasised that according to evidence brought by Carter *et al.* (1988) the majority of requests for wave information in the 1990s are for areas close to land and in the form of short-term site-specific forecasts. This is just the problem area covered by regional satellite oceanography!

The capabilities of SAR satellite imagery for practical engineering- and environmental monitoring-oriented applications were shown to be encouraging, though

> yet, our understanding of the SAR imaging mechanism(s) involved in detecting sandbanks, sandwaves, surface expressions of near-shore currents and patterns of near-shore wave refraction is not on a sound footing. (Werle 1994)

This aspect was already mentioned in Section 4.3.1.

Of particular interest to regional satellite oceanography are the problems related to detection of internal waves. Satellite imagery of the visible band gave many examples of patterns that were interpreted as manifestations of internal waves at the sea surface. One should know that it is actually very difficult to prove these results by sea-truth measurements. Often it is impossible to understand a pattern in a satellite SAR image without *a priori* knowledge of the bottom features that could *possibly* generate this or that detail in the image. An example of a complicated pattern of internal waves registered in the METEOR image of 1 July 1980 (MSU-S scanner, 700–1100 nm channel) in the middle of the Mediterranean Sea was presented by Kazmin and Sklyarov (1981).

According to Fu and Holt (1982) most of the internal waves recorded in SAR satellite imagery (at surface winds exceeding 2–3 m s^{-1}) occur in coastal areas; the waves propagate toward the shore in separate groups of 4–10 crests, with a distance between these groups of several hundred metres to several kilometres; the crests are 10–100 km long. Following Hughes and Gasparovich (1988), Werle (1994) pointed out that the oceanographic information content of satellite SAR imagery of internal waves was considerable. These features may be a source of knowledge on the depth or the density interface which, in turn, 'could be important for planning fishing and military operations ... Also, when internal waves travel over shoals, they tend to break eventually and produce intensive mixing. These areas of influence often have an important effect and impact on the biology and primary production of coastal waters' (Werle 1994).

4.3.7 Ship traffic monitoring

The assessment of the utility of SAR satellite imagery in ship traffic monitoring based on state-of-the-art technologies was presented by Wahl *et al.* (1993). Large quantities of ERS-1 SAR fast delivery images (30 m nominal resolution, pixel size 16 m × 20 m) produced by the Tromso Satellite Station were analysed to assess the capability of this imagery and the relevant algorithms to detect ships of various sizes. Both signals from ships as 'hard' targets and ship wakes were analysed.

In SAR images ships are usually visible as bright targets against the dark background of the sea surface. The actual intensity of a ship pixel is reported to 'depend strongly on aspect angle, ship infrastructure, and the material used' (Wahl *et al.* 1993). The state of the 'background' sea surface, the size of a ship, and the location of the ship in the SAR frame were studied by Wahl *et al.* (1993).

On a calm sea surface at 1–2 m s^{-1} wind speed most ships, even small ones, were detectable, but it was impossible to detect a long ship with a well-known position in a stormy sea (15 m s^{-1} wind speed; 3 m significant wave height). Sometimes it was difficult to detect 100 m long ships at 10–12 m s^{-1} wind speed. Based on the results of the analysis of some 400 SAR images carried out at the Norwegian Defence Research Establishment (NDRE), Wahl et al. (1993) presented the following detection limits, illustrating the ERS-1 SAR detection capability: the smallest ship found in SAR imagery was 11 m long; at 4 m s^{-1} wind speed the lower limit on the length of visible ships was 50 m, at 10 m s^{-1} it was 100 m. Ship wake statistics for ERS-1 SAR showed that dark turbulent wake events were the most frequent cases, with Kelvin arms being in second place (Wahl et al. 1993). The analysis supported the theory that the 'hard' target, the ship itself, is 'a more robust signature than the wake, even at stronger winds' (Wahl et al. 1993). An automated system for detection of ships in SAR imagery was developed at NDRE, in which a full digital terrain map of the Norwegian mainland and islands (90 m grid) was included. The system was capable of finding most of the ships that were found by operators.

The ship detection capability of ERS-1 SAR was better in the very long-range part of the swath than in the near-range part, the major limitation being the steep incidence angle. This means that the use of larger incidence angles should favour ship detection. Future satellites (Radarsat and ENVISAT) will make the use of larger incidence angles possible. In conclusion, Wahl et al. (1993) considered detection of ships more than 120 m long 'feasible ... under most weather conditions'; detection of ships shorter than 50 m was regarded as 'very uncertain'. 'Detection of medium-size ships is feasible much of the time, but problematic at wind speeds above 10 m s^{-1}; ... it is not recommended to use ERS-1 in its standard mode for counting fishing vessels, except for at very low wind speed' (Wahl et al. 1993).

CHAPTER FIVE

Regional satellite oceanography: case study of the Baltic Sea

5.1 Introduction

In Chapter 4 many examples of applications of satellite data to various aspects of regional oceanographic research, monitoring the marine and coastal environments, and management of coastal zones in many seas were presented. These examples showed that satellite oceanography of the seas or 'regional satellite oceanography' is indeed becoming an effective instrument of marine research and operational oceanography.

This chapter will deal with a single sea. An attempt will be made to illustrate the systems analysis approach to the problems of the study of the Baltic Sea based on complex use of the available satellite imagery of various bands and *a priori* data consisting of sea-truth subsatellite data and experts' knowledge of processes and phenomena.

As an object of study by means of remote sensing observations, the Baltic Sea is a very hard nut for the following reasons: (a) the spatial scale of phenomena and processes here is rather small, and generally only high-resolution satellite data are useful; (b) the shape of the sea is very remarkable – it extends from the South to the North, with the Gulf of Bothnia spreading far to the North, and the Gulf of Finland spreading far to the East, thus leading to rather distinct variability of climatic conditions based on mere geographical dimension; (c) the coastal line is not smooth, it is chopped (cut), with a lot of small bays and caps, each playing their role in generating the overall pattern of oceanographic fields; (d) the total number of cloud-free days is not very high (see Chapter 3 for details) which makes it difficult to obtain time series of satellite imagery of visible and IR bands providing the best spatial resolution.

These were probably the reasons that pioneering studies in regional satellite oceanography were carried out elsewhere. Except for studies of ice cover by Swedish and Finnish experts, more or less regular use of satellite data for marine science in the Baltic Sea started as late as in the 1980s when collection of selected images of various areas of the Baltic Sea taken from the NOAA (AVHRR) series of satellites and from 'Nimbus-7' (CZCS) was presented along with plausible explanations of various patterns recorded in the fields of radiative temperature in terms of biological productivity (Horstmann 1983). This publication struck a chord in the Soviet Baltic oceanographic community, and the question was put forward: 'Are there Soviet satellite images of the same high quality?'

Basing on shipborne measurements of the vertical inhomogeneity of chlorophyll concentration Kahru (1986) discussed the feasibility of remotely sensed two-dimensional data on chlorophyll distribution in the Baltic Sea. This topic-oriented study has led to the investigation of interannual variations of surface cyanobacterial accumulations in the Baltic Sea based on AVHRR/NOAA data (Rud and Kahru 1994).

The Baltic was used as a test sea for the development of techniques and data processing procedures by the Soviet remote sensing community. The concept of regional satellite oceanography originated here (Victorov 1988a, 1989b), and long-term experimental satellite monitoring of the Baltic Sea as a whole was performed in the 1980s (Victorov 1986b, 1987, 1988b; Bychkova et al. 1990a). A series of studies has been carried out dealing with various dynamical processes, including oceanographic fronts and upwellings (Bychkova et al. 1982a, 1985a, 1987, 1988a; Bychkova and Victorov 1987), eddies (Bychkova et al. 1985b; Bychkova and Victorov 1988), mushroom-like structures (Brosin et al. 1986, 1988), river discharges and plumes (Bychkova et al. 1990b), as well as seasonal algae blooms and suspended matter distribution (Victorov and Sukhacheva 1984; Victorov et al. 1984, 1987, 1988, 1990a).

Along with those topic-oriented studies some studies on the subregional level, e.g., the south-eastern part of the Sea (Victorov et al. 1986a), or site-oriented studies followed, including complex investigations of the Gulf of Finland (Victorov et al. 1989a; Victorov 1990a; Bychkova et al. 1994), the Neva Bay (Victorov 1991b; Victorov et al. 1991; Victorov and Sukhacheva 1992a,b; Sukhacheva and Victorov 1994; Sukhacheva and Tronin 1994) and the Kurshi Bay (Victorov et al. 1986b). These activities will be reviewed in Sections 5.3–5.5. Three complex oceanographic subsatellite experiments have been carried out (Berestovskij et al. 1984, 1993; Brosin and Victorov 1984; also Chapter 3 here) with the former USSR and the former GDR oceanographers participating.

Co-operative efforts in oceanographic research is a remarkable feature of the Baltic Sea studies. International co-operation here has a long tradition going back to the nineteenth century (Smed 1990; Matthaus 1987; Brosin et al. 1994). Even in the period of the cold war there were joint activities in the Baltic Sea with scientists from the 'West' and the 'East' participating, e.g., the Cooperative Synoptic Investigation of the Baltic in August 1964, the International Baltic Year 1969/70, the 1977 Baltic Open Sea Experiment (Kullenberg 1984), and the 1986 Patchiness Experiment (Dybern and Hansen 1989). Some information on Baltic Sea monitoring activities can be also found in Section 5.2.

Still there was no East–West collaboration in satellite oceanography and it was only in 1988 when, for the first time, an expert from the USSR had the opportunity to attend the 16th Conference of Baltic Oceanographers (CBO) in person (Victorov 1989a), though some papers had been presented 'on behalf' of the Soviet remote sensing community at the 13th (Victorov et al. 1982a) and the 14th CBO (Victorov 1984a).

Among the recommendations of the 16th CBO, giving to the regional oceanographic community the guidelines and research priorities for the next two-year period, along with customary recommendations, a special 'Recommendation 1988–4 on Remote Sensing' was issued:

The 16th Conference of Baltic Oceanographers,

- following the general tendency of modern world oceanography,
- taking notice of the increasing activity in applying the methods of satellite oceanography to the studies of the Baltic Sea, and the positive results already obtained,
- taking further notice of the future development in satellite sensors, data evaluation algorithms, and image processors,
- recommends intensification of studies in regional satellite oceanography especially with application to problems of water dynamics (upwellings, eddies, currents), eutrophication and ice cover studies in the Baltic Sea,
- recommends further that common conception of satellite data evaluation to the needs of the Baltic Sea studies be worked out,
- recommends also that scientists in the countries around the Baltic Sea conduct joint studies in the remote sensing including complex subsatellite experiments, satellite imagery banks, modification of data processing algorithms, etc. (taken from the proceedings 1988).

As a response to this recommendation an attempt was made to create a general framework which could cover many aspects of the monitoring, analysis and assessment of the state of the environment (including marine and coastal) in the Baltic Sea region with adequate involvement of various data sources, among which RSO techniques could find their place. The concept of a super-GIS has been proposed, driven by concern over the current situation in the region where, among the other serious matters, the existing scientific knowledge and operational facilities are not capable of providing adequate monitoring of the pollution of the sea, detection of hazardous spills and forecast of their development, where the current level of confidence between laymen and environmental agencies is not very high, and where there is a strong need for environmental education of the population starting from the children.

5.2 The 'Baltic Europe' geographical information system

5.2.1 General remarks

The following text is based on my concept of a regional 'Baltic Europe' GIS put forward for the first time at the International Workshop on 'Environmental Management in the Baltic Region' held in Leningrad, USSR in 1989. The broad scope of scientific and managerial problems was discussed at this Workshop, organised by the Centre for International Environmental Cooperation (Centre INENCO) including 'environmental problems and socio-economic processes in the Baltic region', 'Monitoring needs and conditions', 'Scientific knowledge, decision-making and public acceptance', and 'Systems analysis and modelling'. We attempted to discuss informally the activities of various international bodies (the Helsinki Commission, the UN Economic Commission for Europe, European Statisticians, WHO/Europe, Baltic Oceanographers) and numerous national institutions in the Baltic region to better understand the aims of each participating part, to become informed of current problems and needs of participants, and to find the intersections and gaps between the programmes. This approach, and the fact that a recently established strictly non-governmental organisation – Centre INENCO – initiated and hosted this Workshop, was absolutely new for the Russia of 1989. (We had problems with publishing the Workshop Proceedings and managed to do it only after three years; see Environmental Management in the Baltic Region 1992.)

Later the concept of the 'Baltic Europe' GIS was discussed at the Work Session on Specific Methodological Issues in Environment Statistics (a separate symposium in the framework of the Conference of European Statisticians held in Stockholm on 5–8 June 1990) as part of my paper dealing with satellite environmental monitoring of the Gulf of Finland (Victorov 1990a). It was recommended that this concept be presented as a separate document. Following this recommendation the text of proposal 'International Project in Regional Environmental Geoinformatics. Part 1. Concept' was written (Victorov 1992b).

This concept was also publicized at the International Conference on a Systems Analysis Approach to Environment, Energy and Natural Resource Management in the Baltic Region (Copenhagen 1991) (see Victorov 1991a).

A version of the text emphasising population health issues and their relation to the state of the environment was presented at the WHO Consultation 'Development of a Health and Environment GIS for the European Region' (Bilthoven, The Netherlands, 1990) (Victorov 1990b).

5.2.2 Objective of the project

The principal objective is to improve the state of environmental geoinformatics in the Baltic Sea region, thus providing for all the participating countries reliable, truthful, comprehensive and consistent information on the state of the environment in the whole region both in operative and non-operative modes (including an element of alarm control for some parameters).

This objective is being naturally pursued in four main approaches:

(1) the development of existing (and, where needed, the establishment of new) information networks with the aim of installing standardised reliable and precise sensors for measuring the necessary parameters of the state of the environment on land, sea and atmosphere;
(2) the development of procedures for the collection and standardisation of routine data treatment, and exchange of information products (based on network data, expeditions and remote sensing data) between the participating countries with the aim of feeding the databases with the bulk of information 'on line';
(3) the establishment of a regional integrated geographic information system (GIS) providing free access for each participating country to all the information concerning the state of the environment in the whole region in operative (or close to) mode and a broad variety of analytical information products upon request;
(4) the establishment within the GIS of an 'alarm structure' providing, through special 'hot-lines', alarm signals of various levels indicating discomfort situations in the environment which might be of potential harm for the neighbouring countries (territories or aquatories).

Thus, the proposed project differs greatly from some regional projects in geoinformatics, for example, the CORINE Project (CORINE 1989), comprising all the steps of collecting information, its processing and use (see figures and tables below).

5.2.3 Area of the project

The project is meant for the Baltic Sea region – 'Baltic Europe'. This term was taken from Serafin and Zalesky (1988) and Zalesky and Wojewodka (1977), and used in connection with regional GIS by Kondratyev and Victorov (1989). The geographical bases of the 'Baltic Europe' concept is the drainage basin of the Baltic Sea. Baltic Europe includes Sweden, Finland, Denmark, Poland, part of the former German Democratic Republic, Schleswig–Holstein in the former Federal Republic of Germany, the republics of Lithuania, Latvia and Estonia, and Leningrad, Novgorod, Pskov and Kaliningrad provinces (oblasts) of Russia. (The reality of today's Russia is that the city of St. Petersburg is the centre of the still existent Leningrad oblast.)

Figure 5.1 shows the boundaries of Baltic Europe. Some basic geographical characteristics of Baltic Europe are presented in Table 5.1.

The 'basin' principle (the sea plus its drainage area) is the natural sequence of the applied systems analysis approach to environmental problems. Baltic Europe is a very complex ecosystem of subcontinental level (Table 5.2) with heavy industrial (Table 5.3), agricultural (Table 5.4) and urban (Table 5.5) load closely connected with the shallow and semi-enclosed Baltic Sea, which integrates the problems of the region.

5.2.4 Scientific and technological background

There is no need to give even a brief review of the present environmental situation in Baltic Europe. It may be characterised as rather serious in general and very serious at some local sites.

The amount of polluting emissions is tremendous. Some data on these emissions for the Baltic Sea are given in Tables 5.6–5.8, compiled from the materials published in the experimental compendium *Environmental Statistics in Europe and North America* (1987). In this book the United Nations Statistical Commission and the Economic Commission for Europe presented a comprehensive review of the available data on the Baltic Sea environment. This 'statistical monograph' (Environmental Statistics in Europe and North America 1987) is of real value for everyone interested in the problem.

A comprehensive review of the present situation for the Baltic Sea is given in the *First Periodic Assessment of the State of the Marine Environment of the Baltic Sea Area, 1980–1985* published by the Helsinki Commission (Baltic Marine Environment Protection Commission 1986). The second Assessment is being prepared. New

Table 5.1 Geography of Baltic Europe (including Kattegat and Danish Straits) (Serafin and Zalesky 1988)

Area of administrative domain (000 km^2)	1652
Area of drainage basin (000 km^2)	1721
Water surface area (000 km^2)	415
Water volume (000 km^3)	21.7
Mean depth (m)	52.3
Maximum depth (m)	459

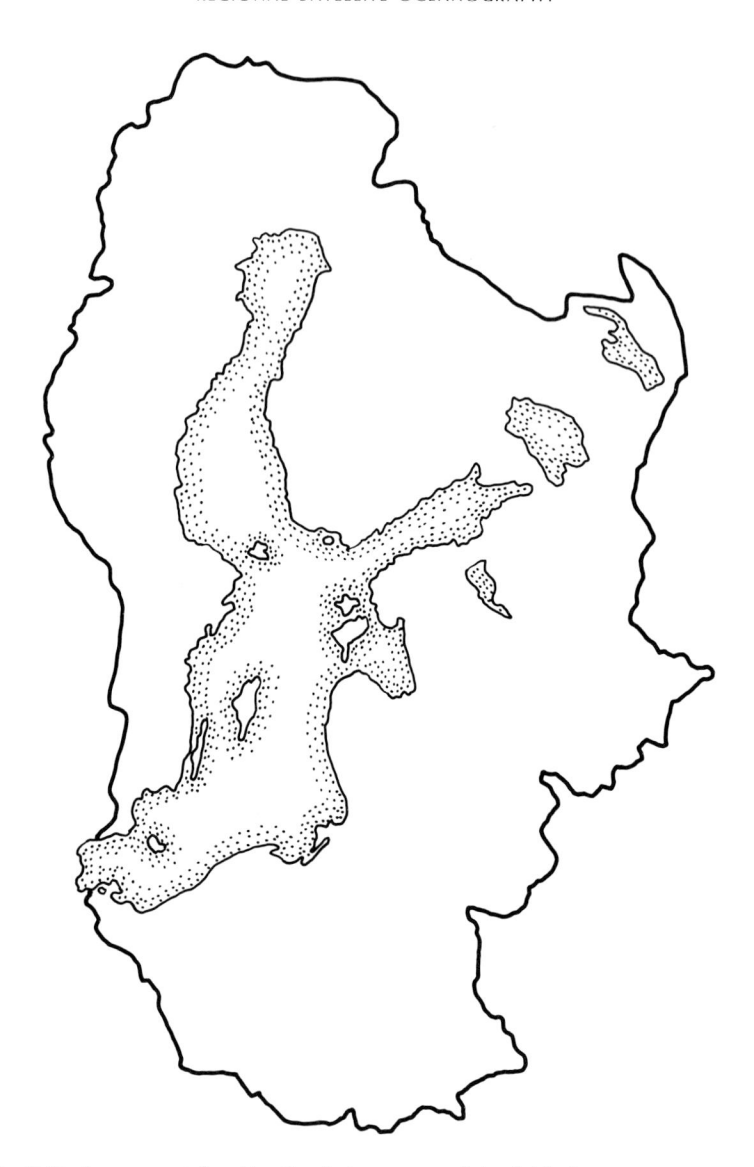

Figure 5.1 Baltic Europe as defined by the drainage area of the Baltic Sea.

results in the studies of Baltic Sea pollution are currently published in the Proceedings of the Conferences of Baltic Oceanographers held every even year (the latest was held in Sopot, Poland, 1994). Below, some remarks on the state-of-the-art of environmental monitoring and GIS in Baltic Europe will be presented.

Bilateral activities of the Baltic countries in marine environmental protection are encouraging. Special symposia on monitoring activities are taking place in cooperation with the Baltic Marine Environment Protection Commission – Helsinki Commission (see, e.g., Baltic Marine Environment Protection Commission 1986).

Long-term monitoring in the Baltic Sea itself started in the beginning of this century within the framework of ICES (International Council for the Exploration of

Table 5.2 Changes in the population of Baltic Europe (Serafin and Zalesky 1988)

Country/region	Area (000 km²)	1850	1900	1950	1984	1984 Density persons/km²
		(000s persons)				
Denmark	43.1	1415	2450[a]	4281	5110	119
Finland	337.0	1637	2656	4030	4860	14
Sweden	449.8	3471	5137	7041	8330	19
Poland	312.7	4852[b,c]	25106	25008	36571	117
GDR	108.2	7400[d]	16700	18388	16699	154
Estonian SSR	45.1	620[d]	1107[e]	1104[f]	1518	33
Lithuanian SSR	65.2	1050[d]	2029[e]	2561[f]	3539	54
Latvian SSR	63.7	910[d]	1845[e]	1954[f]	2584	40
Leningrad oblast	85.9	485	1267	3321	4827	75
Kaliningrad oblast	15.1	106[d]	189[h]	705	632	54
Novgorod oblast	55.3	68[d]	175	698	741	13
Pskov oblast	55.3	80[d]	220	744	842	15
Schleswig–Holstein	15.7	980	1380	2598	2615	156
West Berlin	0.5	380[d]	1600	2085	1820	3640
Baltic Europe	1652.6	23454	61861	74518	90888	56

[a] 1901; [b] Russian partition only; [c] 1854; [d] estimated; [e] 1922–1925; [f] 1951; [g] Konigsberg 1867; [h] Konigsberg 1897.

the Sea) and had been carried out by the bordering countries until World War 1. After World War 2, the Baltic Oceanographers and the Baltic Marine Biologists together with ICES planned joint international research. (The history of the Conferences of Baltic Oceanographers is described by Matthaus 1987.) In 1957 seven countries agreed upon the standard deep sea stations and areas to be observed regularly. Baseline monitoring started in 1972 within the framework of ICES and

Table 5.3 Selected economic indicators for Baltic Europe

Indicators	Year	
Annual population growth rate (%)	1950–1960	0.9
	1970–1981	0.3
Average real annual growth of agriculture (%)	1960	3.4
	1981	1.6
Average real annual growth of manufacturing (%)	1960	6.0
	1981	4.6
Food consumption per capita (calories/day)	1960	3300
	1981	3400
Energy consumption per capita (kg coal equivalent)	1960	334
	1981	6535
Movement of commodities by water (thousands of tonnes)	1968	31105
	1981	160469

Table 5.4 Indicators of agricultural and energy development of Baltic Europe for 1900–1985 (Serafin and Zalesky 1988)

Year	Production of wheat (million tonnes)	Electricity generated (billion kWh)
1900	4.8	1.1
1925	4.5	—
1950	5.7	53.4
1975	10.3	340.0
1985	14.0	402.7

continued from 1979 within the framework of the Baltic Monitoring Programme (BMP) (Guidelines for the Baltic Monitoring Programme 1984).

As was stated by Astok et al. (1986) at the Symposium on Monitoring Activities 'the most serious critical remarks made on the BMP concern the sparseness of its station grid and the low sampling frequencies ...'. The improvement in monitoring quality is attainable in two alternative ways. The first 'is to observe more frequently in time and space. This is the ideal way from the viewpoint of marine sciences'.

Figure 5.2 shows the spectrum of the energy of the Baltic Sea (after Astok et al. 1986). The scale of BMP is shaded and the arrows indicate what the temporal resolution could be if the scale of monitoring activities is extended.

Now Figure 5.2 illustrates the point of view that the main tendency in developing the monitoring is to shift the bar to the right using remote sensing data (where appropriate). Another approach is 'to reduce the number of observations performing monitoring at the space points where and when the variability is minimal' (Astok et al. 1986). Here also remote sensing data could be very helpful in indicating the areas with high and low gradients of oceanographic fields.

It is worth mentioning in this connection the experience gained during the Joint Complex Oceanographic Satellite Experiment performed by scientists from the State Oceanographic Institute, Leningrad and the Institute for Marine Research, Rostock in 1985, when the research vessel was actively guided by means of commands worked out on the basis of on-line analysis of satellite IR imagery revealing the highest gradients (anomalies) of the sea surface temperature field in the south-eastern Baltic (see Section 5.3 for details).

International monitoring of the Baltic Sea is complemented by national investigation programmes for coastal waters in order to survey discharges from land-based sources

Table 5.5 Growth of cities in Baltic Europe (number of inhabitants in thousands) (Serafin and Zalesky 1988)

Year	Copenhagen	Stockholm	Leningrad	Warsaw
1800	101	76	220	100
1850	129	93	485	160
1900	401	301	1267	638
1950	1168	928	3321	601
1982	1372	1402	4779	1635

Table 5.6 Nutrient emissions by source (tonnes per year) (Environmental Statistics in Europe and North America 1987)

Area	Municipalities Total N	Municipalities Total P	Industries Total N	Industries Total P	Rivers Total N	Rivers Total P	Total Total N	Total Total P
Kattegat	6490	1735	590	240	46000	1120	53080	3095
Belt Sea and western bays	14230	3810	4180	2061	39600	2550	58010	8421
Sound	8050	1620	400	610	10100	310	18550	2540
Baltic Proper	24780	3613	1140	4740	140650	13131	166570	21484
Gulf of Riga	438	60	47400	1000	47838	1060	—	—
Gulf of Finland	8500	560	995	212	64500	4080	73995	4852
Archipelago Sea	1160	64	200	10	5800	540	7160	614
Bothnian Sea	1924	199	1678	350	45600	2350	49202	2899
Bothnian Bay	1630	140	2366	174	49500	3240	53496	3554
Total	67202	11801	58949	9397	449588	28381	464313	47459

Table 5.7 Emission of trace metals by source (tonnes per year) (Environmental Statistics in Europe and North America 1987)

Metal	Source	Baltic Sea (total)
Mercury	Municipalities	1093
	Industries	268
	Rivers	3707
	Total	5068
Cadmium	Municipalities	3216
	Industries	9320
	Rivers	46340
	Total	58876
Lead	Municipalities	18
	Industries	8
	Rivers	239
	Total	265
Arsenic	Municipalities	4
	Industries	101
	Rivers	72
	Total	177

from rivers. ICES has coordinated studies of contaminants of fish and shellfish in the Baltic Sea. The monitoring programme today provides data of physical and chemical determinants, harmful substances in selected organisms and biological determinants. Regulations concerning the Baltic Sea area are included in national laws, the Helsinki Convention and other relevant international legal instruments. (Environmental Statistics in Europe and North America 1987)

It is our feeling that in general the situation with co-operation is much more substantial when we consider the exploration of the Baltic Sea than its drainage

Table 5.8 Oil emissions by source (tonnes per year) (Environmental Statistics in Europe and North America 1987)

Area	Municipalities	Industries	Rivers	Total
Kattegat	430	40	2335	2805
Belt Sea and western bays	1013	68	185	1266
Sound	488	78	195	761
Baltic Proper	1881	151	6606	8638
Gulf of Riga	960	—	3000	3960
Gulf of Finland	3700	34	10300	14034
Archipelago Sea	70	5	0	75
Bothnian Sea	215	40	2400	2655
Bothnian Bay	220	160	1000	1380
Total	8977	576	26021	35574

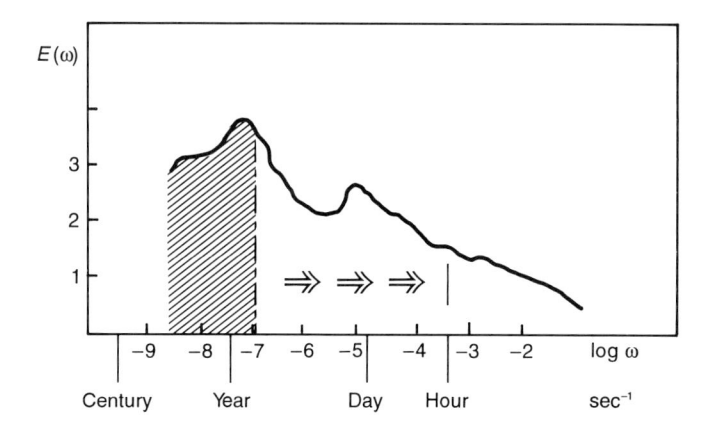

Figure 5.2 Energy spectrum of the Baltic Sea. See the text for explanations.

area. All the bordering countries have their national programmes of ecological monitoring of the atmosphere, land and inland waters. Measurements of various physical, chemical and biological characteristics are being performed on a regular basis. A list of environmental parameters is rather long but different numbers of parameters are regularly determined in various countries. Principles of monitoring activities also differ. Various methods of measurements and data evaluation are used. Technical equipment in various countries differs in its accuracy, technological level and capacity. Hence intercalibration is obviously required. In some cases there is a shortage of specific devices. In each country a large number of various administrative bodies and institutions are participating in the environmental monitoring activities. Some examples of these activities are given by Victorov (1992b).

Analysis of the current situation in technical equipment, methodological regulations and managerial aspects of monitoring activities in Baltic Europe countries is far beyond the scope of this section. Recently a number of international organisations have inspected the situation with environmental issues in the former USSR countries including the monitoring activities, and the interested reader may address the original documents (e.g., the OECD assessments on environmental information systems in Belarus and in Russia). Efforts aimed at improvement of the technical and methodical level of environmental monitoring are being made continuously in the framework of international and bilateral co-operation.

Nevertheless, we think that life requires us to work out a new approach to the problem. The stimulus and the tool for improvement of the quality of monitoring is likely to become the creation of an integrated 'Baltic Europe' GIS based on both the existing components of national databases (or GIS where appropriate) and the new modules. Monitoring of the environment (with the elements of population health and socio-economic issues) in the GIS domain is to be integrated, based on and supported by general cartography, geology, meteorology, industrial, municipal, agricultural and demographical statistics, etc.

Since 1985 the CORINE project has been under development 'with the objective of gathering, co-ordinating, and improving the consistency of information on the state of the environment in the European Communities. The participating countries are: Belgium, Denmark, France, Germany, Greece, Ireland, Italy, Luxembourg, the

Netherlands, Portugal, Spain and United Kingdom' (CORINE 1989). It seems reasonable to start the 'Baltic Europe' GIS project bearing in mind its compatibility with the CORINE GIS, using the experience gained during its development (and, maybe, considering their linking in the years to come).

Besides, in the long run the 'Baltic Europe' GIS project could be used as an effective managerial tool to improve a whole set of issues comprising up-to-date environmental monitoring (in the broad sense of the term) in the Baltic countries with economies in transition.

5.2.5 An outline of the 'Baltic Europe' GIS

General structure and sources of information The subsystems and the general structure of GIS are shown in Figures 5.3 and 5.4. There are two principally different sources of information on the state of the environment and relevant socio-economic issues: (a) obtained from various sensors, and (b) others (non-sensored data).

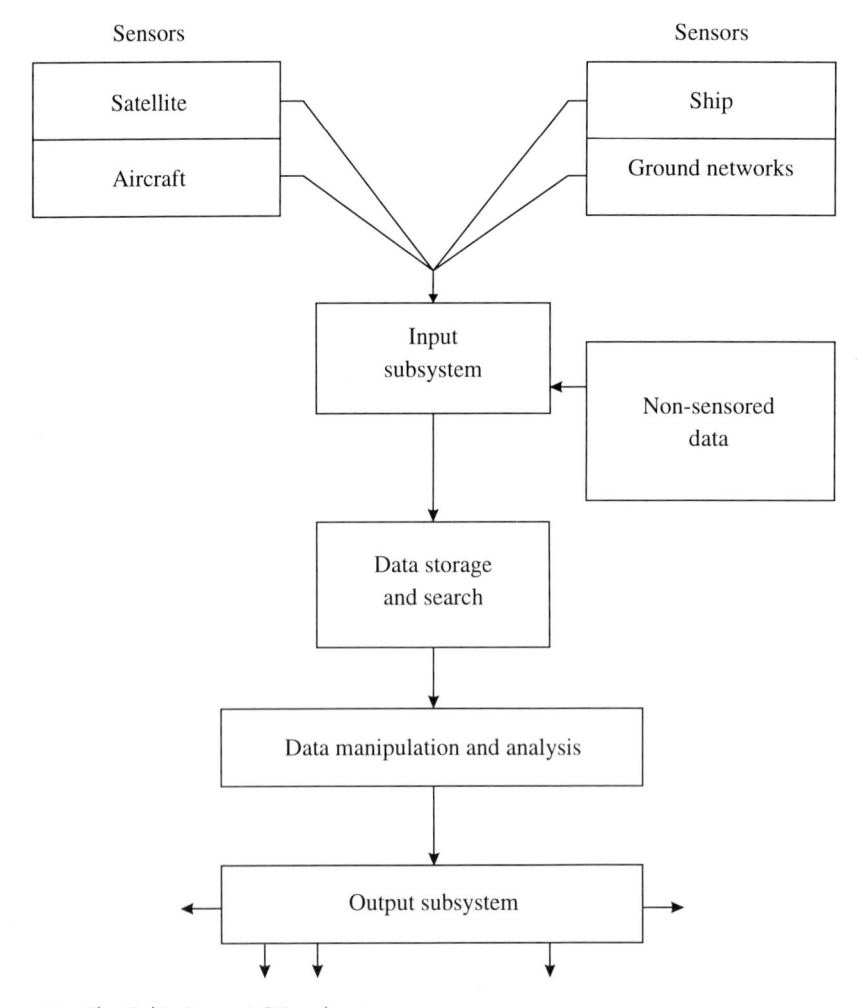

Figure 5.3 The 'Baltic Europe' GIS: subsystems.

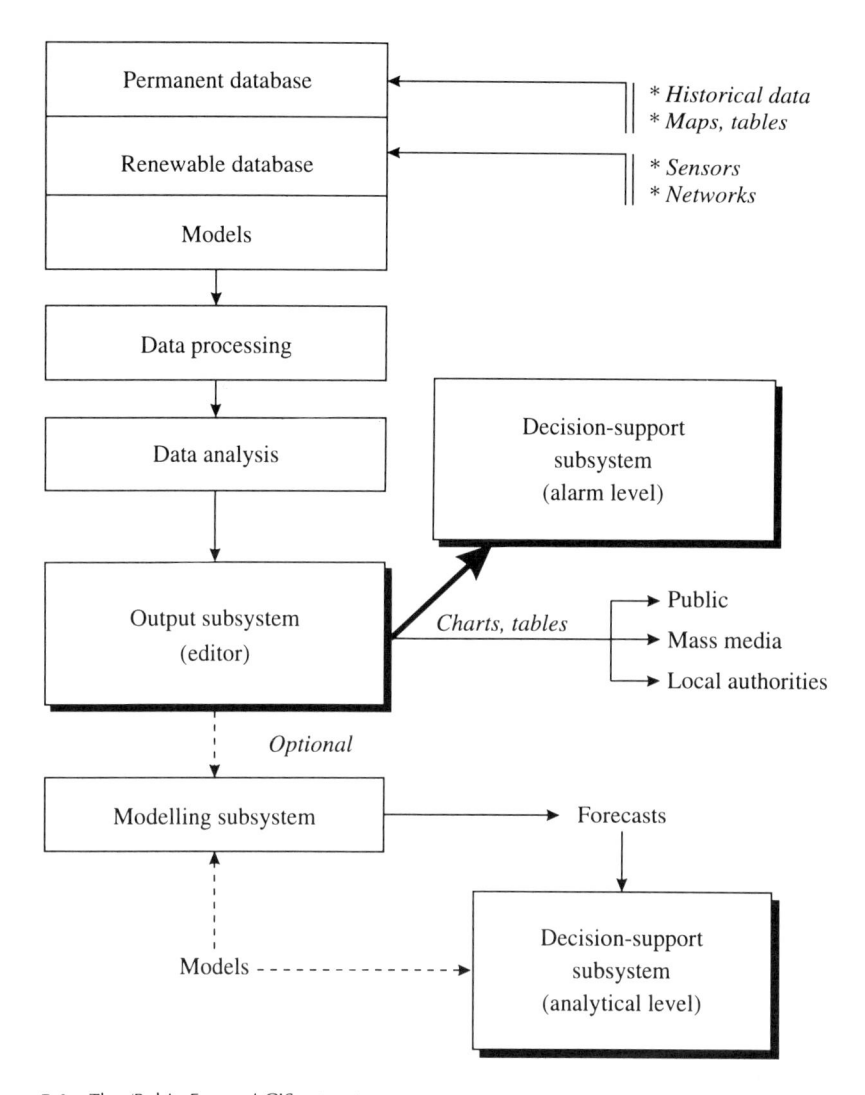

Figure 5.4 The 'Baltic Europe' GIS: structure.

(a) Measured (sensored) information from the following types of sources:
- ground-based networks of all kinds;
- shipborne data from networked routine cruises, research vessels and ships of opportunity;
- airborne information, including routine and special surveys;
- satellite data (raw and processed) from various sensors, including ground networks' data relayed via satellites.

(b) Statistical (non-sensored) information:
- data collected in the forms of various regular reports, censors (demography, population health, etc.);

■ data in the form of urgent reports (epidemic events).

Some types of data to be included in the GIS are shown in Figure 5.5.

An assessment of the quality of the existing informational networks in the region is needed in order to decide whether to include a certain type of data in the GIS on the first, second, ..., stages of the project, and to assess the running and preparation

Databases

- Meteorological network
- Hydrological network
- Water quality control
- Oceanographic data
- Air-pollution control
- Seismic network
- Electromagnetic field data
- Magnetic field data
- Radiation control data
- Sanitary control data
- Population health data
- Demographic data
- Land use data
- Socio-economic data
- Geological (soil) data

Expert subsystem concepts, models

Macroparameters

Complex assessment of the (local) environment and the quality of life

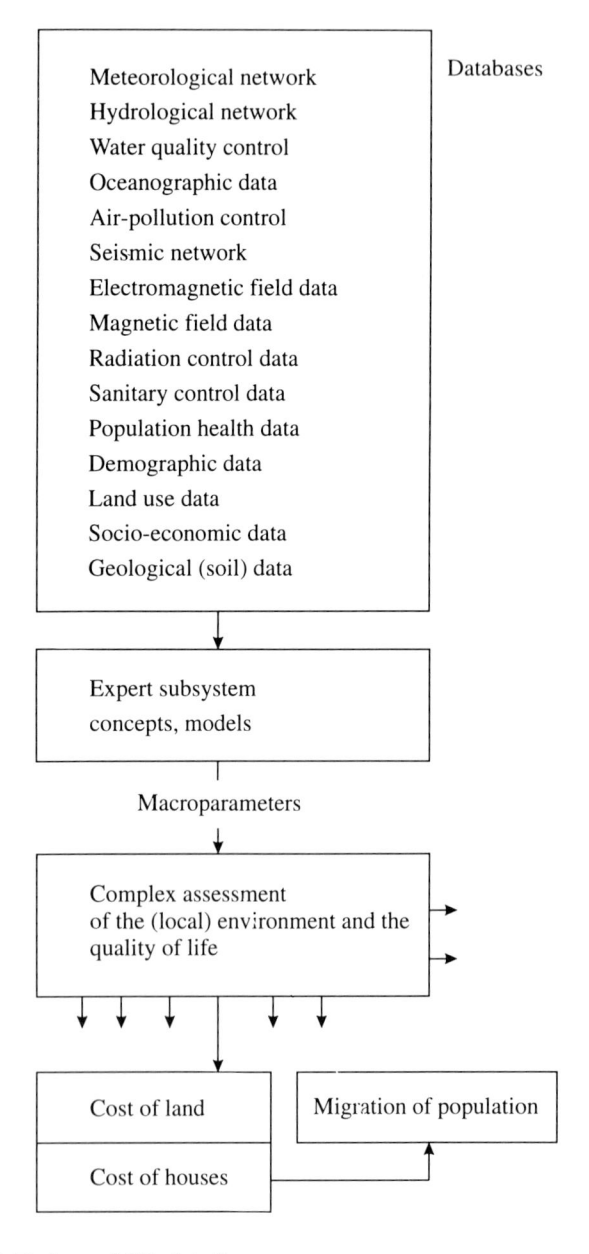

Cost of land

Cost of houses

Migration of population

Figure 5.5 The 'Baltic Europe' GIS: data flow.

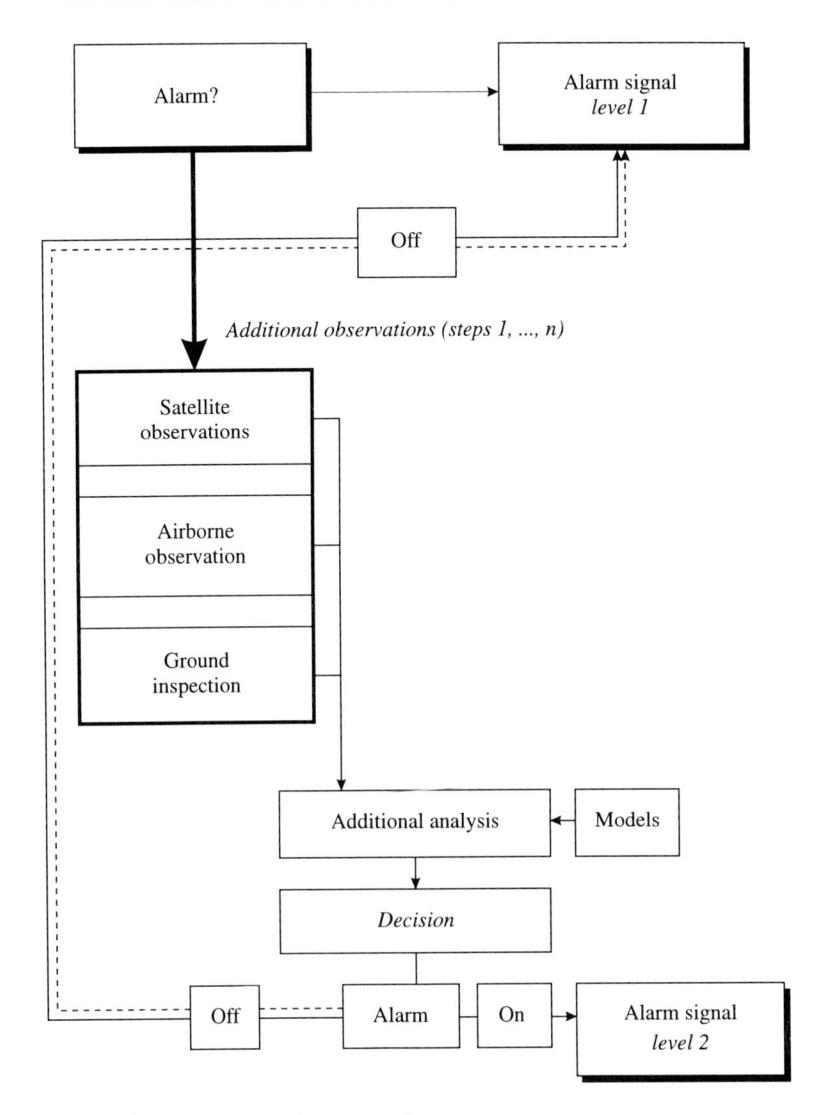

Figure 5.6 The 'Baltic Europe' GIS: alarm scenario.

of a recommendation on the improvement and development of a certain national informational network. A special international group of experts could analyse the situation with the existing national and international networks and provide the necessary recommendations.

5.2.6 Databases

It is suggested that the databases consist of 'permanent' parts and 'operational' or 'renewable' parts. The latter should be fed from informational networks and should become 'permanent' after the next step of their updating procedures (see Figures 5.4 and 5.5).

It is important to note that the databases should include a bank of various types of numerical models for data evaluation, the models that accumulated the known relationships between the parameters of the environment, social and economic processes ('knowledge bases') and special models assimilating current remotely sensed data.

5.2.7 Data manipulation and analysis

The data processing and data analysis subsystems shall comprise the analytical part of the GIS (Figures 5.4 and 5.5). An expert subsystem shall use the data and the models from bases and (possibly through a number of 'macroparameters' constructed from the measured parameters of the environment) could produce complex assessments and/or forecasts on the state of the environment in a certain local site. This could be used to estimate the quality of life at this site. In its turn this information could result in a change in costs of land, houses and even lead to migration of the population. These aspects of regional GIS applications seem to be useful for commercial aims, providing the municipal councils and individuals with chargeable informational products. Some other types of system output products (see Figure 5.4) could also be commercialised.

Provided that operational network data are used, two subsystems of decision-making (or rather decision-support) could be incorporated in the GIS structure: (a) a subsystem of the *analytical* level (as described above), and (b) the subsystem of the *alarm* level, working out the 'level 1 alarm signal' (see Figure 5.6).

Alarm scenario modules in the environmental regional GIS should comprise a reliable checking subsystem preventing 'false alarm' events and subsequent operations. According to an optional variant of such a scenario, the alarm signal of 'level 1' is only a preliminary one, and additional information is required to make a decision on either to generate an alarm signal of 'level 2' or to switch off the alarm subsystem. Figure 5.6 shows that more than one step of additional observations might be needed. These 'ad-observations' shall consist of satellite observations, airborne surveys and/or ground inspections.

During alarm situations the satellite observational subsystem may be operated in a specific regime (Figure 5.7). Following the 'level 1' alarm signal data from the other satellites can be provided, target operations can be performed (in the case of pointable sensors on board the satellite), and the current operational mode of data transmission from sensor(s) can be changed in order to provide information with better spatial and temporal resolution as compared to the routine ('standby') mode. Besides alarm-related operation on board the satellites, priorities of data processing and dissemination in the ground segment can be changed as well.

The need for an *alarm* subsystem in the 'Baltic Europe' GIS is obvious. The concentration of potentially hazardous enterprises in the region is very high, and their technological level is often inadequate, which leads to a high level of environmental risk for this region. A lot of 'chemical time bombs' exist in the Baltic region. The lesson of the Chernobyl atomic plant disaster showed that the alarm control system in the region appeared to be unsophisticated. To the best of my knowledge, current activities in the region on the international level seem to be based on separate topics (e.g., radiation control only) with inadequate involvement of other

```
                    ┌──────────────┐
                    │ Alarm signal │
                    │   Level 1    │
                    └──────┬───────┘
                           │
                           │           Operations in space
                           │        ┌────────────────────────┐
                           ├────────│  Additional satellite(s)│
                           │        └────────────────────────┘
                           │
                           │        ┌────────────────────────┐
                           ├────────│   Targeting operations  │
                           │        └────────────────────────┘
                           │
                           │        ┌────────────────────────┐
                           ├────────│   Sensor operation mode │
                           │        └────────────────────────┘
                           │
                           │        ┌────────────────────────┐
                           ├────────│  Data transmission mode │
                           │        └────────────────────────┘
                           │
                           │         Ground-based operations
                           │        ┌────────────────────────┐
                           ├────────│ Priority of data processing│
                           │        └────────────────────────┘
                           │
                           │        ┌────────────────────────┐
                           └────────│     Priority of data    │
                                    │      dissemination      │
                                    └────────────────────────┘
```

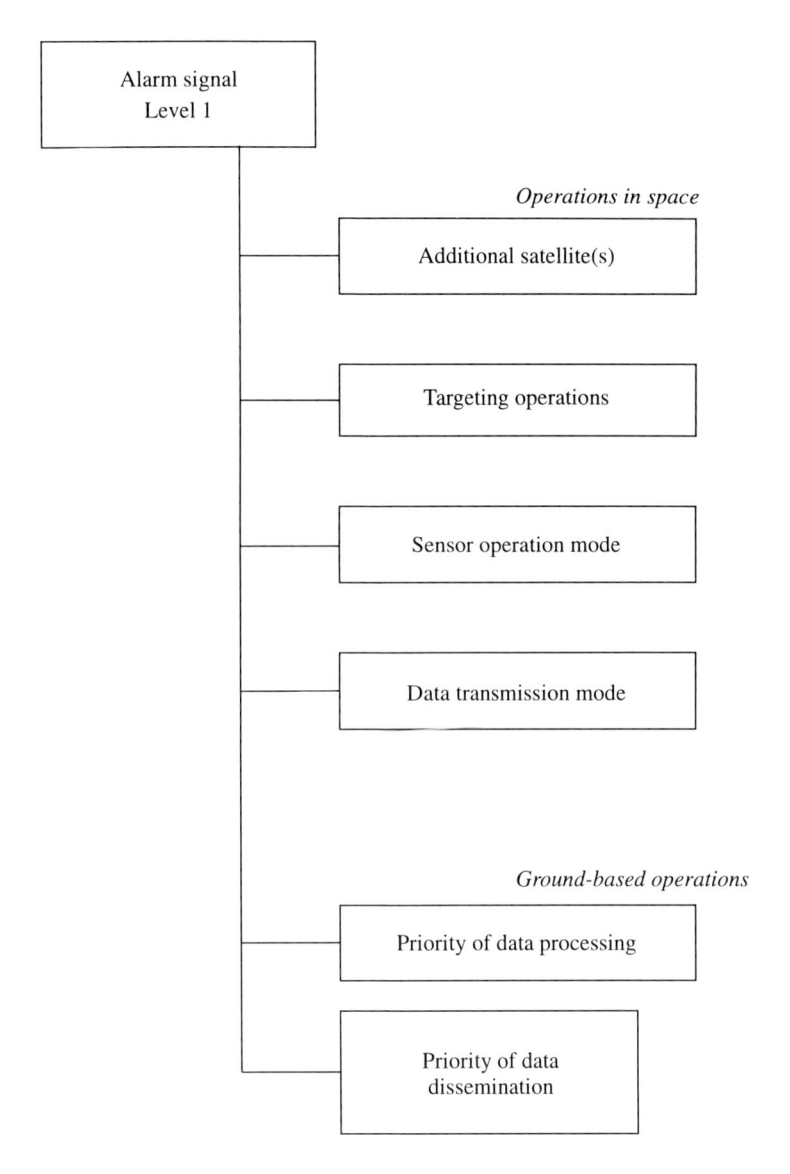

Figure 5.7 The 'Baltic Europe' GIS: alarm operations in the satellite segment.

closely related matters (e.g., meteorology, sea water dynamics, etc.). I would be glad to have been mistaken.

5.2.8 Output subsystems

The routine informational products of the regional GIS shall be maps, charts, tables and brief releases based on the measured parameters of the state of the environment (level 1), various forecasts, based on modelling procedures (level 2), and analytical reviews and assessments (level 3). One stream of these informational products can be produced for the local authorities, the public and mass media on a non-commercial basis. The second stream (mainly, some types of forecasts and analytical value-added products) shall be produced and distributed on a commercial basis (similar to the situation with meteorological information).

In order to provide a reasonable level of confidence between the public and the environmental agencies (especially for the sites where this level is not too high, as is the case in the former USSR), within GIS (or attached to it) a special module has been proposed – ECOLOGIUM (Victorov 1989c). Figure 5.8 shows the possible structure of ECOLOGIUM: a regional informational and analytical public facility centre for 'Environment and Population Health' comprising, along with the scientific unit, educational facilities, a museum of environmental studies, a public library, a press centre, and a public display centre of current (live) environmental information for the region, in the local site and in the neighbouring regions (sites) including aquatories. The project proposal on ECOLOGIUM specified for the St. Petersburg

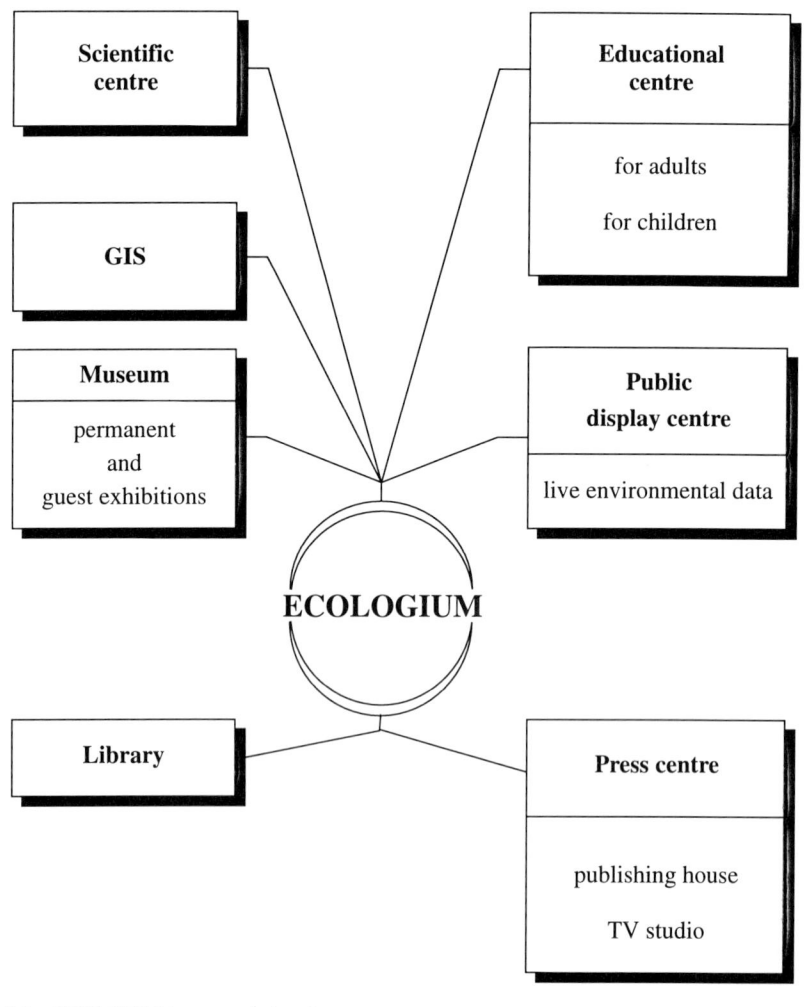

Figure 5.8 ECOLOGIUM: general structure.

metropolitan area – north-west Russia and Baltic Europe – was presented by Victorov (1989c). But it seems to me that the idea of ECOLOGIUM is actually of much broader sounding. This idea was first published in 1989. Since then it has found support among the participants of various international conferences and workshops who recommended the setting up of the ECOLOGIUM (but could not make the necessary contributions to make it a reality). Excellent examples set up by the Danish Science Centre EKSPERIMENTARIUM (Copenhagen) and the Ontario Science Park (Toronto, Canada) give encouragement to this venture.

In my opinion, at a period of general public concern over the state of the environment, some kind of local Ecologium should be set up now in each coastal town to provide public access to the operational data on the state of marine and coastal environments, including the sea water quality, the quality of public beaches, and the level of fish contamination with toxic chemicals and radionuclides. In recreational areas a local mini-Ecologium could also provide – in the most vivid, simple and attractive form – the nowcasts and forecasts of upwelling events and algae bloom events for swimmers.

5.2.9 Too early or too complicated? (Concluding remarks)

All the principal aspects of the 'Baltic Europe' regional GIS project are to be consistent (at a reasonable level) with the existing relevant GIS and databases, current monitoring programmes, and alarm and warning systems (and those under development) in the region. An international group of experts could consider the whole set of problems. This group should include specialists on GIS and environmental statistics, experts in each type of data to be included in the databases, and experts in remote sensing, communications, public relations, etc., together with representatives of the relevant international organisations – ECE, HELCOM, WMO, WHO, UNEP and others.

This review of the principal features of the project is now complete. The general political climate was rather favourable for the 'Baltic Europe' regional GIS. The 'Baltic Sea Declaration' was signed by the representatives of the Baltic states on the governmental level in Ronneby, Sweden in September 1990 in which the participants declared

> their firm determination to:
>
> ...10. Promote additionally, through supportive measures, increasing transfer of knowledge regarding the environment;
>
> ...13. Extend and strengthen the programme of monitoring in order to improve the assessment of the present and future state of the marine environment of the Baltic Sea area and encourage the cooperation between statistical agencies to improve demographic and other statistics relevant to the protection of the Baltic Sea;
>
> ...15. Encourage a strengthening of cooperation and facilitation of human contacts in the region to improve the environment of the Baltic Sea, including *inter alia* participation of *local and regional governments*, governmental and private institutions, industries and *non-governmental organisations* in the field of economy, trade, *science*, culture, information, etc. (Baltic Sea Declaration 1990; emphasis by S. Victorov)

Still it seems that the concept of a systems analysis approach as expressed in the 'Baltic Europe' regional GIS project proposal was ahead of its time and probably too complicated to become a reality. In fact current activities in the region appear to be channelled into topic-oriented informational systems or databases with no adequate concern for their compatability. However, I am pleased to note that some features, very similar (!) to those in the proposed 'Baltic Europe' GIS, can be easily traced in the HELCOM databases, in the informational issues of the current UNEP activities in this region (though for the case of Russia again based on exclusive co-operation with the Moscow-located central governmental agencies rather than with *regional and local governments and non-governmental organisations*, thus totally neglecting the above quoted document and bringing additional tensions in the relations between the central and regional structures), and in the recently launched international project BALTEX dealing with the hydrological cycle in the Baltic region.

The integrated 'state of the environment plus population health' approach, which is a characteristic feature of the 'Baltic Europe' GIS concept, was also used in the project design of a regional monitoring system for the St. Petersburg area (see Kondratyev and Bobylev 1994), carried out by a group of experts with the author's participation.

5.3 Dynamical processes in the Baltic Sea as revealed from satellite imagery

Elements of water dynamics in the Baltic as recorded in satellite images of various bands can be analysed from two viewpoints: as pure oceanographic phenomena, and in the ecosystem context – as the mechanisms by means of which the nutrients, biota, sediments and pollutants are being transported, dispersed and exposed to various driving forces such as solar irradiance or the atmosphere. The marine ecosystem approach needs better understanding of the biophysical and biochemical processes and much more sophisticated models (in terms of their spatial and temporal resolution) than the relevant marine sciences are able to provide at present. One may say that the observational capabilities of remote sensing are in some cases ahead of the marine ecologists' capabilities to assimilate the observations. Thus, though bearing in mind the possible future ecosystem approach, in this section we will mainly discuss the oceanographic meaning of the remotely sensed data.

In the studies described in Section 5.3 we used mainly SST data from AVHRR/NOAA received by our operators at the data acquisition station in the village of Lesnoje (Kaliningrad oblast) on a regular basis in spring–autumn of 1986–1988 (258 days of acquisition) and on a non-regular basis (during COSE and other cruises) in 1983–1985 (30 days). We also used the satellite images of the Baltic Sea for 1980–1982 published by Horstmann (1983) as the source of raw data (78 days). In the period of regular data acquisition at the autonomous station in APT mode the spatial resolution was about 3 km and temperature resolution was about 0.4–0.5 K. About 1000 images were received, of which about 400 were analysed. The reliability of satellite data was verified using sea-truth airborne and shipborne data and also comparing satellite data from the adjacent overpasses. The correlation coefficient of satellite and *in situ* data collected during the AINS (allowed interval of non-synchronicity, see Chapter 3 for details) was 0.8–0.9. The root-mean-square deviation between the satellite SST data and *in situ* shipborne data was 0.4–1.5 K, depending on local conditions of data acquisition, variability of atmospheric effects, etc. (Bychkova *et al.* 1990a).

5.3.1 Seasonal alongshore fronts

These fronts can be observed during the spring heating and the autumn cooling of the water. The peculiar feature of the Baltic Sea water – its low salinity – creates preconditions for the formation of density fronts which are similar to the 'thermobar' phenomenon in large lakes (see Chapter 3). Such density fronts appear as a result of the inhomogeneous heating/cooling of the nearshore and offshore waters in spring/autumn.

In spring the offshore water surface temperature usually does not exceed 2°C, while the nearshore water surface temperature increases quickly. Reaching the point of its maximum density (e.g., at the salinity value of 6–7‰ corresponding to 2.5–2.7°C), the surface water with reduced salinity submerges, warm/cold water comes to its place, and as a result of their mixing a density front appears.

The location of this front at a fixed moment may coincide with the location of a certain isobath. While the sea is heated up, the front is moving offshore, following the next isobath configuration further and further out. The horizontal velocity of the density front extension is known to be dependent on the incoming heat flow and is inversely proportional to the slope of the sea bottom (Bychkova *et al.* 1990a).

Figure 5.9 (Plate 2) shows some examples of thermal fronts during the 'hydrological spring' in the southeastern part of the Baltic Sea. These fronts were also recorded in shipborne *in situ* measurements in May 1985. The lifetime of such fronts is about a month, and the temperature difference across the front-line can reach 3–4°C (Bychkova *et al.* 1990a).

5.3.2 Upwelling events

Upwelling events in the oceans are well known, and the physical nature of this phenomenon was understood many years ago. A strong alongshore wind plays the main role in generating the so-called Ekman transport effect. The Ekman transport in the upper water layer leads to the deviation of the drift current from the wind direction (in the southern hemisphere, to the right). Coastal upwelling events are of importance as the bottom water masses coming to the surface bring nutrients thus affecting the marine environment and biological productivity; upwelling zones are known to be rich in fish.

Baltic oceanographers have been studying upwelling events for many years (see, e.g., Gidhagen 1984), but it is not so easy to study them using routine techniques, as the lifetime of upwellings in the Baltic is short and their dimensions are sometimes too small to be detected by rather scarce coastal hydrometeorological stations. For these reasons the study of coastal upwellings in the Baltic has benefited so much from regional satellite oceanography.

Horstmann (1983), Gidhagen (1984), and Bychkova *et al.* (1985a) have presented satellite images with indications of upwelling events. The first systematisation of the upwelling zones in the Baltic based on satellite data was made by Bychkova and Victorov (1987).

5.3.2.1 *Tasks of study*

We set the following tasks:

- to determine the zones of upwelling events;
- to estimate the appearance frequency of upwellings in each zone;
- to find out the spatial dimensions of the zones and the direction of fronts;
- to estimate the intensity of upwellings (the temperature difference between the upwelling zone and the neighbouring waters, and the value of the horizontal gradient of temperature within the zone);
- to study the role of the bottom topography and the coast-line configuration in the generation of upwellings;
- to describe the development of an upwelling in time and to determine its main phases and the 'inertial period' (the delay in time between the setting of a favourable wind and the moment of cold water arrival at the surface).

5.3.2.2 *Development of coastal upwellings*

An upwelling event along the eastern coast of the Baltic Proper was recorded in the satellite image of 12 July 1983. Figure 5.10 shows three stages of its development. Airborne measurements of SST were carried out on 9 and 10 July and recorded the initial stage of the upwelling event. In a narrow band near Liepaya a minimal value of 15.4°C was registered, while near Gotland Island it was 21°C. The upwelling event at its maximal stage was recorded in the satellite image of 14 July in a northwest wind. The upwelling zone was about 250 km long from the Kurshi Bay in the south to the Irben Strait in the north and about 20 km wide on the average (its width was only 6 km near the Kurshi Bay). A sharp front could be seen near Liepaya, stretched in the south-east direction at a distance about 60 km. There were about 30 'patchy structures' (possibly eddies) 6–10 km in diameter at the front of the upwelling zone. The averaged values of the temperature difference between the upwelling zone and the surrounding sea was about 4–5 K, with the maximum value of 7–8 K in the central 'core' near Liepaya on 14 July. Horizontal gradients of SST were about 2–4 K per 3 km (one element of spatial resolution). The structural function of the two-dimensional SST field in the region of the upwelling zone had a maximum corresponding to 24 km, which is close to the width of the zone. On 16 July the wind direction changed to the west, and relaxation of the upwelling began. The satellite image of 18 July showed no upwelling in this region. This analysis was confirmed by *in situ* measurements at 38 stations and the results of a dispersion analysis. The total lifetime of this upwelling was five days. A wind speed of 6–7 m s^{-1} measured at the coast corresponded to a depth of the Ekman transfer of 20 m which exceeded the depth of the thermocline at this season (Bychkova *et al.* 1985a).

A time series of satellite images showing the development of the upwelling near the eastern coast of the Gulf of Riga in May 1985 is shown in Figure 5.11 (Plate 3). The upwelling event was recorded in AVHRR imagery for the first time on 13 May near Skulte (see also Figure 3.10 in Chapter 3) in northeastern winds. On 15 May

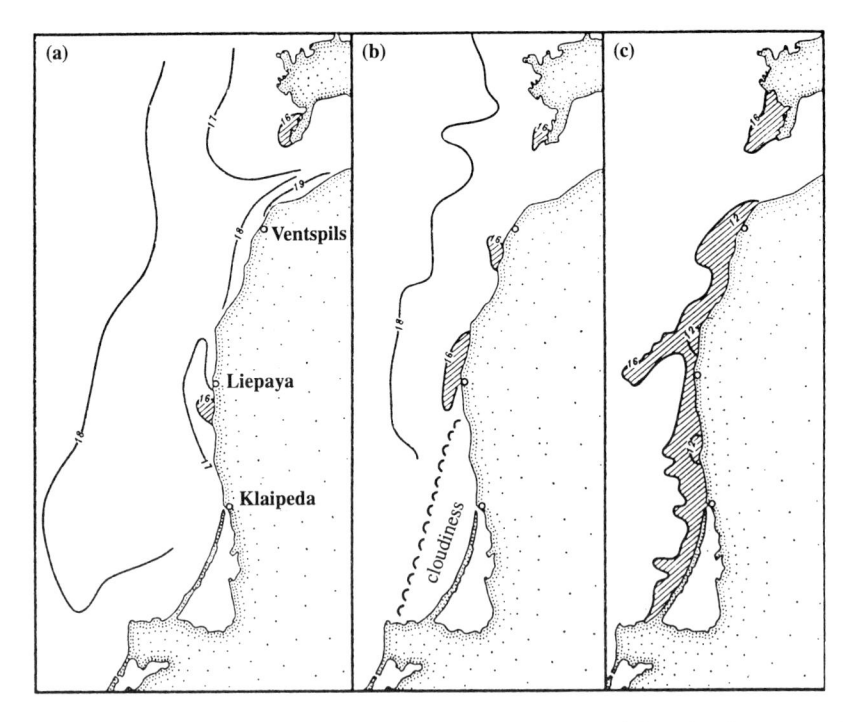

Figure 5.10 Development of the upwelling along the eastern coast of the Baltic Sea as revealed from remotely sensed data. SST charts (figures show SST in °C) based on: (a) airborne measurement of 10 July 1983; (b) NOAA-7 image of 12 July 1983; (c) NOAA-7 image of 14 July 1983 (Bychkova et al. 1985a).

the upwelling waters spread along the coast for about 75 km and for about 15–20 km offshore. The coolest water (the centre of the upwelling) was recorded as a patch less than 10 km in diameter near Skulte. On 18 May the 'tongue' of cold water located near Tuya spread at a distance of 40 km from the shore, and on 19 May the upwelling water joined the cold core waters in the northern part of the Gulf. The cold patch in this part of the Gulf has been registered from aircraft and from satellite many times during each year; probably it is not connected with the wind conditions and may be caused by a water exchange process with the central (colder) part of the Sea and may also be connected with some features of the bottom topography. On 19 May the satellite image showed the maximal phase of the upwelling with cold waters spread at a distance of about 80 km. On 20 May the relaxation of the upwelling front began, caused by a change of the wind direction to the southwest. The upwelling zone was divided into two separate zones which continued to relax. On 23 May the upwelling could be seen in satellite imagery for the last time. On 25 May the region was covered with clouds, and no upwelling was recorded in the satellite image.

Satellite data detected the upwelling from 13 to 23 May, with a minimal temperature in its zone of 3°C. Local networked coastal stations recorded this upwelling event from 12 till 23 May. On 23 May the water temperature at the eastern coast was 6.4°C, and on 24 May it was as much as 11.8°C.

The development of another upwelling – at the southern coast of the Gulf of Finland in May 1984 in a cloudiness situation – was presented in Section 3.7.3 dealing with data processing in these conditions.

It is obvious that the two-dimensional patterns of the development of coastal upwellings presented above could have been studied only by means of satellite imagery. The regional satellite oceanography methods became a tool of long-term studies of the Baltic upwellings.

5.3.2.3 *Summary of upwelling events*

Bychkova *et al.* (1985a) suggested the technique for elucidation of coastal upwellings in satellite imagery of the IR band (in SST patterns). The reliability of the detection of upwellings had to be checked by using the *in situ* subsatellite sea-truth measurements. These datasets had to be analysed by means of a dispersion analysis tech-

Table 5.9 Parameters of the Baltic Sea upwelling zones (except the Gulf of Bothnia) as revealed from satellite imagery (Bychkova and Victorov 1987)

No.	Area	Length (km)	Width (km)	Wind at the coast	Lifetime, temperature gradient
1	NW of Rügen	50–100	20–40	Change from W to E	4–6 days
2	Polish coast	150–200	10–50	E, SE	6–7 days
3	Near Hel peninsula	50–100	10–30	E, SE	6–7 days; $\Delta T = 3$ K
4	Lithunian and Latvian coast	250	6–20	N	$\Delta T = 4$–8 K
5	East coast of Gulf of Riga	75–100	10–30	SE	0.5–10 days
6	West coast of Saarema	55	5–30	N	0.5–1 K km^{-1}
7	South coast of Gulf of Finland	20–30–40 (chopped)	5–40	E, NE	0.5–1 K km^{-1}; $\Delta T = 6$–8 K; 7–8 days
8	Northern coast of Gulf of Finland	100–300	30–40	SW–W	0.5–1 K km^{-1}; $\Delta T = 2$–3
9	Along Swedish east coast	160	10–50	SW	$\Delta T < 10$ K
10	West coast of Gotland	80	1–5	NE	
11	East coast of Öland	130	5–10	SW	
12	East coast of Gotland	30	5–10	SW	
13	Hanö Bight	100	5–15	NW, W	7 K (9 km)$^{-1}$; $\Delta T < 10$ K
14	South coast of Sweden	60	5–40	SW, NW, W	

nique with Tukey classification (this method has already been described in Chapter 3). Expert criteria based on the knowledge of regional peculiarities of coastal upwelling events in the Baltic Sea was suggested, namely:

- the decrease of sea water temperature in the spring–summer period of 3°C or more during 1–2 days at some coastal hydrometeorological stations or at some stations during ship cruises; and simultaneously
- the absence of this effect at the neighbouring stations located in the areas with quite different configuration of coast-line.

These standard criteria were used in the analysis of bulk satellite imagery described at the beginning of Section 5.3. About 100 upwelling events were analysed.

Bearing in mind the tasks which we set for the study (Section 5.3.2.1), we determined 14 zones of coastal upwellings in the Baltic Sea (without the Gulf of Bothnia). Basic characteristics of these zones are given in Table 5.9. The last column gives (where possible) the lifetime of the event, the temperature difference T between the upwelling zone and the surrounding waters, and the horizontal gradient in units of Kelvin degree per km.

Figure 5.12 shows the location of the upwelling zones listed in Table 5.9.

Table 5.9 indicates the wind conditions during the lifetime of the upwelling in a certain area. All the cases shown in Table 5.9 (with the exception of the upwelling near Rügen Island) were recorded in the strong alongshore wind accompanied by the Ekman transport effect. The upwelling phenomenon near Rügen, according to Horstmann (1983), is connected with the bottom topography features, namely the interruption of the outflow in deeper water layers by the Darss Sill (e.g., based on numerical calculations by Kielmann 1982), or in relation with the low-frequency waves radiated seaward from the shore.

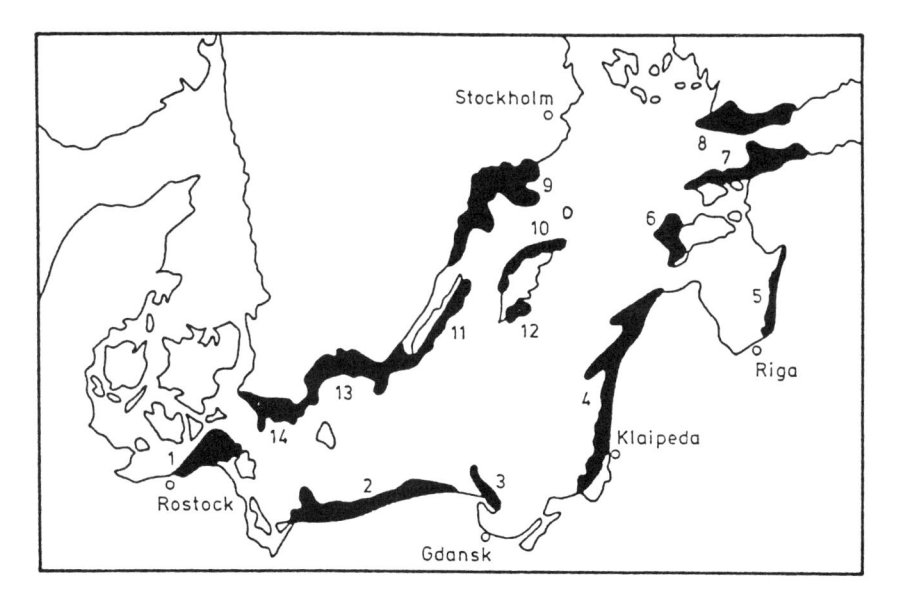

Figure 5.12 Schematic chart of the Baltic Sea coastal upwelling zones (without the Gulf of Bothnia) as revealed from satellite imagery of 1980–1984 (Bychkova and Victorov 1987).

Figure 5.13 shows typical shapes of the fronts of the upwelling zones as recorded in satellite imagery in various areas of the Baltic Sea with various configurations of the coast-line and in various wind conditions.

Four cases are shown in Figure 5.13a–e:

- superposition of the upwelling fronts near Rügen Island (a) for 12 July 1982 (1), 11 June 1980 (2), 14 May 1980 (3);
- upwelling along the smoothed shore with several capes in the winds along the coast-line (b) and in the 'blow off' winds from the coast (c);
- upwelling along very smoothed shore (d),
- upwelling along cropped (cut) shore (e).

The 'inertial' period for the generation of the upwellings was estimated to be 0.5–1 day which was confirmed by satellite observations. The lifetime of the upwelling events depends on the duration of a favourable wind situation. In general the horizontal dimensions of the upwelling zones appeared to be about 100 km in length along the shore and about 10–20 km in width. Temperature gradients at the sea surface reach 0.5–1 K km^{-1} (sometimes up to 4 K km^{-1}); the temperature difference between the upwelling waters and the surrounding waters is from 2 to 10 K. (The events when the temperature decreased by more than 15 K during 3–4 days were reported by Altshuler and Shumakher (1985). On 25–28 June 1966 the water temperature at the eastern coast of the Gulf of Riga decreased by 16 K (Bychkova et al. 1988a).)

In some images the movement of an upwelling front can be traced with a velocity of 10–15 km per day, which may be caused when the waves are 'trapped' by the coast (Gidhagen 1984; Horstmann 1983). In some cases the upwelling zones gener-

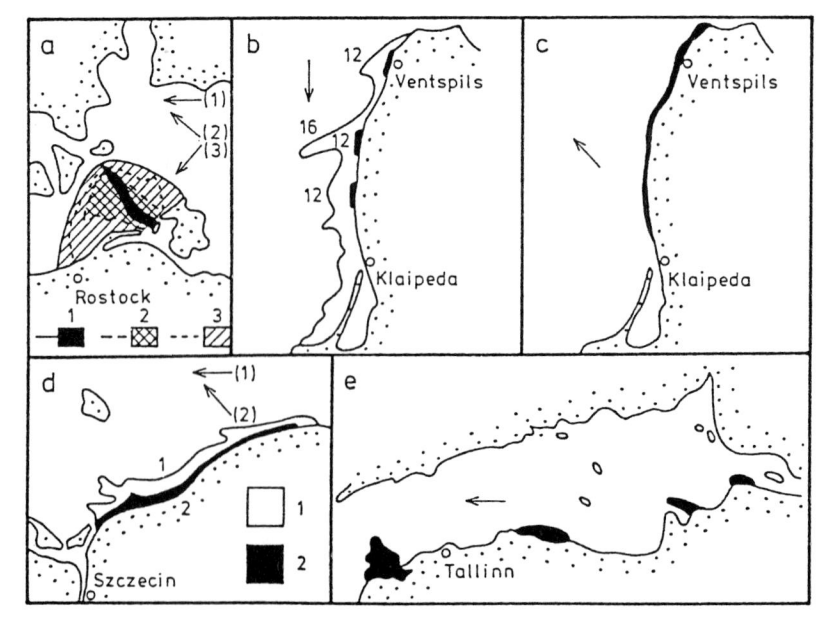

Figure 5.13 Typical shapes of the upwelling zone fronts in the Baltic Sea as recorded in satellite imagery. Arrows show wind direction. For explanation see the text. (Bychkova and Victorov 1987.)

ate cold water filaments, recorded, for example, at the southern coast of Sweden in the Carlscruna–Oland island area (directed to the Bornholm Basin); near Liepaya and Visby the filaments were about 100 km long.

The lifetime of the coastal upwellings in the Baltic Sea does not exceed 7 days in 94% of the observed events. Short-term upwelling events with a lifetime of one day or less were observed when the thermocline was close to the surface and the dominating wind direction was changing fast. In this case one could observe small-scale upwelling events with a typical length of tens of kilometres and sharp temperature gradients (Bychkova et al. 1990a).

These results of coastal upwelling studies in the Baltic Sea based on satellite imagery have brought a new sounding to this topic. They have led to further investigation of this phenomenon by means of satellite data and have also triggered the analysis of other aspects of Baltic upwellings.

5.3.2.4 More on the coastal upwellings in the Baltic Sea

Thorough analysis of the time series of satellite SST AVHRR NOAA images for 1980–1988 provided more detailed information on the upwelling events in the Baltic Sea. Some 20 zones (and subzones) of the upwelling events have been elucidated in the sea as a whole (Bychkova et al.1988a). But the short time series of satellite data does not give enough information on the appearance frequency of the upwelling events. Other sources of data have to be used.

Satellite data were complemented with all the available historical records of the sea temperature in which the information on the upwelling events could have been contained. Records of cruises since 1908 (Lebedintsev 1910), data from light vessels and coastal stations since 1936 (Hydrographical Observations 1953–1961; Rocznik 1953–1972; Surface Temperature 1954, etc.) and also the charts from regular airborne surveys for 1965–1986 (carried out by the personnel of the North-Western Regional Administration of the USSR Hydrometeorological Service/Committee) have been analysed (Bychkova et al. 1988a).

The same methodical approach and criteria (see Bychkova et al. 1985a) have been applied. Data of the sea temperature measured at about 20 stations during 1940–1961 have been found consistent and selected for detailed analysis of the upwelling situations (some 1300 upwelling events). Tables 5.10 and 5.11 based on lightvessels and coastal stations data give an idea of the lifetime and appearance frequency of the coastal upwelling events.

The appearance frequency of the coastal upwellings in each zone varies considerably during the spring–autumn period. In some months in a certain site more than five drops in water temperature in the coastal zone occurred. For example, in August 1962 in the Gulf of Riga, eight events were recorded when water masses as cool as less than 10°C appeared at the sea surface. In May–June coastal upwellings are frequent at the eastern coast of the Baltic Sea and along the southern coast of the Gulf of Finland; in July, at the coast of Finland; in August, at the coasts of Estonia and Sweden; and in September, near the Swedish coast. In some sites the frequency appearance of coastal upwelling exceeds 20%.

The results of the analysis of durable time series of observations showed that the intensity of the upwellings in the coastal zone depends also on such factors as the close location of river estuaries, bottom topography and the configuration of coastline. Satellite data showed that the most intensive upwellings occur at the wind side

Table 5.10 Lifetime of coastal upwellings in the Baltic Sea (Bychkova et al. 1988a)

Number of events (%)	Lifetime (days)
39	1
22	2
14	3
7	4
6	5
3	6
3	7
1	8
1	9
1	10
3	15
0.2	20

Table 5.11 Appearance frequency of coastal upwellings (%) (Bychkova et al. 1988a)

Site	Duration of observations (years)	May	June	July	August	September
Ulkokalla	22	0	4	23	13	3
Tankar	22	1	6	16	12	3
Saapi	22	6	17	16	12	4
Tvarmin	16	4	13	21	7	9
Harmajo	22	1	9	20	10	5
Sidostbrotten	13	0	2	9	7	5
Svenska Bjorn	18	0	6	9	16	4
Havringe	16	0	8	5	14	10
Falsterbjorev	19	0	3	5	6	1
Tallinn	14	2	14	6	12	7
Liepaya	14	7	17	5	2	6
Klaipeda	14	8	24	3	3	3
Ristna	14	8	22	12	6	8
Melno	19	2	15	9	12	4
Vladislavovo	19	1	16	6	10	6
Hel	14	4	12	2	2	0
Gdyunya	19	2	11	5	2	2

of capes which is in agreement with theory (Crepon et al. 1984). The bottom deeps and gutters near the shore are the favourable factor for coastal upwellings in the winds from the coast.

5.3.2.5 Upwellings and large-scale atmospheric circulation

As a result of joint analysis of satellite imagery and all the sources of *in situ* data mentioned in Section 5.3.2.4 an updated (and sometimes slightly corrected) version

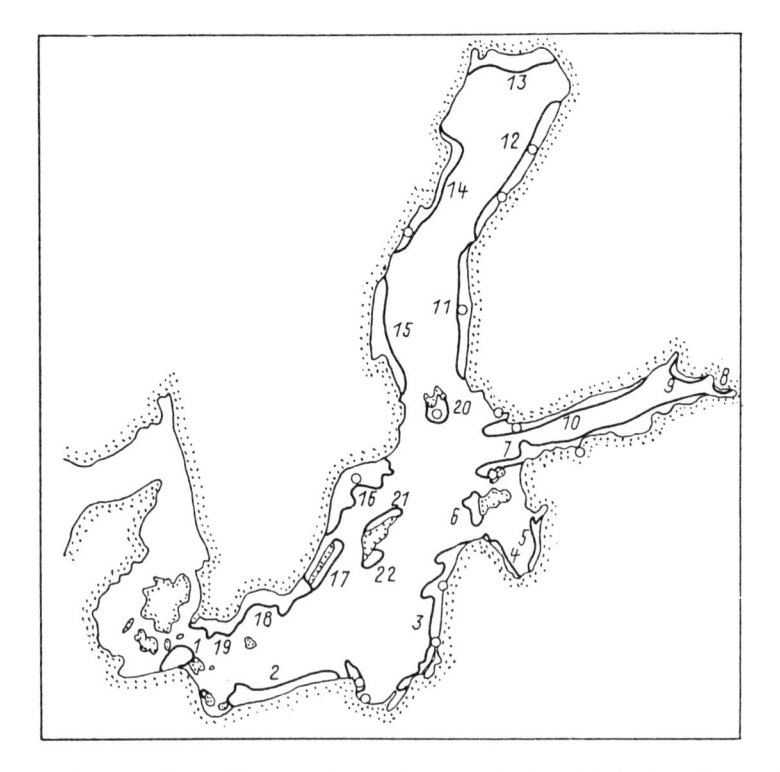

Figure 5.14 Schematic chart of the coastal upwelling zones in the Baltic Sea based on satellite and *in situ* data (Bychkova *et al.* 1988a).

of the list of coastal upwelling zones in the Baltic Sea (including the Gulf of Bothnia) has been obtained (Table 5.12) and these are shown in Figure 5.14.

Bychkova *et al.* (1988a) related the coastal upwellings in the Baltic Sea with large-scale atmospheric circulation in north-west Europe, of which the Baltic region is a constituent part. Following Tiuryakov *et al.* (1983) all of the variability of the synoptic situation causing the upwellings was represented by 11 principal synoptic situations (PSS) shown in Figure 5.15.

In Figure 5.15 the PSS are labelled I–XI, and the last column in Table 5.12 indicates the type of PSS when the coastal upwelling events occur in a certain zone.

5.3.2.6 Forecasts of coastal upwelling events

The results of investigations of coastal upwelling events in the Baltic Sea have made it possible to consider two approaches to the complicated problem of the forecast of these events.

Based on the Ekman theory of upwelling, one can suggest the following scheme of forecast. The velocity W of vertical transport of water caused by the Ekman divergence can be estimated by the equation:

$$W = t/p_{\text{water}} f L,$$

Table 5.12 Coastal upwellings in the Baltic Sea as revealed from remotely sensed and in situ data (Bychkova et al. 1988a)

No.	Area	Length (km)	Width (km)	Intensity (K)	PSS
1.	NE of Rügen Island	50–100	30	4–6	Change VII to III
2.	Along coast of Poland	100–300	30	4–6	IV, V, VI, X, XI
3.	Along coast of Lithuania and Latvia	100–300	30	6–7	I, II, III, XI
	Gulf of Riga				
4.	Western coast	50–100	15	6–7	IV, V, X
5.	Eastern coast	50–100	15	7–8	I, II, III, X, XI
6.	Western coast of Saaremaa Island	50–75	20	7–8	I, II, III, XI
	Gulf of Finland				
7.	Southern coast	100–400	40	6–8	II, III, X
8.	Neva Bay	20–25	3	7–8	II, III, VIII
9.	Northern coast/USSR	50	10	3–5	I, II, III
10.	Northern coast/Finland	300–350	30	6–7	VI, VII, VIII, X
	Gulf of Bothnia				
11.	Eastern coast	200	30	6–7	VI, VII, X
12.	North-eastern coast	200–250	30	6–7	VII, VII, X
13.	Northern coast	100–150	40	5–7	VI, VII, X
14.	North-western coast	200–350	30	5–7	VI, VII, IX, X
15.	South-western coast	100–150	30	5–7	VI, VII, X
16.	Eastern coast of Sweden	200–300	40	6–8	VI, VII, IX, X
17.	Eastern coast of Oland Island	100	20	7–8	VI, VII, IX, X
18.	Hano bukten	100–150	40	7–8	V, VI, VII, X
19.	Southern coast of Sweden	60	30	4–5	V, VI, VII, X
20.	Southern Aland archipelago	50–100	50	5–7	VI, VII, VIII, IX, X
21.	Western coast of Gotland Island	50	5	3–5	II, III
22.	Eastern coast of Gotland Island	50	15	4–5	VI, VII, X

where t is the alongshore wind tension; $f = 1.2 \times 10^{-4}$ s^{-1} is the Coriolis parameter; $p_{water} = 10^3$ kg m^{-3} is mean water density; and L is the averaged width of a certain coastal upwelling zone. The value of L can be obtained from a series of satellite images. The value of t can be calculated using another simple equation:

$$t = 1.3 \times 10^{-3} p_{air} V,$$

were V is the wind speed (m s^{-1}), and $p_{air} = 1.29$ kg m^{-3} is the air density. The value of the wind speed can be obtained from the nearest meteorological station, or from satellite data of the microwave band (see Chapter 2). We tested this scheme in 1992 (though in a limited number of cases) and got encouraging results.

Another approach is based on the current analysis of synoptic charts and a one-day forecast of atmospheric pressure. Using the library of charts shown in Figure 5.15, one can find a similar pattern, thus choosing the type of synoptic situation (among the PSS from I to XI) which is likely to occur the next day. Then,

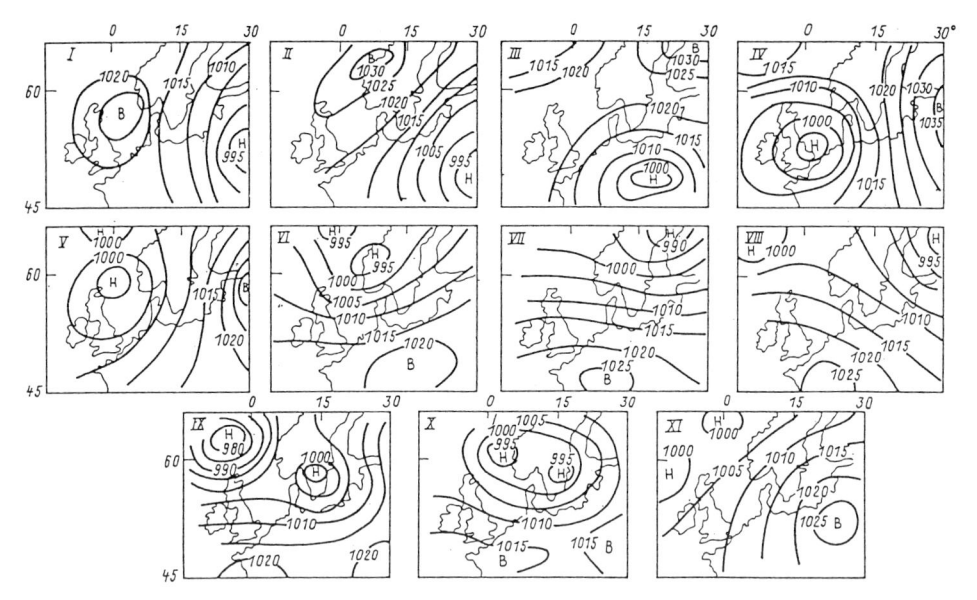

Figure 5.15 Principal synoptic situations (PSS) in the Baltic region. Atmospheric pressure is indicated in units of millibar. Letter H stands for low pressure, letter B stands for high pressure (Bychkova et al. 1988a).

using Table 5.12, one can determine the areas where the upwelling may take place, provided that this synoptic situation remains for at least two days. Using, further, the knowledge base on the bottom topography and the coast-line configuration, one can provide a forecast of an upwelling event. This scheme was tested using the *in situ* data for 1961 (the most comprehensive and easily accessible databases were available for this year). In the period of 1 June–10 September, 60 upwelling events were forecasted in various areas of the Baltic Sea. The forecasts were confirmed in 55 cases. Besides, four upwelling events actually occurred but were not forecast, probably due to the fact that not all the synoptic situations were covered by Table 5.12.

5.3.3 Eddies

5.3.3.1 On visualisation of water motion in remotely sensed two-dimensional data

Development of satellite oceanography techniques in general, and the facilities enabling us to display and manipulate the two-dimensional (2D) remotely sensed data in particular, provide a new approach to the problem of the study of dynamical processes in the seas. Of great importance in this context is the visualisation of real natural processes of any spatial scale. Visualisation of fine structure of flows makes it possible to present complex dynamical processes as simple patterns of currents (Countwell 1984).

Thus the process of 'drawing pictures' using satellite data becomes a constituent part of the study of turbulence in natural conditions. The analysis of satellite images enables us to distinguish between the principal types of water motion, while various scales of images provide the necessary level of generalisation of the currents structure. This is important because in the majority of turbulent shift currents the transfer processes are driven by large-scale non-random vortex motions (Countwell 1984). Digital presentation of satellite data supports the implementation of the techniques of classical hydrodynamics in satellite oceanography. In turn, this makes it possible to check theoretical equations on different scales or to extrapolate the well-known mathematical models of the natural processes.

Natural and artificial tracers can be used in the visualisation of motion in the upper layer of the sea. Artificial tracers are widely used in aircraft surveys. In regional satellite oceanography, natural tracers are used: the sea surface temperature (SST), ice, suspended substances, algae. SST data are most useful, as thermal fronts can be studied all the year round, but the fine structure of thermal patterns can be visible only when the temperature gradients are of considerable value.

Blue–green algae drift freely with sea water and often concentrate at convergence lines. There are two periods of seasonal blooms in the Baltic Sea, but the time, intensity and location of bloom vary from year to year depending on a number of factors.

Ice can be used as a seasonal tracer, though large ice fields cannot trace small-scale water motions and may even lead to mistakes showing just surface wind-driven motion of ice fields. Ice as a tracer in the ocean was studied by Nazirov (1982). Ice as a tracer in shallow sea with very complicated bottom topography (which is the case for the Baltic Sea) needs to be studied (Bychkova and Victorov 1988).

5.3.3.2 Internal structure of deep water and its manifestations at the sea surface

Here we will briefly discuss the question of whether the two-dimensional patterns recorded in satellite imagery correlate with the processes in the depth, and what do they actually show. There was a strong opinion that the remotely sensed SST field cannot provide information on the water layer because IR radiation does not penetrate into the water more than a few tens of microns. In June 1983 Bychkova *et al.* (1985b) managed to collect data relevant to this issue in the central part of the Baltic Sea (see Figure 5.16) during the second Complex Oceanographic Subsatellite Experiment (see Section 3.4 for details).

Before the analysis of satellite data we applied the technique of estimation of the AINS (allowed interval of non-synchronocity) described in Section 3.5. The estimation showed that the standard *in situ* shipborne point measurements of the sea water temperature could be used within 12 hours of the satellite overpass. For the satellite–aircraft system at the test area the AINS value was estimated to be three hours. Further, only the data matching those AINS criteria were analysed.

There is a statistical relationship between the radiative temperature, T_r, data measured from satellite and aircraft, and the *in situ* thermodynamical temperature, T_o, data.

Datasets were collected on 6 June (rough sea, wind speed 10–12 m s^{-1}) and on 8 June (still sea, wind speed 2–3 m s^{-1}).

A correlation analysis technique was used to study the relationship between T_r and T_o. For the test area, the *selected parity correlation coefficient r* was calculated

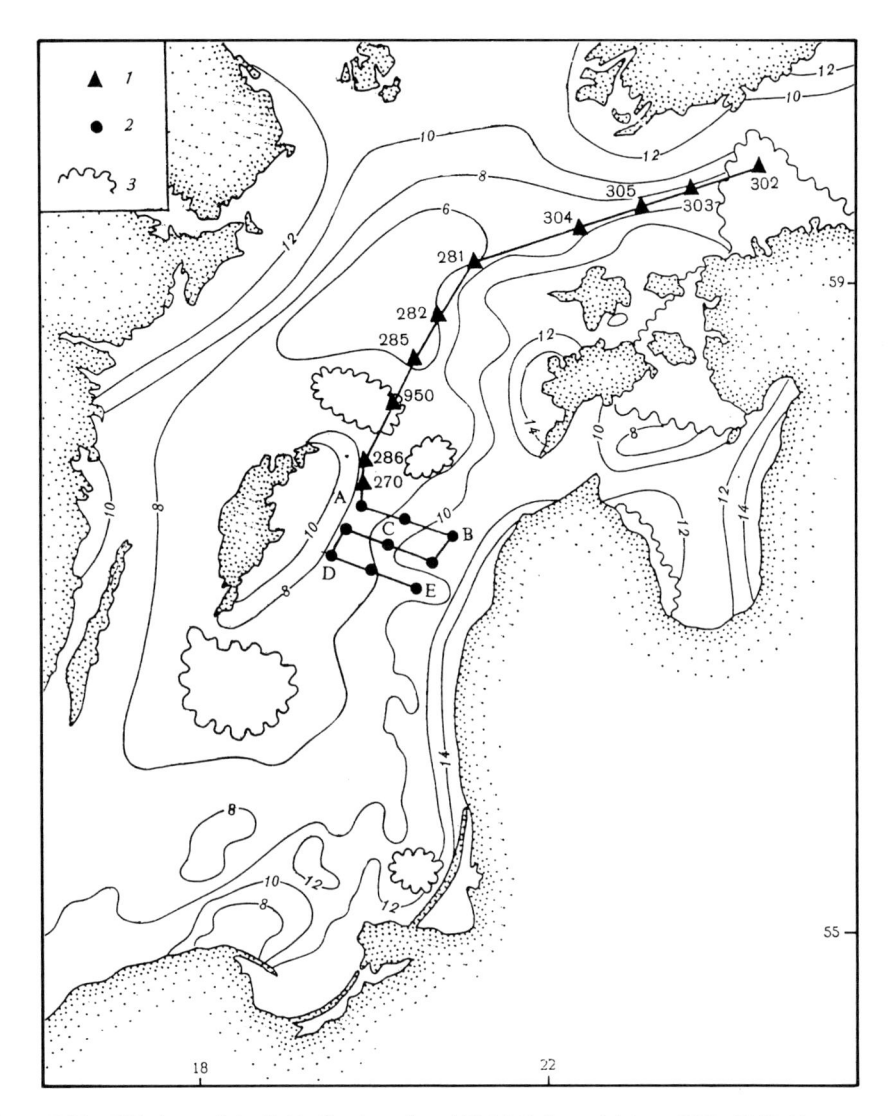

Figure 5.16 SST chart of the Baltic Sea based on NOAA-7 data of 1 June 1983. 15:52 Moscow time. Temperature in °C. (1) Oceanographic stations of research vessel *Alexander von Humboldt* on its way to test area; (2) its stations at the second COSE test area (centre is designated with C and the four corners are designated as A, B, D, E); (3) clouds. (Bychkova et al. 1985b.)

by the equation:

$$r(h) = \frac{\sum_{i=1}^{n}(T_{ri} - \bar{T}_r)[T_{oi}(h) - \bar{T}(h)]}{nS_r S(h)}$$

where n is the number of stations, $T_{oi}(h)$ is the temperature value at depth h at each station, $\bar{T}(h)$ is the mean temperature value at depth h at all the stations within the test area, T_{ri} is the radiative temperature value measured from satellite or aircraft at the points corresponding to the location of the stations, $\bar{T}_r$ is the mean value of the radiative temperature at the test site, and S_r^2 and $S_o^2(h)$ are the dispersions for the sets of radiative temperature data and *in situ* data at depth h, respectively.

The results of calculations are shown in Figure 5.17. We omit the further routine procedure and come to the conclusion that, at a value of r exceeding 0.7, there is a correlation between the two datasets. This means that, irrespective of wind conditions, a correlation exists between remotely sensed sea surface temperature data (both satellite and airborne sets) and shipborne *in situ* measurements of the sea temperature, from the surface to a depth of 17 m. At a depth of 35–40 m the correlation coefficient becomes negative, which corresponds to the generation of a current with opposite direction in the Ekman layer. The calculated value of the 'friction depth' was about 30 m. The conclusion is that satellite SST data can be used to study drift currents in the Baltic Sea (Bychkova *et al.* 1985b).

Figure 5.18 shows the three-dimensional (3D) structure of the test site on 6 June based on satellite and shipborne data. Satellite images recorded an S-shaped warm eddy located within the second COSE test site with maximal temperature near the centre of the test site. The shipborne measurements recorded characteristic shapes of isothermal lines indicating warmer waters at the surface.

Taking into account that (a) during the second COSE the dynamical processes in the upper water layer were governed mainly by the variability of wind-driven (drift) currents and (b) that a strong correlation between satellite and shipborne data was shown in the layer 0–17 m, one may conclude that the eddy recorded in the satellite image of the IR band was generated by instabilities in the drift currents. The SST satellite data recorded the following parameters of the eddy temperature front: the maximum temperature gradient was 0.5°C km^{-1}, the temperature difference was

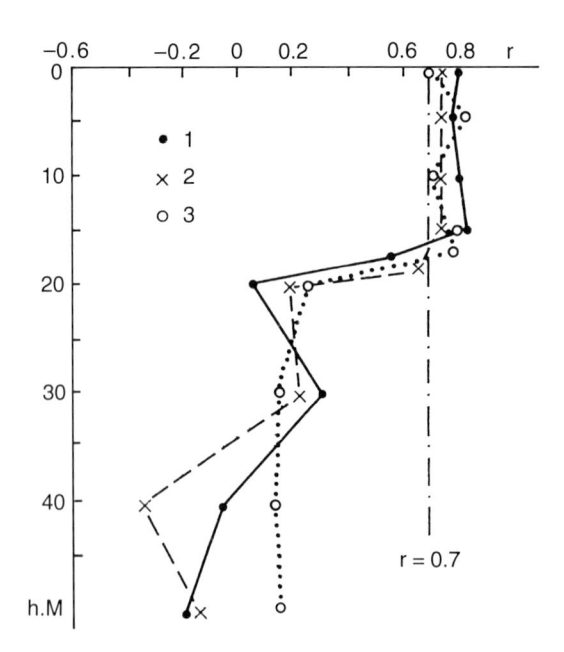

Figure 5.17 Correlation between radiative temperature at the test area and *in situ* measurements of water temperature at various depths: (1) satellite data of 6 June 1983; (2) airborne data of 6 June 1983; (3) airborne data of 8 June 1983 (Bychkova *et al.* 1985b).

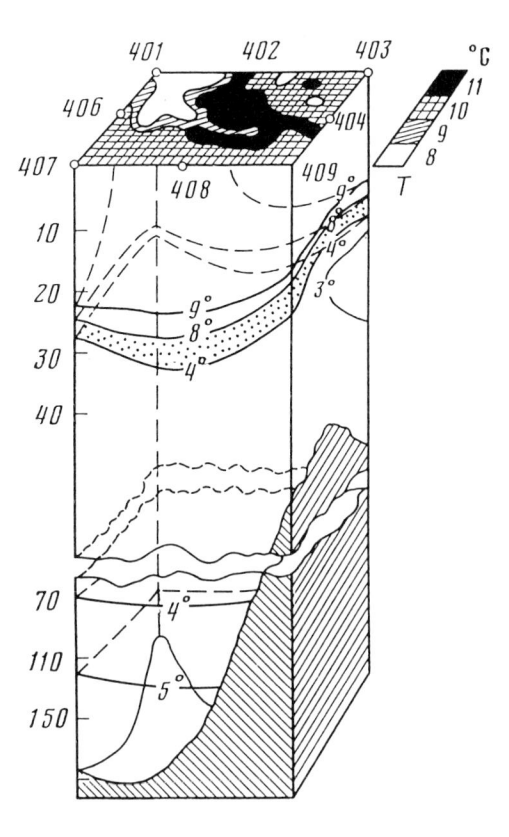

Figure 5.18 Schematic view of 3D water temperature field at the test area of the second COSE (Bychkova et al. 1985b).

3°C between the eddy core and the water outside the eddy, the length of the front was 40 km, and the frontal zone width was from 3 to 15 km (Bychkova et al. 1985b). According to Nelepo et al. (1983) characteristic S-shaped isotherm lines within the eddy zone and an S-shaped temperature front are the indicative features of a synoptic eddy.

In some cases the eddies in the Baltic Sea actually cause distortion of the halocline and show anomalies in the distribution of chemical elements. For example, in April 1981 and June 1892, a 10–30 m dome-like uprising of the halocline near the station located at 55° 05′ N; 15° 35′ E was detected from the ship with more than a double column chlorophyll value in the eddy core, and also maxima of SiO_2 at 40 m, of PO_4 at 50 m, of NH_3 at 60 m and minima of O_2 at 40 m and 50 m (Elken et al. 1984).

Taking into account the possibly important (though not yet estimated quantitatively) role of eddies in energy transfer and mass transport in relation to the general circulation of water masses in the semi-enclosed Baltic Sea, and also in the context of ecosystem studies, the following tasks were set (Bychkova and Victorov 1988) in the beginning of the 1980s:

- to analyse all the satellite datasets where the eddies (and the 'mushroom-like' eddy structures) could have been recorded;

- to determine the principal parameters of these dynamical structures; and
- to classify them by the area of origin and by the mechanism of origin.

The results presented in Sections 5.3.3.3.–5.3.3.6 and 5.3.4.1 were based on our analysis of 78 satellite images for the April–October period of 1980–1982 from NOAA and 'Nimbus-7' satellites published by Horstmann (1983), 82 images for the April–September period of 1983–1985 from NOAA satellites received at the autonomous data acquisition stations by our operators in Leningrad and Lesnoye, and the images from the USSR 'METEOR-Priroda' satellites for 1980–1985 in the database of the Laboratory for Satellite Oceanography (Bychkova and Victorov 1988).

5.3.3.3 Frontal eddies

Eddies of this type are known to be caused by baroclinic-barotripic instabilities of currents and can be traced at the upwelling fronts and at the fronts separating the coastal zone from the open sea. Strictly speaking, there are no permanent 'classical' currents in the Baltic Sea. The water motion, which (on average) is directed towards the North Sea, is called the Baltic current. The Coriolis force intensifies this current near the Swedish coast. The eddies can be traced in the zone of the Baltic current by their cores with well-pronounced temperature gradients. Eddies can be visible in satellite SST fields which enables us to study their structure – the core parameters and the thermal gradients in the core and in the peripheral parts. Figure 5.19 presents some examples of various types of eddies in the Baltic Sea as recorded in satellite imagery of IR and visible bands.

In Figure 5.19(a) an eddy in the zone of the Baltic current (indicated with an arrow) with a warm core is shown. Frontal eddies in this zone are usually 5–20 km in diameter. According to Horstmann (1983) frontal eddies can move 7–15 km during one day.

In the Danish Straits the eddies can be generated at the fronts of thermohaline intrusions, as the Straits connect water masses with sharply different values of salinity. Generation of frontal eddies in the Baltic Sea is affected by the bottom topography, each feature of the bottom structure acting as a catalyser of hydrodynamical instabilities at the front (Blatov *et al.* 1983; Nelepo *et al.* 1984). In this connection another type of eddy, the 'topogenic' eddy, may be considered as a separate group. They originate when the currents meet some bottom features such as banks, sills, trenches, etc. Bearing in mind the very complicated bottom topography of the Baltic Sea, one could expect a lot of topogenic eddies here, but they can be seen only in shallow-water areas, because the current instabilities originating near the bottom may not manifest themselves at the sea surface. Dimensions of topogenic eddies are of the order of the bottom features. Topogenic eddies may be in a stationary position over the bottom features, or they may escape. Eddies over shallow-water banks are mainly anticyclonic (Blatov *et al.* 1983).

In Figure 5.19(b) a fragment of an SST chart of the Baltic Sea is shown with the shallow-water Slupsk Bank (contoured) in the middle. A cold water eddy is located near the bank.

In offshore regions and at slopes, the synoptic phenomena of the dual wave and eddy nature can be seen; these may be called 'topographically captured (trapped) waves' (Nelepo *et al.* 1984). Topographic Rossby waves in the Baltic may cause eddies of 30–40 km in diameter moving along isobaths (Horstmann 1983; Aitsam

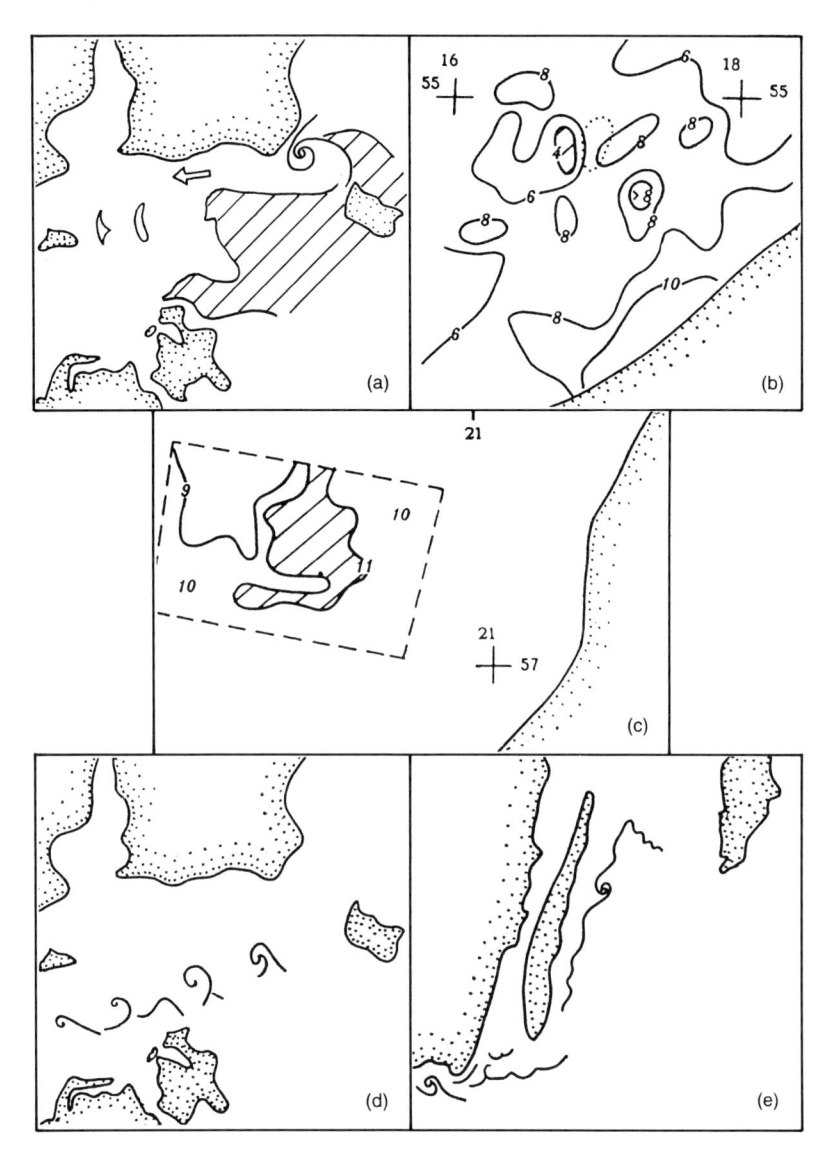

Figure 5.19 Showing eddy structures in the Baltic Sea as recorded in satellite imagery: (a) frontal eddy in the zone of the Baltic current, 2 June 1982, NOAA (tracer – SST); (b) SST chart in the vicinity of the Slupsk Bank (contoured); (c) advective eddy in SST field, 6 June 1983, NOAA; (d) friction eddies (tracer – blue–green algae), Nimbus-7; (e) eddy chain along the coast of Oland Island (tracer – blue–green algae), Nimbus-7 (Bychkova and Victorov 1988).

and Elken 1982). Inhomogeneities of the offshore and the slope relief are the favourable factors in generating these eddies.

5.3.3.4 *Advective eddies*

In the open Baltic Sea one can often observe wind-driven drift currents and gradient currents caused by the difference in the levels of various areas of the sea as a result of wind stress. These currents are very variable, their hydrodynamical instability leads to generation of the frontal eddies described above. Still, taking into account the short time period when the processes generating eddies cause the advection of waters, we considered it worthwhile to single out the advective eddies within a class of frontal eddies (Bychkova and Victorov 1988). Figure 5.19(c) shows one of these eddies studied during the second COSE (Bychkova et al. 1985b). Characteristics of the advective eddy recorded in the satellite image of 6 June 1983 are given in Section 5.3.3.2.

5.3.3.5 Friction eddies

Friction eddies are generated when a velocity shift occurs in the jet flow and are usually observed in the coastal zone and at the boundary of two currents with opposite directions. The horizontal dimension of these eddies is 1–10 km, the vertical dimension is about 10 m, and their lifetime is 1–7 days. These parameters were obtained by means of *in situ* measurements (Blatov et al. 1983; Nelepo et al. 1984). Figure 5.19(d) shows a chain of eddies westward of Bornholm Island; blue–green algae were used to trace the eddies in the satellite image of the visible band. Horstmann (1983) writes that a shift in stability of the current caused this dynamical phenomenon. In the same scene recorded in the IR band, only one eddy could be seen, the rest of them are hardly recognisable, which means that the temperature gradients in their frontal zones are low.

While analysing satellite images of the Baltic Sea it is often difficult to distinguish between topogenic and friction factors in the processes of eddy generation. Sometimes one needs additional *in situ* data to decide which 'classical' criteria a certain eddy matches.

An example of tackling the satellite and additional shipborne data relevant to an eddy follows. Let us consider in detail the dynamic patterns traced in the satellite image of the visible band using blue–green algae as a tracer (Victorov et al. 1988). In the satellite image of 22 August 1983 the algae formed a spiral feature curled clockwise – the anticyclonic eddy at the entrance to the Gulf of Finland north-east of Hiiumaa Island (Figure 5.20).

The distance between the two peripherical parts of the spiral was about 35 km. The eddy was recorded near the coast where the bottom depth was changing sharply; the feature was slightly asymmetric, being compressed from the land. Though further observation of this feature was complicated due to cloudiness, the eddy was recorded in the satellite images of 23 and 27 August. Thus its lifetime was at least five days. During this period its centre moved about 25 km north-east.

The set of satellite images showed that the eddy travelled at the southern periphery of the current, entering the Gulf of Finland with a mean velocity of 5–6 cm s^{-1}, which is 1–2 cm s^{-1} less than the integral velocity which could be estimated using the mean wind field for the period 20–27 August.

The Ekman equation:

$$V = 0.0127W/\sqrt{\sin \alpha},$$

where: V is the velocity of the surface current; W is the velocity of the surface wind; and α is the geographical latitude; gives 6–8 cm s^{-1} for the velocity of the surface current. Nearly the same value of the integral velocity of the surface layer, 7–8 cm

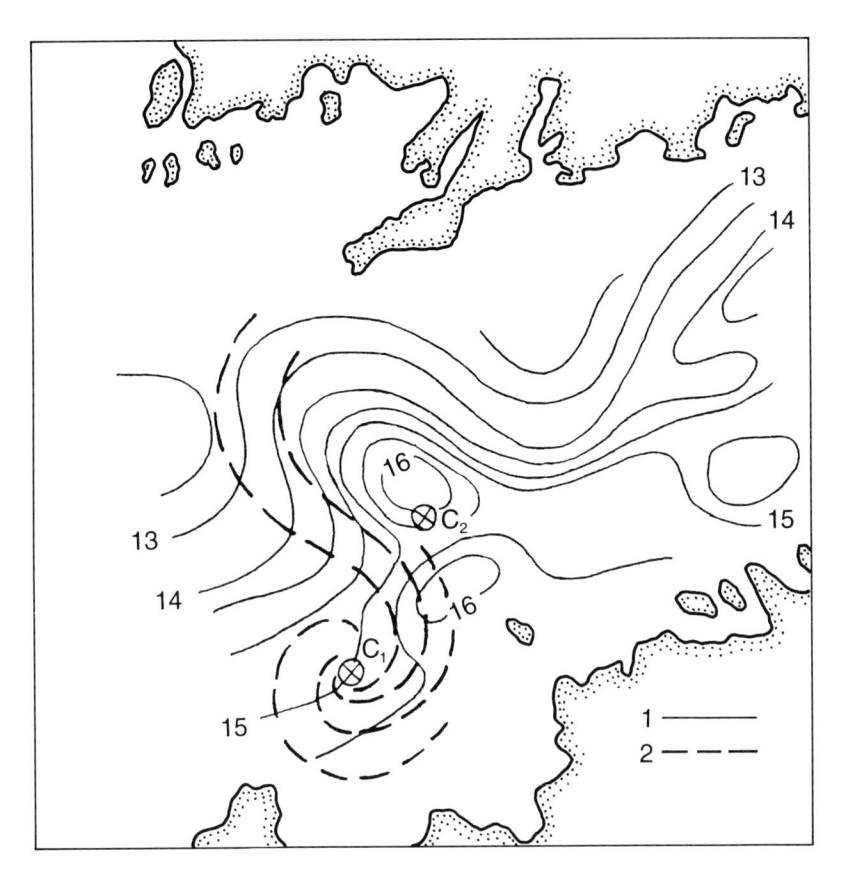

REGIONAL SATELLITE OCEANOGRAPHY: CASE STUDY OF THE BALTIC SEA 191

Figure 5.20 Observation of friction eddy movement at the entrance to the Gulf of Finland: (1) surface temperature distribution (°C), *in situ* measurements from research vessel *Rudolf Samoilovich* on 24–28 August 1983; (2) contours of eddy structure recorded in the image of 22 August 1983 (MSU-S scanner on METEOR satellite). C_1 and C_2 indicate the location of the eddy centre in satellite images of 22 and 27 August 1983 respectively (Victorov *et al.* 1989a).

s^{-1}, was obtained from a set of satellite images of 20–23 August, in which the bloom fields were recorded to the south and west of Hiiumaa Island.

The above data could be used to determine the type of this eddy. According to genetic systematisation (Bulatov and Tuzhilin 1980), the observed eddy was of friction type. This was confirmed by:

(a) its characteristic spiral shape, known as a 'spin-off' eddy;
(b) the place of origin – at the border of a current flow;
(c) the trajectory of its movement – along the current flow at its periphery;
(d) the asymmetry of the eddy;
(e) its horizontal dimension of 35 km, which is more than the Rossby internal radius (12 km).

For this region the Rossby radius, L_R, can be estimated by the equation:

$L_R = \bar{N}H/f,$

where $\bar{N} = 2 \times 10^{-2}$ s^{-1} (the characteristic value of the Vaisel–Brent frequency averaged in depth); $H = 80$ m (the mean depth); and $f = 1.3 \times 10^{-4}$ s^{-1} (the Coriolis parameter).

The friction eddies are most intensive in the zone of maximal current velocities, in our case in the surface layer. The orbital velocities of particles in the eddy are approximately equal to its phase velocity. Eddies are not participating in the horizontal transfer of water mass; they just affect the structure of hydrological fields, practically not changing the existing temperature–salinity relation.

To confirm the results of the analysis presented above along with satellite observations of this eddy the additional *in situ* quasi-synchronous cruise data from research vessel *Rudolf Samoilovich* of 25–28 August 1983 were used. The comparison of satellite images of 22 and 27 August (showing the eddy patterns as traced by algae) with the shipborne charts of the sea surface temperature (Figure 5.20) showed that the warm anomaly in the chart corresponds to the centre of the anticyclonic eddy. Horizontal temperature gradients are small. In the relatively warm core the intensified thermocline is deeper for, on the average, about 10 m (Figure 5.21). Unfortunately large distances between ship routes (an example of the inflexibility of the routine cruise programme; see Section 3.4.3 for a comparison with the COSE programme with the possible guidance of the vessel using satellite imagery) did not allow us to observe with appropriate accuracy the spatial variations of hydrophysical parameters in the eddy zone, as well as the transformation of the eddy itself.

Still, the shipborne data complement the satellite data with crucial details. From the available shipborne data the chart of the depth distribution of the isopicnic surface, $\sigma_t = 4.5$, provides the most impressive information (Figure 5.22).

An asymmetry of the eddy is revealed in its intensification to the south – from the coastal side. A more pronounced asymmetry is revealed along the route between stations 79 and 85. Figure 5.23 presents the results of the calculation of currents with the dynamical method. The velocities were calculated with respect to the 30 decibar level, allowing us to estimate the general circulation of the surface layer.

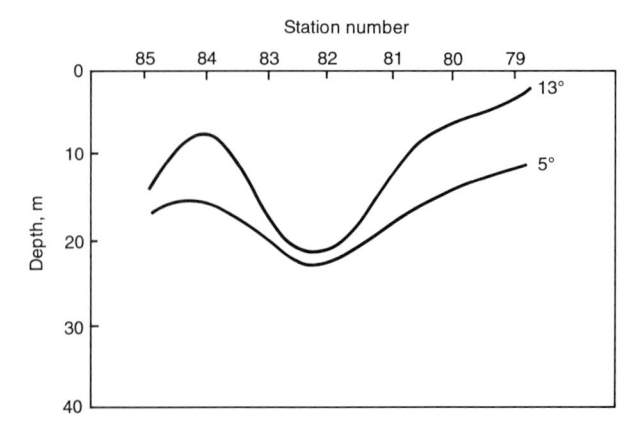

Figure 5.21 Depth (m) of thermocline on the route across the eddy. Numbers of stations of research vessel *Rudolf Samoilovich* are indicated (Victorov et al. 1988).

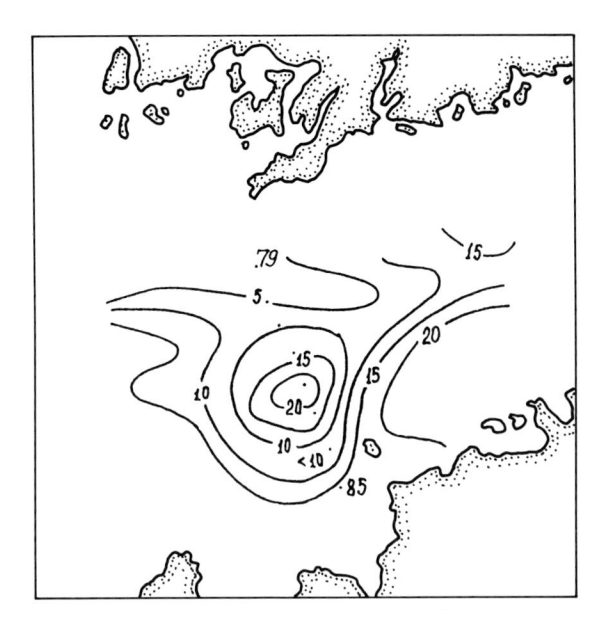

REGIONAL SATELLITE OCEANOGRAPHY: CASE STUDY OF THE BALTIC SEA   193

Figure 5.22 Depth (m) of isopicnic surface $\sigma_t = 4.5$. Research vessel *Rudolf Samoilovich*, 24–28 August 1983 (Victorov et al. 1988).

(Positive values of velocity in Figure 5.23 stand for the currents directed from the figure.)

In full agreement with the results of the analysis of satellite imagery, on this route an eddy was observed which was anticyclonic in its rotation, asymmetric in shape, and of friction type in origin; the eddy was formed at the southern periphery of the entering current. Almost identical temperature–salinity relations show the lack of

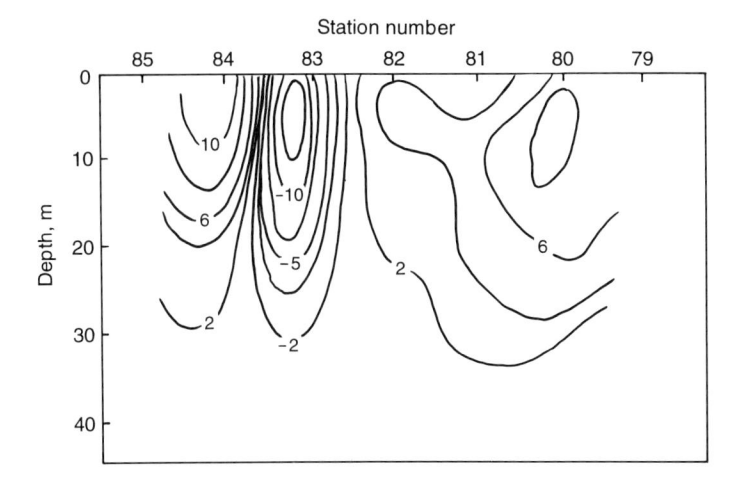

Figure 5.23 Isotaches of relative velocity of the current on the route of research vessel *Rudolf Samoilovich* across the eddy for $P = 30$ decibars (Victorov et al. 1988).

mass transport by the eddy. Thus the shipborne data confirmed the initial assessment based on satellite data.

5.3.3.6 *Peculiar features of eddies in the Baltic Sea*

Table 5.13 gives some examples of eddies recorded in various parts of the Baltic Sea. Figure 5.24 shows the satellite imagery-derived location of some eddies of various shapes (the digits in Figure 5.23 do not correspond to the list in Table 5.13) in the

Table 5.13 Parameters of some eddy structures identified in satellite imagery of the Baltic Sea (Victorov 1988b)

Area of observations	Diameter (km)	Type of eddy[a]
1. Western part of Gulf of Finland	15	C
2.	10	C
3.	35–40	A, S-shaped
4.	20	A
5.	20	C
6.	25	C
7. Landsort Deep	15	C
8.	10	C
9.	30	A
10. Gotland Deep	30	A, S-shaped
11.	10	C
12.	40	A, S-shaped
13. Gdansk Deep	25	A
14.	35	C
15. Slupsk	10	C
16. Bornholm Deep	10	C
17.	10	C
18.	10	C
19.	45	C, S-shaped
20. Hanobukten	15	C
21.	10	C
22. Arkona Basin	10	C
23.	10	C
24.	10	C
25.	5	C
26.	15	C
27.	10	C
28. Kadet-Rennen Strait	10	C
29.	10	C
30.	10	C
31. Mecklenburg Bight	10	C
32. Small Belt Strait	10	C
33. Kattegat Strait	5	A
34.	10	C
35. Kalmarsunn Strait	5	C

[a] C, cyclonic eddy; A, anticyclonic eddy.

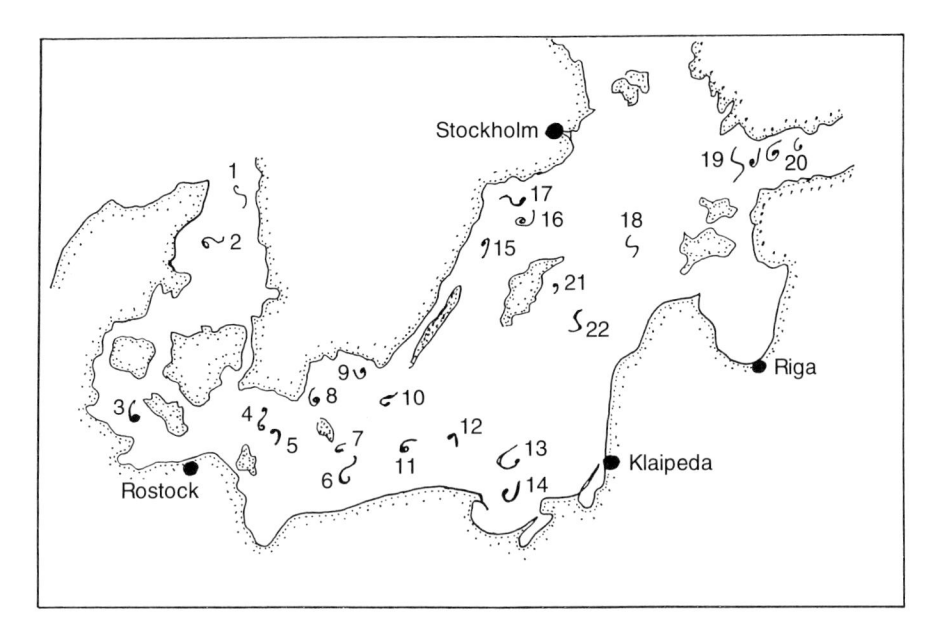

REGIONAL SATELLITE OCEANOGRAPHY: CASE STUDY OF THE BALTIC SEA   195

Figure 5.24 Examples of eddy structures in the Baltic Proper as revealed from satellite imagery.

Baltic Proper. Some eddy structures in the Gulf of Finland are shown in Figure 5.25.

The major part of the eddies recorded in satellite imagery in the Baltic Sea (see Countwell 1984; Kagan 1978; Horstmann 1983) are of cyclonic type. This corresponds to the general cyclonic circulation of waters in the Baltic Sea, with the existing quasi-permanent currents being close to the coast. Besides, in the satellite imagery of the IR band, so-called 'mushroom-like' dynamical eddy structures are

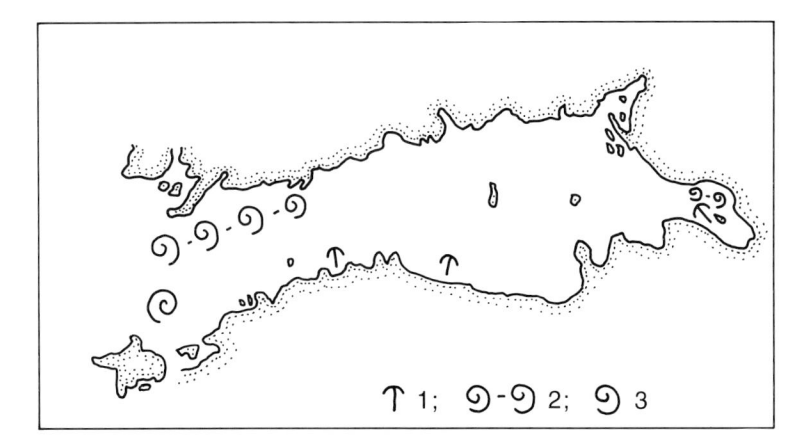

Figure 5.25 Chart of principal areas of generation of medium-scale eddies in the Gulf of Finland: (1) 'mushroom-like' structure; (2) chain of eddies; (3) single eddy (Victorov et al. 1989a).

recorded (see Section 5.3.4 below), actually consisting of two eddies, cyclonic and anticyclonic. Sometimes only the cyclonic part could be seen in the satellite images of the IR band (Ginzburg and Fedorov 1984a), thus leading to a bias in calculations.

Another peculiar feature of the eddy structures in the Baltic Sea is the generation of long chains of eddies. They are recorded in satellite imagery in various parts of the sea (see Figures 5.19(e) and 5.25).

It should be noted that with the development of satellite sensors for sea surface observation the still finer spatial structure of hydrophysical fields such as sea surface temperature, sea roughness, or patchiness in the visible band, can be seen and analysed. One of the characteristic features of the spatial 'order' of these fields is the eddy patterns which represent the universal type of the distribution of matter in Nature on scales from the Milky Way to microscopic viruses.

Hence, when moving step by step from an AVHRR image in APT mode (spatial resolution of 3–4 km) to an AVHRR image in HRPT mode (1.1 km), then to a LANDSAT TM image (120 m in the IR band), to MSU-E (30 m) and to satellite photography (several metres), one can see in the Baltic a still more complicated pattern of interconnected eddy structures of various types, shapes and dimensions. This pattern, resembling astrakhan fur, seems to be a specific indicator of a water mass with internal structure (see, e.g., Figure 5.9). One can see more and more eddy structures of smaller and smaller size at each step of the development of satellite sensors. So those 'good old times' when the detection of each eddy was considered to be worth announcing in a separate article (see, e.g., Elken et al. 1984) have gone forever.

This is why in regional satellite oceanography it seems more reasonable to discuss the feasibility of a study not of separate eddy structures but of a local eddy field in time, that is, of a kind of 'climatology of eddy fields'. Bearing in mind these remarks, one should regard the dimensions of eddy structures (listed in Table 5.13 and mentioned in the cited sources) as a certain approach to a future, more comprehensive, description of these complicated phenomena.

5.3.4 'Mushroom-like' structures

5.3.4.1 *'Mushroom-like' structures recorded in satellite imagery*

The analysis of satellite imagery of the sea surface in visible band (Ginzburg and Fedorov 1984a,b) gave clues to another form of movement of water – the so-called 'mushroom-like' currents. Similar structures named 'dipole eddies' or 't-structures' were also described by Ikeda and Emery (1984) and Ikeda et al. (1984).

Figure 5.26 shows a 'mushroom-like' structure with a 'foot' and a 'hat'. The letter A indicates the 'centre' of the 'mushroom'. Characteristic dimensions are also indicated.

The generation mechanism of 'mushroom-like' dynamical structures (MLDS) seems to be closely connected with the retardation of a jet stream taking place in a thin near-surface layer under the influence of a local impulse. The sources of the local impulse were discussed by Ginzburg and Fedorov (1984a,b). Possible mechanisms of the generation of 'mushroom-like' structures were also discussed by Kozlov and Makarov (1985).

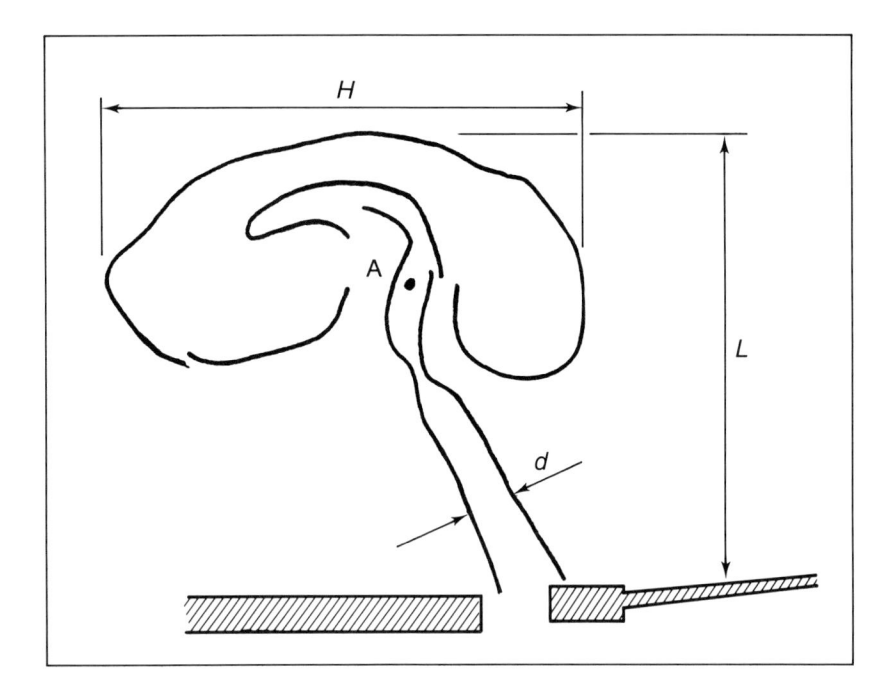

Figure 5.26 'Mushroom-like' eddy structure with characteristic dimensions (see the text for explanations). This particular pattern was recorded in an airborne image taken in the vicinity of the St. Petersburg flood barrier on 7 September 1987. One of its water gates generated this 'mushroom-like' structure. The tracer was suspended matter (Victorov et al. 1989a).

According to the model calculations of Ikeda and Emery (1984) and Ikeda *et al.* (1984) the mechanism of baroclinic instability is involved in the development of these structures.

The tasks for studying MLDS in the Baltic Sea were decided, similarly to those relating the eddies (Section 5.3.3). The same database was used to detect and analyse the MLDS. (The detection of MLDS in the images published by Horstmann (1983) was carried out and the classification of MLDS by their origin was performed by Bychkova and Victorov 1988). Some examples of MLDS, caused by various mechanisms of generation, follow.

MLDS caused by inshore water circulation and local wind conditions Figure 5.27(a) shows MLDS in the eastern part of the Gotland Basin as recorded on the AVHRR NOAA image of 22 May 1981. The mutual location of these structures corresponds to the general cyclonic circulation of waters in this basin. The wind field at that period was also favourable for the cyclonic circulation. A similar pattern was recorded in the satellite image of the visible band (tracer was blue–green algae) on 7 August 1981 in the western part of the Gotland Basin.

MLDS related to river discharges MLDS can often be seen near river estuaries in satellite images of the visible band. Suspended matter is the tracer in this case. MLDS of this type were recorded many times near the River Neman (Nemunas) Delta in the

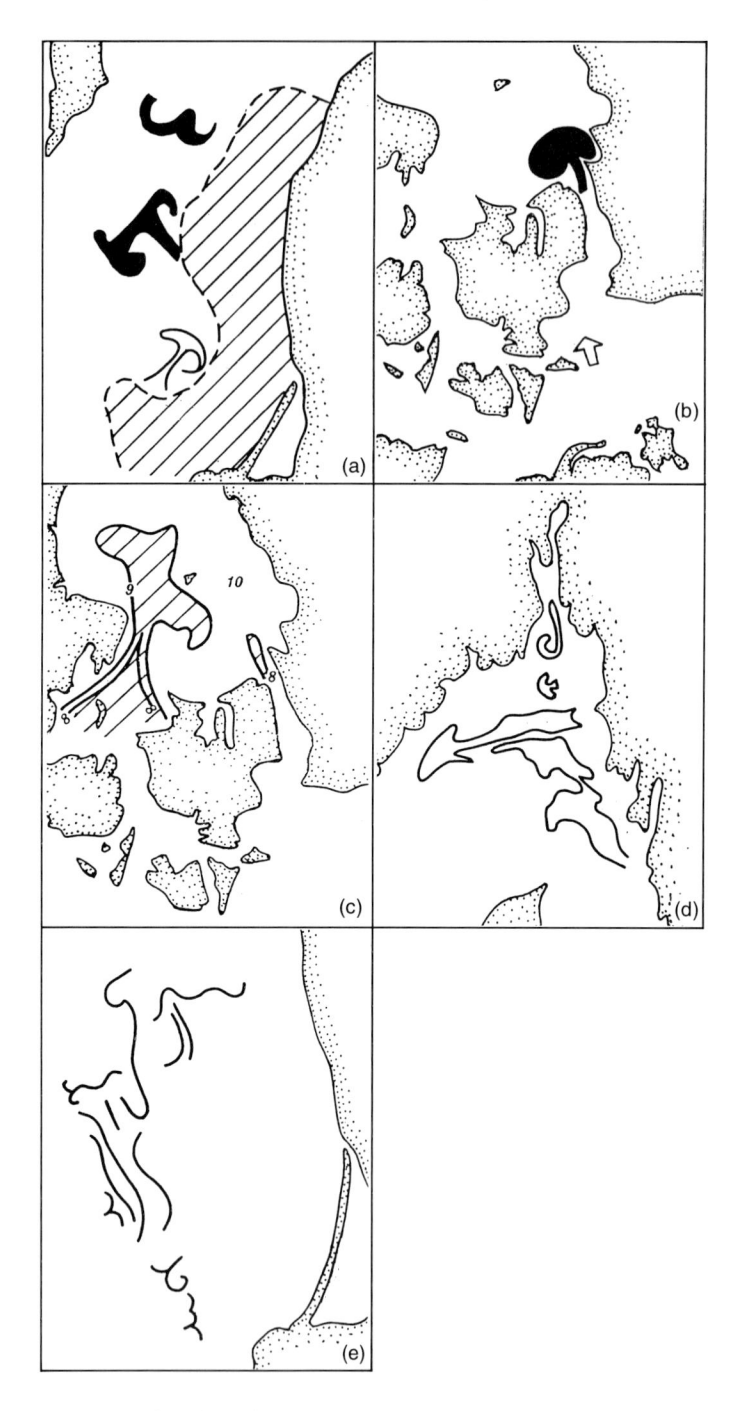

Figure 5.27 Some examples of 'mushroom-like' dynamical structures in the Baltic Sea as recorded in satellite imagery: (a) in the eastern part of the Gotland Basin, tracer – SST, NOAA, 22 May 1981; (b) at the exit from the Oresunn Strait, tracer – SST, NOAA, 12 July 1982; (c) at the exit from the Great Belt Strait (temperature is indicated in °C), 19 May 1985; (d) in Skagerrak Strait, tracer – ice, METEOR-Priroda satellite, (e) packages of MLDS, tracer – blue–green algae, METEOR-Priroda satellite, 7 August 1981 (Bychkova and Victorov 1988).

Kurshi Bay (Victorov et al. 1987) in METEOR imagery of the 0.7–1.1 m km band. Figure 5.28 shows MLDS in the Kunda Bay (Estonian coast of the Gulf of Finland).

MLDS caused by water exchange variability in the Danish Straits Figure 5.27(b) shows MLDS recorded at a period of a general change of direction of water flow. On 10 July 1982 the current direction was 270° (the flow through the Kadet–Rennen Strait), while on 11 and 12 July it was 320° (the main flow through the Oresunn Strait). On 11 July the flow at the exit from Oresunn generated an MLDS with a 'hat' dimension of 35 km, the next day it was 50 km (with anticyclonic part being less than the cyclonic part, probably due to restriction from the coast).

Figure 5.27(c) shows the MLDS at the exit from the Great Belt and Oresunn Straits on 19 May 1985. MLDS here are probably caused by the decrease in viscosity of the Baltic waters current velocity in these narrow straits. MLDS in the Kadet–Rennen Strait corresponds to the turn of the outflow to Oresunn, which means a change of the local impulse direction.

MLDS related to the instabilities of fronts and currents MLDS of this type are recorded in satellite images where sharp high-gradient fronts occur in the Baltic current and in the transition zone between the current and the neighbouring waters. In transition zones, with their high stratification, the sea surface temperature and the drifting ice fields can be used as tracers (Figure 5.27(d)).

Figure 5.29 shows some of the MLDS recorded in the Baltic Sea satellite imagery. (Digits indicate the MLDS in the list, which is not presented here.) It should be noted that the MLDS recorded in satellite images could be generated by a mixture of various driving forces: the local wind, general circulation, etc. In some cases the satellite images show a very complicated pattern of currents with packages of MLDS (see Figure 5.27(e)).

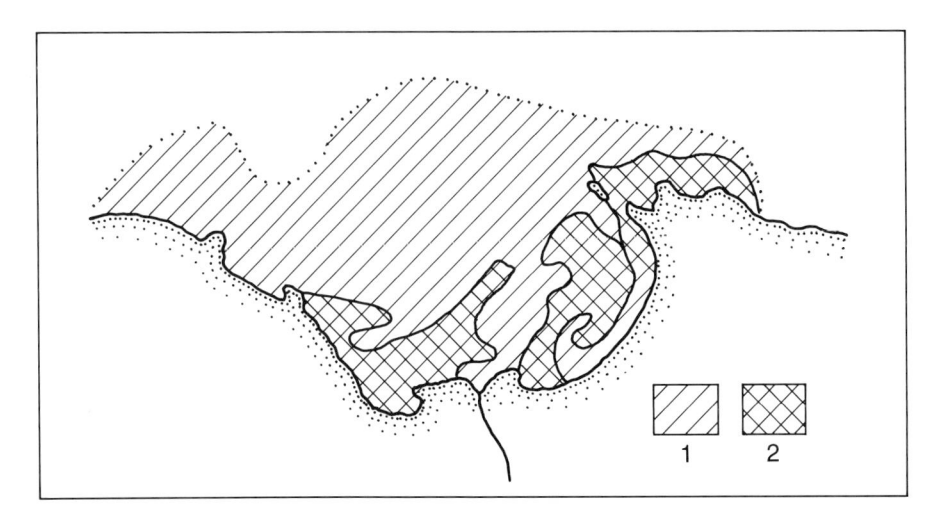

Figure 5.28 MLDS in the Kunda Bay, Gulf of Finland. Tracer was suspended matter. Based on MSU-E data from KOSMOS-1939 satellite of 23 September 1988. Digits 1 and 2 show zones with various concentrations of suspended matter (Victorov et al. 1989a).

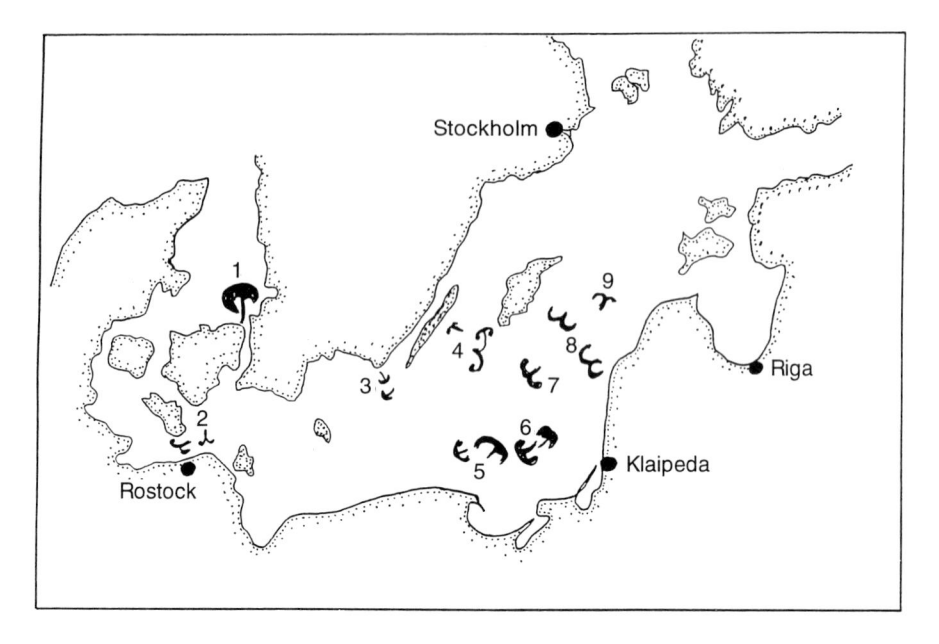

Figure 5.29 Some MLDS in the Baltic Sea as revealed from satellite imagery.

In general the characteristic dimensions of the MLDS recorded in satellite imagery in the Baltic Sea are as follows: 15–45 km in length, with a 'hat' dimension of 20–50 km, and with a 'foot' diameter of 2–8 km. Some examples of MLDS with their location and direction are given in Table 5.14.

There were no *in situ* observations of MLDS in the Baltic Sea before 1985. The first ever complex study of 'mushroom-like' structures in the Baltic was carried out during the third international COSE in May 1985 (Brosin *et al.* 1986, 1988) when the NOAA (in operative mode) and METEOR satellite data were used alongside intensive *in situ* measurements performed from the research vessel *Alexander von Humboldt*.

5.3.4.2 *Complex study of a 'mushroom-like' structure based on satellite and shipborne measurements*

In the second half of May 1985 the meteorological situation in the Baltic region was dominated by stable high-pressure conditions with high insolation. An averaged daily increase in the water surface temperature was 0.3 K for the period 17–19 May and 0.37 K for 19–25 May.

Intensive insolation, along with the spreading of heated and less saline waters from the Kurshi Bay, caused the formation of a large thermal anomaly in the south-eastern part of the Baltic Sea. (This anomaly could often be traced in the satellite images of this region in the 'hydrological spring' period.) Within this anomaly two patches of warm water located to the north-west (up to about 70 km from the Kurshi Bay mouth) and to the west-south-west (in a more elongated form) of the Kurshi spit, were recorded in satellite images of the IR band at 14 h on 15 May and

Table 5.14 Parameters of some MLDS in the Baltic Sea identified in satellite imagery (Victorov 1988b)

Coordinates of (·) A (°)		L	H	d	Direction
Longitude	Latitude	(km)	(km)	(km)	of current
19.6	55.7	30	45	6	NE
19.0	56.2	30	40	4	SW
19.6	57.8	25	45	8	SW
20.3	57.3	15	30	4	N
11.8	54.5	35	35	4	E
12.3	54.5	15	30	4	S
19.0	55.3	30	45	4	SW
12.0	56.2	35	35	4	N
12.5	56.2	35	50	4	N
18.5	55.3	30	60	4	NE
17.5	56.3	30	40	6	SW
16.0	56.0	15	20	6	SE
16.1	55.8	38	25	4	SW
17.3	56.6	12	55	4	E
17.1	56.6	15	25	2	W
17.3	56.8	30	30	4	NW
20.3	56.5	45	45	4	W

18 h on 18 May. Figure 5.30 (Plate 4) shows the anomaly as recorded in the satellite image of 19 May.

There was a significant frontal zone between the inshore warm, less saline water and the cooler offshore water. Shipborne measurements recorded temperature gradients at the frontal zones of the anomaly of up to 2 K km^{-1} and salinity gradients of 0.2×10^{-3} km^{-1}. The spreading of coastal water took place in a near-surface layer only 15 m thick at the most. In the following days the frontal zone moved westward at first with a velocity of 6.6 cm s^{-1}, and later more slowly at about 1.5 cms^{-1}.

The coastal water was not only traceable in the temperature field but also in the optical parameter datasets and in the distribution of suspended matter. At inshore station 25 (located 35 km in front of the Kurshi Bay mouth) the chlorophyll 'a' and phaeopigments concentration reached, at the upper 10 m layer, the value of 11.1 mg m^{-3}, which was 3 or 4 times the pigment concentrations measured at the adjacent stations. (Hence the spreading of the water from the Kurshi Bay could be seen even in a non-calibrated image in the visible band taken from METEOR satellite No. 30 on 20 May.)

By 18 May the winds from the north had shifted the southern part of the anomaly southward with a sharp increase of temperature gradients. Shipborne meteorological observations at distances between 25 km and 100 km off the shore showed an increase in the wind velocity from 3.5 m s^{-1} to 5–7.8 m s^{-1} since 13 h, 17 May. On 18–19 May a sharp change in the direction of the wind took place, e.g., the coastal station in Nida at the Kurshi Spit recorded a change from 360° (from the north) to 90° (from the east).

The wind from the east, along with the instability in the frontal zone of the anomaly, generated a strong westward jet flow with an MLDS in its head; the 'hat' and the 'foot' were both about 70 km each (see Figures 5.30 and 5.31).

The development of the MLDS took place on 19 and 20 May from the maximal phase (as recorded in the satellite image at 7 h, 19 May) through the relaxation phase (as traced in satellite images at 13 h on 19 May and 7 h on 20 May) to its disappearance in the satellite images of 21 May.

On 19 May the Steering Board of the third Complex Oceanographic Subsatellite Experiment (see Section 3.4 for details) decided to recommend that the research vessel *Alexander von Humboldt* rush to the point A (see Figure 5.31) in order to perform intensive *in situ* studies of the MLDS. The coordinates of point A were calculated, taking into consideration the time of the ship run, the delay in transmission of the recommendation, and the estimated development of the MLDS.

The jet flows in the Baltic are known to have a speed of about 0.2 m s^{-1}. The faster relaxation in the MLDS occurs in their cyclonic parts. These considerations led to the guidance of the research vessel to the forecasted position of the centre of

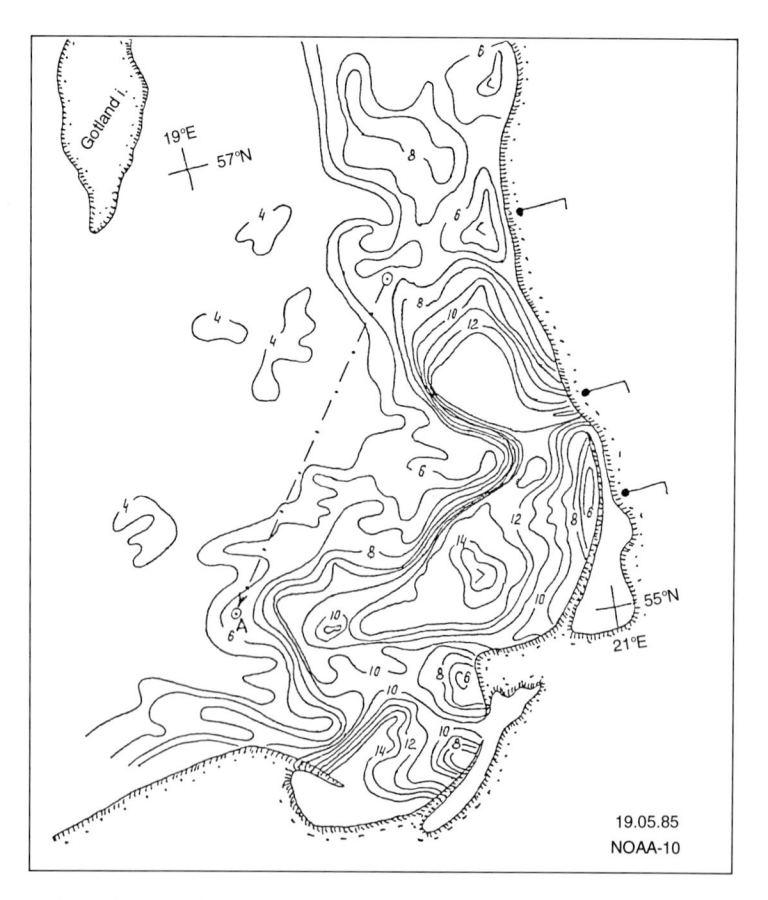

Figure 5.31 Chart of SST on the south-eastern part of the Baltic Sea revealed from the satellite image of 19 May 1985 (relative temperature units). Dashed line shows the route of research vessel *Alexander von Humboldt* guided to point A by the Steering Board (Bychkova et al. 1990a).

the anticyclonic part of the MLDS by the evening of 20 May: 55° 20' N; 18° 30' E (point A in Figure 5.31).

As estimated the research vessel reached point A and performed the first set of measurements (station at point A) at 20 h, 20 May; the second set was obtained at the same station at 3 h, 21 May. Actually the shipborne measurements could be considered as the study of the 'after-effect' (post-effect) of the MLDS in the area around point A.

Characteristics of the thermohaline structure at point A and the data on the hydrostatic stability of the water layers are presented in Table 5.15. The estimated values of hydrostatic stability were carried out using the difference between the measured and the adiabatic density gradients. On 20 May a sequence of hydrostatically unstable layers on a vertical scale of 4 m was determined. Seven hours later a considerable change in stability of the layer between 0 m and 16 m was observed.

The peculiar features of the stratification at point A during the first set of shipborne measurements could be caused by the anticyclonic part of the MLDS dipole. The results of both sets of *in situ* measurements could be considered as a specific indicator of the MLDS relaxation phase.

The data confirmed the existence of a shallow near-surface layer approximately 15 m thick with a temperature of 1.5–2.5°C higher and a salinity of 0.1–0.2×10^{-3} less than those of the surrounding waters. Obviously in the surface layer an intensive vertical exchange took place, manifesting itself also in the distribution of the density versus the stability. The baroclinic Rossby radius within the structure amounted to only 3–4 km (in the inshore area, to only 2 km). Figure 5.32 shows the three-dimensional structure of the temperature field in the vicinity of point A.

Warm and less saline waters could be traced for more than 170 km from the source (the Kurshi Bay mouth), the width of this anomaly being 15–20 km, and the length amounted to about 50 km (Figure 5.33).

The dome-like shape of the isotherm lines at the southern margin of the structure down to 60 m depth could be interpreted as an indication of a cyclonic eddy about

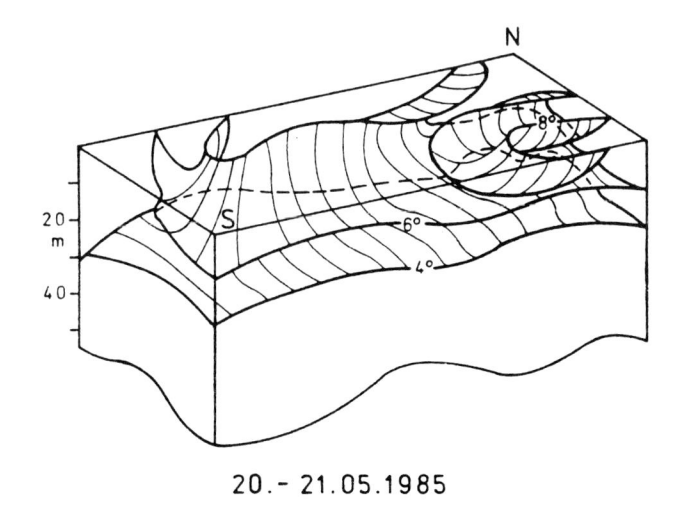

Figure 5.32 3D sea water temperature distribution in the surface layers around point A (°C) (Brosin et al. 1986).

Table 5.15 Oceanographic parameters characterising hydrostatic stability of water layers at point A (Brosin et al. 1988)

	20.5.1985				21.5.1985 03.15 h UTC				21.5.1985 14.15 h UTC			
		20.15 h UTC										
Z (m)	T_w (°C)	S (10^{-3})	σ_t	E (10^{-8} m^{-1})	T_w (°C)	S (10^{-3})	σ_t	E (10^{-8} m^{-1})	T_w (°C)	S (10^{-3})	σ_t	E (10^{-8} m^{-1})
2	7.03	7.63	5.95	0	8.56	7.57	5.78	0	9.03	7.59	5.75	−994
4	7.04	7.63	5.95	−497	8.58	7.57	5.78	0	9.12	7.57	5.73	−497
6	7.13	7.62	5.94	−994	8.59	7.57	5.78	0	9.24	7.57	5.72	−1492
8	7.17	7.61	5.92	994	8.59	7.57	5.78	0	9.41	7.56	5.69	7558
10	7.04	7.62	5.94	497	8.26	7.61	5.84	2982	7.79	7.58	5.85	3976
12	7.08	7.63	5.95	−1988	7.01	7.62	5.94	4970	7.07	7.61	5.93	1491
14	7.06	7.58	5.91	1988	6.92	7.62	5.95	497	6.95	7.63	5.96	497
16	7.01	7.63	5.95	−497	6.82	7.62	5.96	497	6.86	7.64	5.97	1988
18	6.82	7.62	5.94	994	4.37	7.75	6.18	10932	6.55	7.66	6.01	7951
20	6.04	7.60	5.96	10932	4.35	7.75	6.18	0	4.71	7.75	6.17	994
22	4.16	7.74	6.18		4.31	7.75	6.18	0	4.17	7.76	6.19	

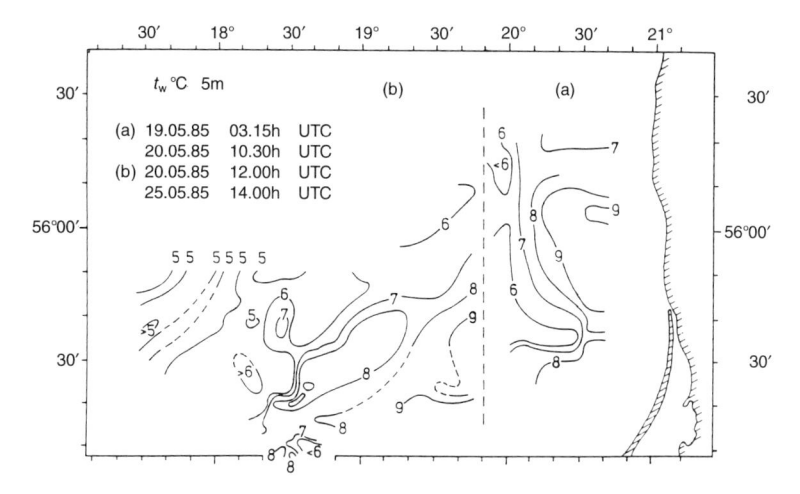

Figure 5.33 2D sea water temperature distribution (°C) at a 5 m depth for the periods of 19–20 May (a) and 20–25 May (b). Shipborne data (Brosin et al. 1986).

10 km in diameter. As for the southern margin the measurements were not conducted far enough southward. Within the anomaly a high variability of the water temperature in the upper layer was recorded. The repetition of the same station in the warm core (patch) confirmed a strong advective temperature increase of more than 0.2 K per hour. (Most of these measurements were taken under calm or light wind conditions.)

Based on satellite imagery and *in situ* shipborne measurements the lifetime of this particular MLDS could be estimated as four days at the most. (Actually it could be detected in satellite imagery of the IR band for about 1.5 days; its 'after-effects' could be traced in shipborne measurements for another couple of days.) During the final period of this investigation both the satellite and shipborne data showed that the two-core pattern in the SST field showing the spreading of the warm water from the Kurshi Bay into the Baltic Proper remained.

Figure 5.34 shows a three-dimensional presentation of the water temperature distribution resulted from shipborne measurements on 19–20 and 24–26 May. Shipborne measurements recorded the sea water temperature gradients in the upper layer of 0.5–1.0 K within a distance of only 0.4 km at the front of the anomaly. The satellite image of 27 May (Figure 5.9) shows the SST pattern which was formed by the warm waters from the Kurshi Bay with an addition of warm waters from the Vistula Bay in the course of the development of the pattern recorded on 19 May (Figure 5.30).

As a result of this complex oceanographic study with the involvement of satellite data and the use of shipborne data collected from a guided research vessel, the three-dimensional characteristics of the MLDS were obtained (though at the stage of its relaxation) bringing information about the details of the generation and development of MLDS (Brosin *et al.* 1986, 1988).

The estimated lifetime of an MLDS in the Baltic Sea based on the results of the third COSE made it possible to formulate the requirements for further studies of MLDS (or oceanographic phenomena with similar characteristics). Namely, satellite

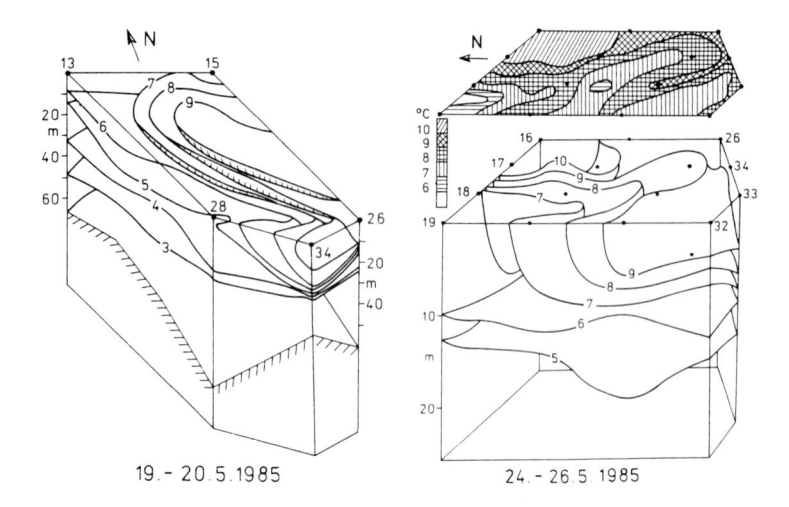

Figure 5.34 3D sea water temperature distribution (°C) for the periods 19–20 May (a) and 24–26 May (b); shipborne data. Numbers of stations are indicated (Brosin et al. 1986).

monitoring of the suspected area should be performed at least twice daily, and the research vessel should be at 'stand-by' mode in a position not exceeding the six hours run towards the forecasted area of MLDS generation. This area could be predicted using the weather forecast and the relevant knowledge base.

5.3.5 Other dynamical phenomena

5.3.5.1 River plumes

Among several initial sources of different water masses that provide 'raw construction material' which will then be used by various dynamical mechanisms to create a complicated multiscale pattern of the spatial and temporal variability (or 'patchiness') of hydrophysical, hydrochemical and hydrobiological fields in the Baltic Sea, the direct river discharges into the open sea and the indirect river discharges through their estuaries should be considered. In the Baltic Sea some shallow-water bays may be regarded as the estuaries of the rivers. For example, the Kurshi Bay may be considered as the extended estuary of the River Nemunas; and the Vistula Bay as an estuary of the River Vistula. Both bays are connected with the open sea by narrow straits acting as a nozzle generating jets, eddies, 'mushroom-like' structures, plumes and separate lenses.

Discharge processes influence significantly the regime of adjacent aquatories and are accompanied by transport of sediments, pollutants, and suspended and dissolved contaminants at great distances from the source. In this section, by 'source' we shall mean the mouth of a certain bay, or the point where the strait meets the open sea. (Though in its broadest meaning the 'source' is actually the whole catchment area of a certain river – see Section 5.2 on 'Baltic Europe' defined from this position.)

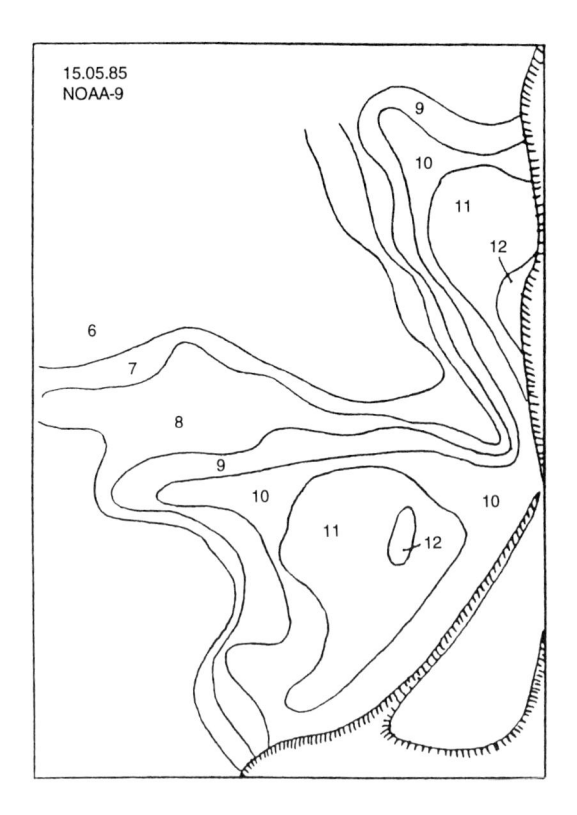

Figure 5.35 Chart of SST in the vicinity of the Kurshi Bay. Relative temperature units. NOAA, 15 May 1985, 14 h (Bychkova et al. 1990b).

River water discharges from the estuaries are known to form well-pronounced plumes. They are recorded in satellite imagery of various bands because their waters differ from the adjacent waters in temperature, and/or in turbidity, salinity, colour, surface roughness, etc., thus having sufficient contrasts to be detected.

River plumes can be traced at considerable distances offshore in satellite images of visible and IR bands. Remotely sensed data allow us to study plumes and to monitor their dynamics. Horstmann et al. (1986) studied the influence of river water on the south-eastern part of the Baltic using CZCS data from the Nimbus-7 satellite. During the period of 1985–1986 Bychkova et al. (1990a) analysed more than 50 satellite images of the river plumes spreading from the Kurshi Bay and about 20 images of Vistula River plumes. Some observations of river plumes from the Rivers Odra, Daugava, Narva and Dejma were also performed.

A study of the Kurshi Bay and its influence on the adjacent Baltic waters based on satellite, airborne and in situ data has been carried out as a co-operative effort of experts from Leningrad and Klaipeda (Victorov et al. 1986a,b). According to Dubra (1978) the mean outflow from the Bay in May for the period 1959–1968 amounted to 896 cubic m s^{-1}. In correlation with the River Nemunas discharge the maximum outflow from the Bay occurs in April. This information was taken into consideration when the international COSE (see Section 3.4) were being planned for the spring period (April, May).

The river waters discharged from the estuaries form zones of mixed waters located in the upper layers of the real sea waters in the form of jets. The direction, the distance of spreading of discharge jets and their thickness depend greatly on the wind conditions and the volume of water to be discharged. The thickness of the fresh water layer may reach 10 m or more (Processes of Sedimentation in the Gdansk Basin 1987). Long-term *in situ* data (Processes of Sedimentation 1987) and satellite imagery (see, e.g., Figures 5.35–5.37) show that the zone of river discharge mixture has a right-side asymmetry caused by the Coriolis force and, partly, by the dominating western winds in this region.

Durable winds of a northern direction in the region of the Gdansk Bay essentially decrease the distance of the spreading of the Vistula waters. In the region of the Klaipeda Strait such winds can lock up temporarily the river waters within the Kurshi Bay. In this case the discharge plume outside the Bay cannot be observed. It is ascertained that the wind influences greatly the location of the mixture boundary of the sea waters and the river waters. The time of response of the jet flow to the change of the wind conditions is proportional to the discharge volume and it usually lasts for several hours.

The width of a jet is a steadier parameter than the direction of the jet axis. There are estimations of a relationship between the river jet width, B, and the barocline local Rossby radius of deformation, R, for local discharge fronts. For example,

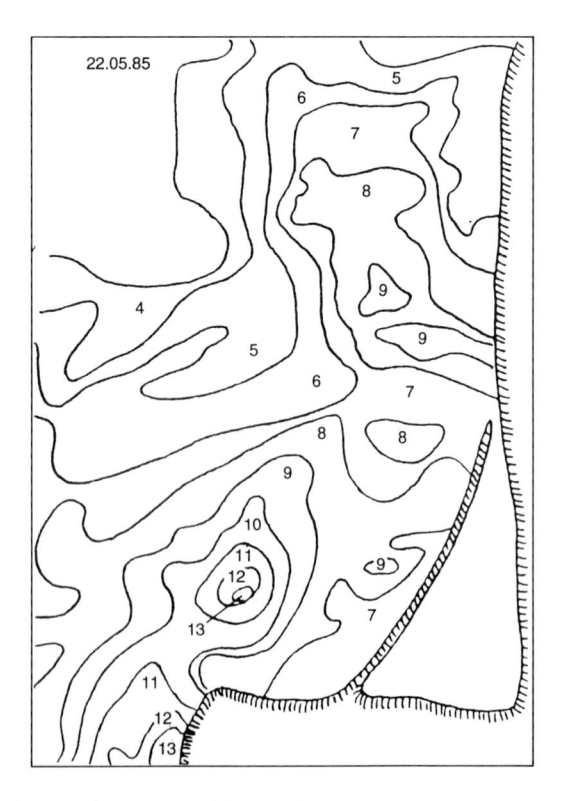

Figure 5.36 Chart of SST in the vicinity of the Kurshi Bay. Relative temperature units. NOAA, 22 May 1985, 13 h (Bychkova et al.1990b).

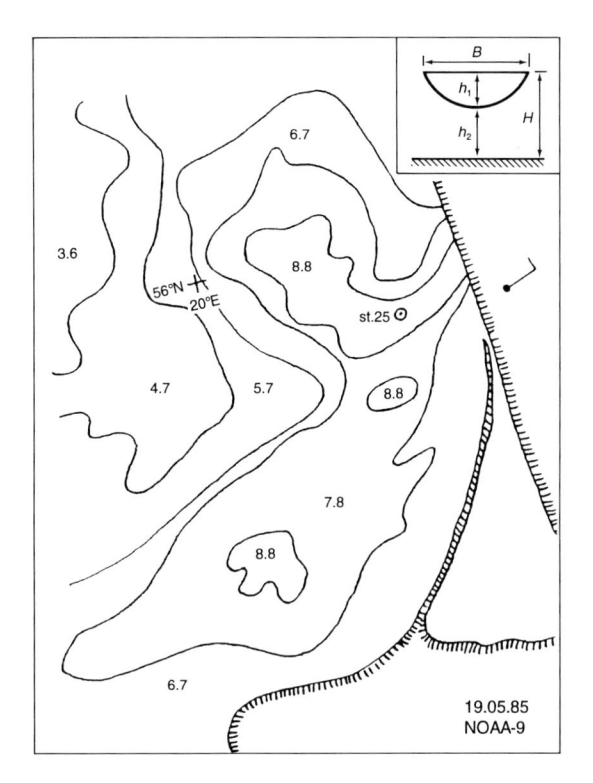

Figure 5.37 The river discharge spreading from the Kurshi Bay as recorded in a fragment of the chart of SST. NOAA, 19 May 1985, 13 h. Station 25 is indicated. Insert: schematic view of cross-section of the jet (Bychkova et al. 1990b).

according to Carstens et al. (1986) for the region of the Nordic Stream, $B = 2R$; for the Onega Lake $B = R$ (Boyarinov 1985).

When studying river plumes and trying to use the data on their appearance frequency to make an estimation of the amount of material released in the open sea, one faces the following problem. The remotely sensed data provide information on the amount of plumes and on their horizontal dimensions, which makes it possible to estimate the area covered with plumes. But one cannot estimate the volumetric characteristics without using relevant *in situ* data.

Bychkova et al. (1990b) have made an attempt to determine the thickness of the river plume flow (for a specific case, based on certain simplified assumptions), without using *in situ* information. Data collected during the third COSE, when the MLDS was studied near the Kurshi Bay, were used to develop a 'rough' three-dimensional model of a river plume flow.

The approach for the estimation of the thickness of the discharge plume using the remotely sensed data only was based on the assumption of the geostrophical balance at the discharge front. On the condition of impenetrability of the jet front and on the condition of relative immobility of the surrounding waters, the slope of a discharge front can be described on the basis of the Margulis equation as

$$\tan \alpha = fU/g, \qquad (5.1)$$

where: α is the angle of the discharge front slope; U is the jet velocity; f is the Coriolis parameter; and g is the gravitational parameter. The scheme inserted in Figure 5.37 shows the cross-section of a jet.

It follows from the geometric relations that

$$\tan \alpha = nh/B, \tag{5.2}$$

where: h is the jet thickness; B is the jet width at the surface; and n is the shape coefficient, closely connected with the geometry of the cross-section.

It is clear that for the alongshore 'cline-like' jet when the angle of bottom slope is much greater than the angle α, the shape coefficient $n = 1$; if both angles are nearly of the same value, then $n = 2$.

When a jet is spreading offshore, as a first approximation its cross-section can be presented as a circular segment with cord B, a line tangent to the front boundary at the sea surface. In this case

$$n = 2 \tan \alpha \sin \alpha/(1 - \cos \alpha). \tag{5.3}$$

It follows from (5.1) and (5.2) that

$$h = Bf U/ng. \tag{5.4}$$

The value of the local Rossby radius of deformation, R, for a mixed jet of discharge stream can be estimated as

$$R = 1/f(gh)^{1/2}, \tag{5.5}$$

or, using a more realistic two-layer model (Csanady 1971; Bowden 1988), as

$$R = 1/f\left(g \frac{h_1 h_2}{H}\right)^{1/2}, \tag{5.6}$$

where: h_1 and h_2 are the thicknesses of the upper and lower layers, and $H = h_1 + h_2$. There are quite forcible reasons to suggest a functional relationship between B and R which, as a first approximation, can be written as

$$B = mR, \tag{5.7}$$

where m is a numerical factor typical for a certain region of discharge front generation.

From (5.4), (5.6), (5.7) it follows that:

$$h_1 = \frac{H}{m^2 U}(m^2 U - nBf). \tag{5.8}$$

One should note that equation (5.8) could be used only in light wind conditions.

The estimation of the shape coefficient value according to (5.3) gives $n = 4$ for a rather wide diapazon of α values (0.03–10). The value of m firstly should be estimated from *in situ* measurements, or may be taken from other sources. The value of B can be determined very precisely from the satellite images (or using the thermal advection equation). The value of H can be obtained from navigational charts.

The verification of the suggested model has been carried out using the satellite and *in situ* data collected in May 1985 near the Kurshi Bay. The satellite data gave $B = 10$ km, $U = 35$ cm s^{-1}. *In situ* shipborne data collected at station 25 (its location was indicated in Figure 5.37) where the sea was 40 m deep, gave a value of 7 m

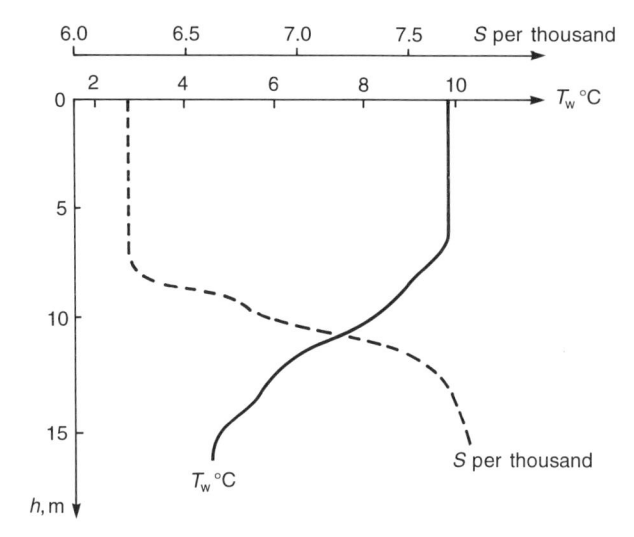

Figure 5.38 Vertical distribution of sea water temperature and salinity at station 25 indicating the thickness of the jet (Bychkova et al. 1990b).

for the jet depth (as seen in Figure 5.38), which was in reasonable agreement with the estimations according to (5.8) for $n = 4$. (For details see Bychkova et al. 1990b.)

Thus, using two-dimensional satellite imagery one can obtain some estimation of a three-dimensional frontal structure of a river water discharge plume provided that the assumptions made above are valid, and on condition that one is able to tune the suggested scheme using local peculiarities of this oceanographic dynamical phenomenon.

5.3.5.2 Discharge lenses

When the upper-layer water masses with low salinity from the estuaries form well-marked plumes, their further development and destruction may generate separated discharge lenses of various spatial scales. These lenses can also be studied using satellite imagery of the IR band if there are thermal contrasts between the river (then estuarine) and the sea waters.

In May 1988 within the framework of the fourth (short-term) COSE, complex remote sensing and *in situ* measurements were carried out in discharge lenses located about 150 km south-west of the Kurshi Bay mouth (Bychkova et al. 1990a). Shipborne measurements were made from the research vessel *Professor Albrecht Penk*. The guidance of this vessel to the SST anomalies was performed using operative SST charts based on AVHRR NOAA-9 and -10 satellite data.

The major findings were as follows. The horizontal scales of such lenses were about 10–30 km, their vertical extension (thickness) varied from 7 to 11 m. Temperature and salinity differences between discharge lenses and surrounding waters reached 1 K and 0.6 promille, respectively. On the basis of consecutive measurements, estimations of horizontal heat flow on the boundaries of the anomalies were carried out, and the lifetime of lenses was also estimated (see Table 5.16 in Section

5.3.6). The calculated value of the horizontal turbulent diffusion coefficient was 240 m^2 s^{-1} (Bychkova et al. 1990a).

5.3.6 Summary of results

Since 1982 satellite monitoring of the Baltic Sea has been carried out as a pilot project (in a non-operative mode) by the Laboratory for Satellite Oceanography, State Oceanographic Institute, St. Petersburg. The imagery from the METEOR, KOSMOS and NOAA series of satellites in the visible and infrared bands, as well as satellite radar imagery, were used to study the distribution of sea surface temperature, fields of algae bloom and suspended sediments with an emphasis on the estuarine and coastal areas. Dynamical phenomena – upwelling events, eddies and 'mushroom-like' structures – have been detected and systematised by the mechanism of origin.

Some hundreds of satellite images have been processed and analysed; the vast majority of NOAA imagery has been received and recorded by the staff members of the Laboratory at the autonomous satellite data acquisition stations. The interpretation and analyses of the satellite imagery were supported by the relevant synchronous and quasi-synchronous *in situ* measurements of oceanographic and meteorological parameters carried out from ships and aircraft. A series of joint Soviet–German Complex Oceanographic Subsatellite Experiments was a constituent part of the work.

The study of dynamical phenomena in the Baltic Sea was described above, and some results of this research are summarised in Table 5.16. It contains data on typical spatial characteristics, lifetime, and peculiar features of some dynamical phenomena recorded in satellite imagery; the tracers used are also indicated. (Figure 5.39 shows an example of a 'still zone' in the Gulf of Finland; this phenomenon was not discussed in this section.)

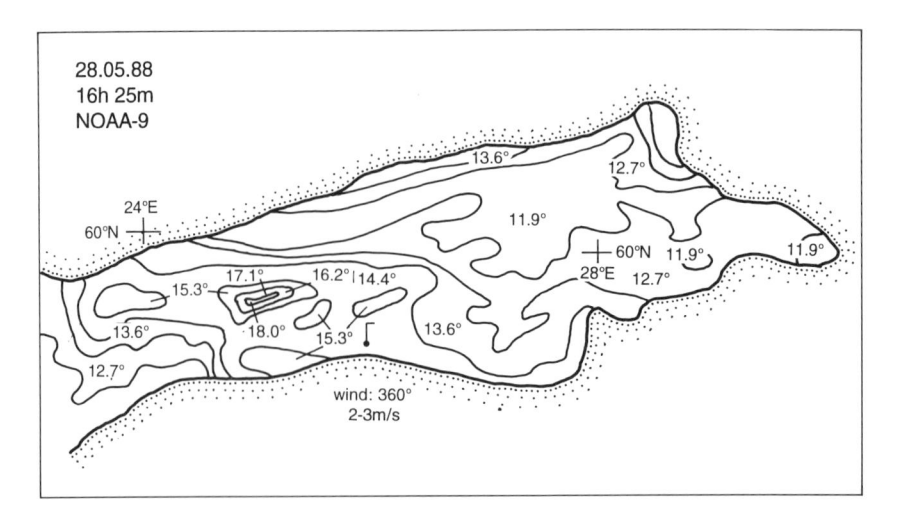

Figure 5.39 'Still zone' in the centre of the Gulf of Finland with much higher SST (17–18°C) as compared to the surrounding waters. Wind conditions are indicated. SST chart based on AVHRR NOAA image of 28 May 1988, 13 h (Bychkova et al. 1990a).

Table 5.16 Characteristics of dynamical oceanographic phenomena in the Baltic Sea as revealed from satellite imagery obtained during the period of experimental satellite monitoring in 1982–1989 (Bychkova et al. 1990a)

(a)

No. (1)	Phenomenon (2)	Tracer (3)	Period of observation (months) (4)	Typical dimensions (km) (5)
1.	Upwellings	SST	4–10	10–20 (2) 200–300 (1)
2.	Discharge plumes	SST, suspended matter	4–11	10–100 (2) 10–30 (1)
3.	Eddy structures	SST, suspended matter, ice, algae	Round the year	1–50 (3)
4.	'Mushroom'-like structures	SST, suspended matter, ice, algae	Round the year	20–50 (1) 2–8 (4)
5.	Discharge lenses	SST, suspended matter	4–11	10–30 (3)
6.	Water exchange through the straits (inflow–outflow)	SST	Round the year	10–20 (1)
7.	Still water zones (6)	SST	4–8	10–100 (3)
8.	Coastal fronts	SST, suspended matter	4–8	10–20 (2)

(b)

No. (1)	Temperature difference (K) (6)	Appearance frequency (7)	Structure features in SST (or brightness) fields (8)	Lifetime (days) (9)
1	2–10	Up to 30% per month	Jets up to 100 km (1) Cores about 10 km (3)	0.5–14
2	2–4	Up to 100% per month (5)	Right-hand asymmetry	Seasonal phenomenon
3	0.5–1	Accidentally	Core and peripheric parts. Mainly cyclonic rotation. Eddy cascades	1–7
4	0.5–1	Accidentally	Pair of eddies	1–3
5	About 1	Accidentally	Area of lens increases with distance from source	1
6	1–2	Continuously	'Mushroom'-like structures may occur	Connected with synoptic processes
7	3–4	Accidentally	Warm core	1
8	1–4	Continuously	Meanders on the front	Seasonal phenomenon

Notes: (1) length, (2) width, (3) diameter, (4) thickness (of the 'foot'), (5) depends on wind situation and river flow, (6) see Figure 5.39.

Another aspect of experimental satellite monitoring of the Baltic Sea – the study of seasonal bloom and the transport of suspended matter – will be presented in Sections 5.4 and 5.5.

5.4 Satellite monitoring of biological phenomena and pollution in the Baltic Sea

5.4.1 Seasonal bloom

5.4.1.1 Introductory remarks

It was shown in Section 5.3 that algae can be used as a tracer to detect and study dynamical phenomena in the Baltic Sea. The algae are of still greater importance as a biological indicator of the state of the marine environment (see Shubert 1984), the state of the entire marine ecosystem of the Baltic Sea. Remotely sensed concentrations of phytoplankton give information on the eutrophication of a certain part of the sea. Hupfer (1982) stated that in the 1970s the phytoplankton bloom intensified considerably, which is the evidence that the nutrition income has already caused the eutrophication of the Baltic. The role of algae as a biological indicator is based, *inter alia*, on the fact that their photosynthetic activity is strongly dependent on external factors. Unfavourable external conditions may depress photosynthetic activity which lead to serious ecological consequences. Some of the external factors can be simulated in the laboratory; e.g., simulation of temperature rise on the phytoplankton growth was described in Shubert (1984).

Both the concentration of phytoplankton and its type are indicators of the state of the marine ecosystem. Some years ago the diatomic algae were dominant in the Baltic (an example of their detection in Kiel Bight using remotely sensed data was presented by Ulbricht 1984). In recent years blue–green algae have pushed out the diatomic algae and caused changes in the whole trophic chain (Horstmann *et al.* 1978; Tsiban 1981; Rinne *et al.* 1981). The diatomic algae are preferable for the 'health' of the sea, but the now-dominant blue–green algae *Aphanizomenon flos aquae* and *Nodularia spumigena* are more accommodative to external factors and are able to fix molecular nitrogen directly from the air and transform it into organic substances. Thus they play a considerable role in the balance of nitrogen in the Baltic Sea. Nielsen and Hansen (1983) calculated that each seasonal bloom of the Baltic Sea produces 2000 tonnes of nitrogen which amounts to 20% of its annual income with the discharge of Swedish rivers. Estimations of specific fixation of nitrogen by blue–green algae were made by Rinne *et al.* (1981), who wrote that the satellite data on the distribution of blue–green algae in the surface layer are required for the quantitative estimation of the total fixation of nitrogen.

Blue–green algae are of micrometric dimension, but they tend to form clusters shaped in long filamentary structures, which makes them detectable from aircraft and satellites. The analysis of multispectral satellite images of the visible band recorded in favourable atmospheric conditions, aside from the Sun glint, enables us to determine optical inhomogeneities of the upper layer of the sea caused by considerable concentrations of suspended substances of mineral and organic origin, including algae.

It is important to discuss some peculiar features of the study of phytoplankton (algae) fields in the Baltic Sea using satellite images. The use of satellite data in the study of algae is based on the effect of visualisation of plants at a certain stage of their development. Strictly speaking, rather an uncertain 'bio-optical parameter' is being registered in satellite imagery. This term reflects many uncertainties in the process of detection and the analysis of the data.

The appearance of blue–green algae at the surface itself is an uncertain phenomenon. Gas cavities of multicellular blue–green algae allow them to accumulate in the upper layer in calm weather, fields being formed whose dimensions and optical properties enable them to be detected from satellites. The process of floating or sinking is governed by biological factors. It is impossible to detect algae if they are located too deep even if their abundance is very high. On the other hand, when the concentration of algae is very high, their clusters may have positive buoyancy and, as was stated by Nielsen and Hansen (1983), 'the upper surface of algae is located above the water and it is almost dry'. But in this case there are no remote sensing techniques at all to determine dry algae concentration! Even in the case of 'wet' algae, there is uncertainty in the vertical distribution of algae which in principle makes it impossible to determine their concentrations on a quantitative level. This problem was studied by Kahru (1981a,b). In general, as already discussed in Chapter 3, the problem of quantitative analysis of satellite imagery in the Baltic Sea is a difficult one due to specific optical properties of the sea water. Hence, here regional satellite oceanography faces, in the Baltic Sea, two separate problems: (1) the detection of algae fields and determination of their instant visible dimensions as a basis for the study of their temporal and spatial variability on various scales, and (2) determination of plankton concentration and assessment of its biomass as a basis for the study of its variability. There is another uncertainty which is a bottleneck for both types of studies: the cloudiness which may impose a bias on any time series of satellite data. It is obvious that only routine standardised technology of satellite data acquisition, processing and analysis can make it possible to study interannual variability of blooms (though with the restrictions imposed by the above-mentioned uncertainties).

5.4.1.2 Bio-optical patterns in satellite images of the Baltic Sea

With these discouraging thoughts in mind, let us turn to the realities of regional satellite oceanography as applied to the study of bio-optical characteristics of the Baltic Sea. The phenomenon of seasonal bloom in the Baltic Sea has been observed accidentally in satellite imagery since 1973, the first source being LANDSAT followed by Nimbus-7, NOAA and most recently METEOR-Priroda (Ostrom 1976; Horstmann *et al.* 1978; Horstmann and Hardtke 1981; Ulbricht 1983, 1984; Nielsen and Hansen 1983; Victorov and Sukhacheva 1984; Victorov *et al.* 1987, 1988).

Horstmann and Hardtke (1981) analysed a series of three cloudless LANDSAT scenes of the south-western Baltic for 2, 3 and 4 March 1976 at the period of spring bloom in this area. Nimbus-7 CZCS images for 1980–1982 with blue–green in the surface layer of the Baltic Sea were presented by Horstmann (1983). In 1980 in CZCS imagery the bloom was recorded on 25 and 31 July, 16 and 17 August; in 1981, on 7 August; in 1982, only on 28 July and 1 August.

In METEOR imagery the blue–green blooms were recorded in 1981, on 7 August (the data coincided with those mentioned by Horstmann 1983); in 1982, on 2 August; in 1983 several scenes with bloom were received (see below); in 1984 one image of 31 July was received (Victorov *et al.* 1987).

216 REGIONAL SATELLITE OCEANOGRAPHY

Before 1983 we could not obtain, in the available satellite imagery, a time series of scenes with reliable records of blue–green bloom. We were lucky to have that series in July–August along with shipborne biological *in situ* measurements. Fields of algae could be traced in satellite imagery from METEOR-Priroda of 26 July; 7, 8, 18 August; of consecutive days 20–23 August; and also of 27 August. It seems that in the last decade since August 1983 hydrometeorological conditions were favourable for the accumulation of blue–green algae in the surface layer of the sea (Victorov *et al.* 1987, 1988).

Most remarkable was a series of images of 20 August (MSU-M), 21 August (MSU-M), 22 August (MSU-S) and 23 August (MSU-M). Only one image of the four (from the medium resolution scanner) was more or less comparable with the previous images of blooms in the Baltic Sea published by Horstmann (1983). Figure 5.40 shows a fragment of an MSU-S image; for comparison a fragment of CZCS is inserted. The quality of the MSU-S image is lower (partly due to the fact that the registration of the signal from the sea surface was not performed properly, see Chapter 3 for details). So the arrows are used to indicate the position of bloom

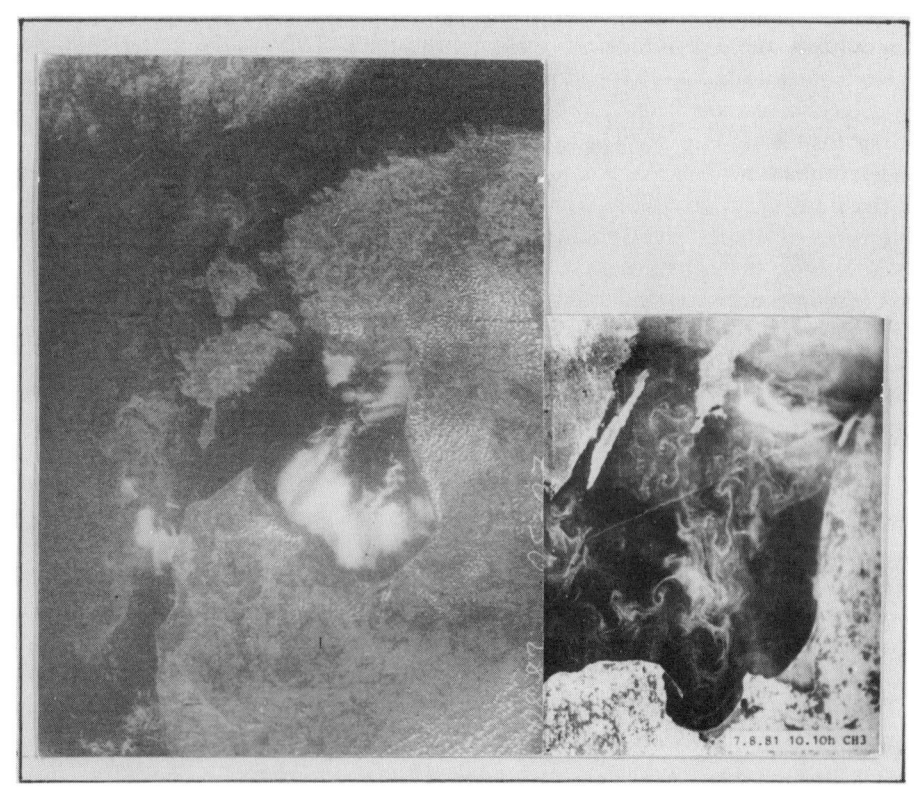

Figure 5.40 Blue–green algae forming an eddy at the entrance to the Gulf of Finland (upper arrow) and filamentary structures to the west (middle arrow) and to the south (lower arrow) of Hiiumaa and Saaremaa islands. METEOR-Priroda satellite image of 22 August 1983. MSU-S scanner, 500–700 nm channel (Victorov *et al.* 1988). For comparison a fragment of a Nimbus-7 CZCS image of 7 August 1981 showing algae fields in the Baltic Proper (Horstmann 1983) is inserted. Composition by the author.

structures, otherwise hardly recognisable. (A set of five images for 20–23 August was published by Victorov *et al.* 1988.)

Only in 1992 did we receive from the Russian satellite an image of the seasonal bloom in the Baltic Sea comparable, in contrast, with those from LANDSAT or Nimbus-7 (see Figure 5.41).

In Figure 5.40 the upper arrow shows the eddy structure (traced by algae) which was described in detail in Section 5.3. Here we will briefly discuss the motion of the algae fields as traced in satellite imagery of 20–23 August; also the relevant *in situ* data will be presented.

According to near-surface synoptic analysis on those days the lower atmosphere in the central and northern parts of the Baltic Proper was characterised by a low-gradient thermobaric field. The weather was mainly cloudless, with high-level clouds in places; the wind velocity was 1–7 m s^{-1}.

In the satellite image of 20 and 21 August the blue–green algae formed a wavy belt along the coast oriented north–south; its length was 180 km and width was 20–40 km. It is interesting to note that a similar configuration of algae field was recorded in the satellite image of 18 July (MSU-S, 500–700 nm channel). On the subsequent days of 22 and 23 August the shape and the structure of the algae field changed considerably. Besides, possibly due to the increase of the algae abundance in the upper layer, the contrasts of brightness increased and some new features could be resolved.

The analysis of these images together with synoptic charts enabled us to relate the movement of the algae field, as revealed from subsequent satellite images, with total wind drift. During four days the phytoplankton field drifted north-north-east for a distance of 30–40 km. The analysis of the satellite image of 27 August showed that by that time a dissipation of the observed algae field had occurred – in any of three channels of the MSU-M scanner one could identify only separate chaotic filamentary features covering practically all the sea surface between the northern part of Gotland Island and the Moonsund.

The results of the analysis of satellite imagery were complemented and confirmed by quasi-synchronous shipborne measurements of the biomass and chlorophyll 'a' concentration, the abundance and types of phytoplankton population, and also by the available data on spatial-temporal variations of their distribution in the period of summer bloom in the Baltic Sea in 1983. The data of hydrobiological observations carried out from the research vessel *Strelets* showed that intensive growth of blue–green algae of *Aphanizomenon flosaquae* and *Nodularia spumigena* was observed during the whole period of cruise research. Considerable spatial and temporal variability of the phytoplankton biomass was determined. During the first and second parts of the observations the concentration of biomass in the layer 0–20 m varied from 1 to 140 mg m^{-3} and from 60 to 380 mg m^{-3} respectively, with the maximum value exceeding 1200 mg m^{-3}. Chlorophyll 'a' concentration in August in the region where the satellite data showed intensive phytoplankton bloom varied from 1 to 5 mg m^{-3} (Victorov *et al.* 1988).

Figure 5.42 shows spatial distribution of chlorophyll 'a' in the layer 0–10 m as revealed from shipborne *in situ* measurements. The high dispersion of the shipborne data can be explained by taking into consideration the multiscale patchy structure of algae bloom fields and the short-term periodic variations of biomass concentration in the upper layer including the diurnal variations. Thus the use of random *in*

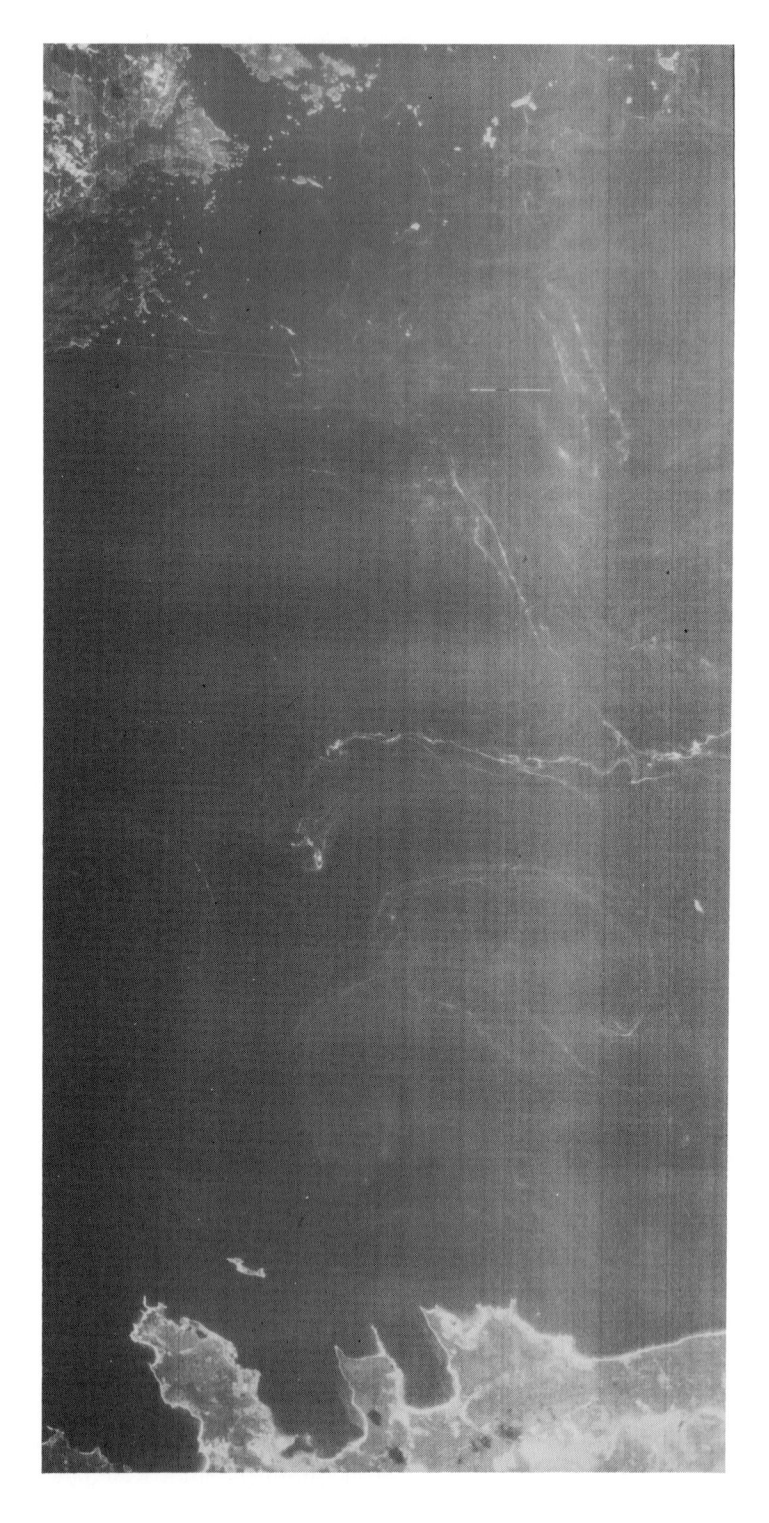

Figure 5.41 Filaments of blue–green algae in the central part of the Gulf of Finland. The frame is located across the Gulf to the east of Tallinn. MSU-E image of 6 June 1992.

Figure 5.42 Chlorophyll 'a' distribution (mg m^{-3}) in the upper layer (0–10 m) in the northern Baltic Proper. *In situ* measurements from research vessel *Zvezda Baltiki* (Baltic star) (Victorov et al. 1988).

situ shipborne measurements of phytoplankton concentration in general seems to be of limited value for studies of patchiness. For effective study of the patchiness of biological fields in the Baltic Sea, complex satellite, airborne and shipborne observations are required.

The scales of inhomogeneities of hydrobiological fields recorded in satellite imagery are limited by the spatial resolution of sensors (e.g., 240 m for MSU-S and 1000 m for MSU-M in our case). Long filamentary structures of algae bloom are detectable in the MSU-M imagery, which means that the filaments are wider than 1 km (though lineaments can be seen in satellite imagery even if their width is less than the spatial resolution of a sensor). Using airborne and shipborne spectrometers and TV cameras one can study the patchiness structures of smaller scales. The fields of blue–green algae recorded as filaments in satellite images can be seen from the ship as parallel stripes of grey–green colour, tens of metres wide and kilometres long (Figure 5.43; Plate 5).

Based on the analysis of the published satellite imagery of the Baltic Sea for 1980–1984 (including a limited amount of imagery from the USSR satellites) Victorov et al. (1987, 1988) noted some peculiar features of the spatial and temporal variability of blue–green algae bloom and compiled a schematic chart of algae distribution in the Baltic Sea (Figure 5.44).

'Analysis of a series of satellite images and sea-truth *in situ* shipborne data collected during the periods of summer bloom in the Baltic Sea for several years enables us to study the year-to-year variability of this phenomenon – appearance frequency of regions of bloom, their duration and intensity.' This long-term task was formulated and set in 1987 (Victorov et al. 1987, 1988).

This was partly done by Rud and Kahru (1994) who got access to multi-year sets of digital satellite data and sophisticated technical facilities at Stockholm University.

220   REGIONAL SATELLITE OCEANOGRAPHY

Figure 5.44 Schematic integrated chart of blue–green patterns in the Baltic Sea as recorded on satellite imagery of 1980–1984 (Victorov et al. 1987, 1988). In the right lower corner a fragment of the algae distribution charts for four indicated years (as presented by Rud and Kahru 1994) is inserted. Composition by the author.

They used 135 archived preselected sufficiently cloud-free AVHRR NOAA scenes recorded in HRPT mode (1 km resolution) for the summer (30 June–24 August) period of the years 1982–1993. For each year a chart of the late summer bloom of nitrogen fixing filamentous cyanobacteria (dominated by the species *Nodularia spumigena* and *Aphanizomenon flosaquae*) was compiled; the total area covered by accumulations was also determined.

The following standardised technique was applied to satellite imagery to detect algae patterns. AVHRR channels 1, 2, 4 and 5 data were involved. Channel 1 data were analysed and three clusters were formed: pure water, cyanobacterial accumulations (with empirically determined albedo range between 2.3% and 4%) and clouds. To account for the internal structure (texture) of filamentary fields of algae, a 3 × 3 pixel digital filter was applied. At the next step, pixels with albedo in channel 2 exceeding the albedo in channel 1 by 0.2% were expelled. Additional threshold filtering was applied using channels 4 and 5 (thermal data) to account for clouds (Rud and Kahru 1994).

It is obvious that the algorithm of algae bloom patterns retrieval should have been tuned: it had to provide detection of bloom features with results comparable to those obtained by an expert manually. Texture analysis could have been used for this purpose.

One should note that the charts of blooms for the years 1982, 1983 and 1984 presented by Rud and Kahru (1994) are in good agreement with the integrated chart of blooms for 1980–1984 presented by Victorov et al. (1987, 1988), which is not a surprise. Rud and Kahru (1994) performed important work, they computerised the procedure of superimposition of single algae bloom patterns, used by Victorov et al. (1987, 1988) manually, and applied their software to a whole set of images in a standard way. Nowadays this is a routine procedure in a sophisticated integrated GIS software package.

It is difficult to say whether an average of 10 satellite images per year is a representative amount of data. CZCS archived data are being analysed in a similar way within the framework of Project OCEAN (Ocean Colour European Archive Network). The OCEAN project will probably yield additional information. Current results obtained by Rud and Kahru (1994) showed that there were no shifts in the period of bloom during 1982–1994. The interannual variations of surface cyanobacterial accumulations were considerable with two peaks in 1982–1984 and in 1990 onwards. AVHRR data showed surprisingly no blooms at all during the summer period of 1987 and 1988; only tiny areas of blooms were recorded in places in 1985 and 1986. The reasons for these variations are not clear.

When considering seasonal algae bloom, one more interesting scientific problem arises. The sea surface temperature field and the algae bloom fields show correlation but SST anomalies (in AVHRR thermal imagery) are usually located offshore compared with the similar anomalies of the algae fields (detected in AVHRR, Nimbus-7 or LANDSAT visible band images). There are different viewpoints on the problem – either the higher albedo of algae as compared to the surrounding water leads to heating of water, or the algae prefer to live in the already heated waters. The discussion of this problem that took place at the Sixteenth Conference of Baltic Oceanographers between U. Horstmann and the author (see Victorov 1989a) showed that the problem was far from being solved. Kahru et al. (1993) seemed to present an example that it was still the blooms that caused heating of the sea surface by up to 1.5°C in 1992.

5.4.2 Transport of sediments and pollutants

River discharges spreading in the open sea containing suspended and dissolved substances can be traced far from the source. The River Nemunas–Kurshi Bay turbid waters are in some cases visible for a hundred kilometres to the north along the Latvian coast as a 70 km wide band. Turbid waters from the source located near St. Petersburg (for details see Section 5.5) can be traced in the Gulf of Finland at a distance of 60–100 km (Figure 5.45).

In some regions of the Baltic Sea even harmless (at first sight) suspended sediments may carry contaminations. In the Neva Bay of the Gulf of Finland, for

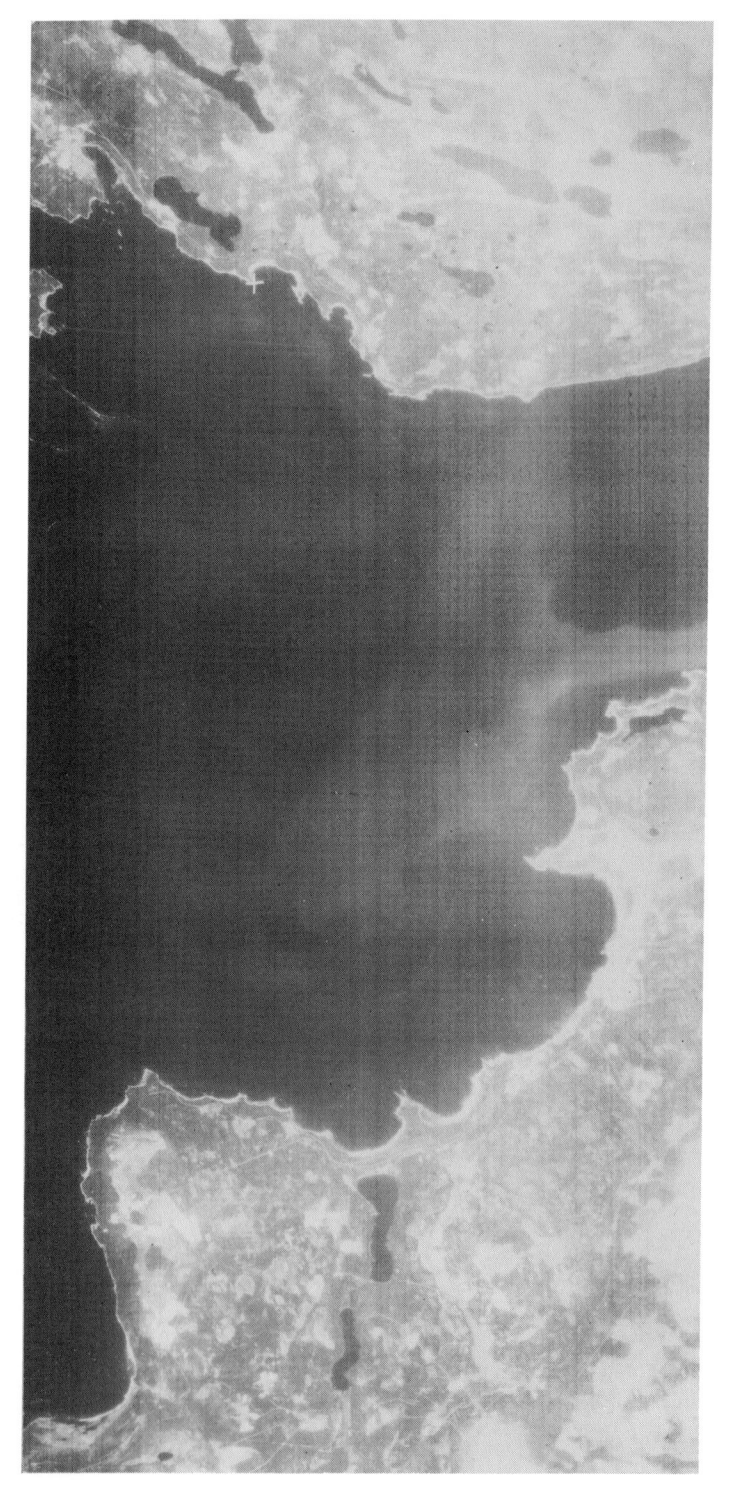

Figure 5.45 Satellite image of the Gulf of Finland of 5 June 1992 showing the maximum distance of spreading of suspended matter from the Neva Bay (from the right, beyond the scene) into the Gulf. The MSU-E frame is located across the Gulf with Koporskaya Bay in the centre.

example, there is a correlation between concentration of suspended sediments and water pollution with heavy metals, chlororganic substances and bacteria (Sukhacheva and Victorov 1994). So while tracing the transport of suspended matter one is able actually to obtain some information on the transport of pollutants.

The local marine and coastal environment may be also affected by 'thermal pollution' from powerful man-made sources of heat. The cooling systems of atomic power stations produce water with a temperature of 6–13°C higher than that of surrounding sea water and require about 150 cubic metres of water per second (Hupfer 1982). There are several atomic power stations in the Baltic region; some of them are located on the shore. There are two nuclear power stations on the coasts of the Gulf of Finland. Hari (1984) presented the results of a study of the marine environment in the vicinity of Loviisa power plant. They show that there was a considerable thermal anomaly in the waters within a radius of only a few kilometres from the plant. These anomalies can be detected in summer time using SST information. In winter time the anomalies in the ice cover caused by thermal outlets can be traced in satellite imagery of the visible band.

Figure 5.46 shows a satellite image of the Gulf of Finland and the Gulf of Riga covered with ice. In favourable conditions in similar images it is possible to detect

Figure 5.46 Satellite image of 22 March 1994 (MSU-S) showing ice cover in the Gulf of Finland and the Gulf of Riga, with Ladoga Lake in the right upper corner.

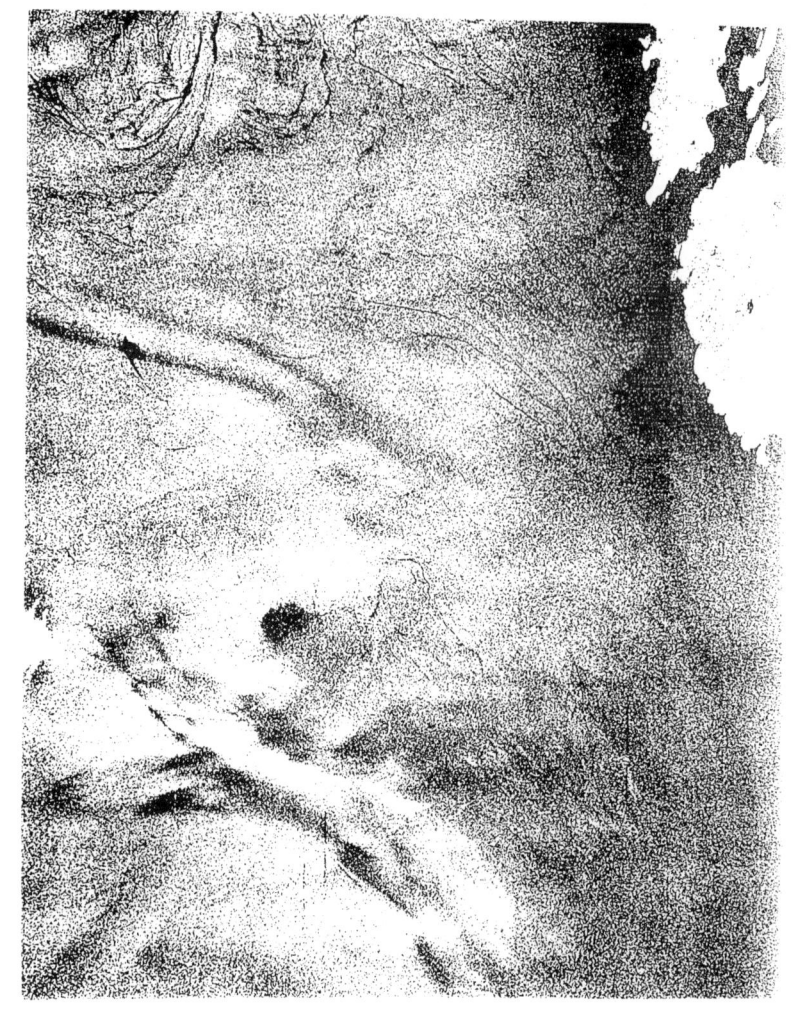

Figure 5.47 Eastern part of the Gulf of Finland. High-resolution synthetic aperture radar (SAR) image.

the thermal outlets of the power plants into the sea, including the nuclear power plant located in Sosnovij Bor (60 km from St. Petersburg, at the coast of Koporskaya Bay) by local anomalies in the ice cover (though more reliable results were obtained during airborne thermal surveys when the IR radiometer showed an increase of up to 10°C in the distribution of SST near the outlet of the power plant in Koporskaya Bay).

In terms of radiation safety near nuclear power stations and monitoring of the environment in the Baltic region, detailed studies of the sea water dynamics in the vicinity of potentially dangerous objects seems to be important. Figure 5.47 shows patterns of dynamical phenomena in the Gulf of Finland near Koporskaya Bay, as recorded in high-resolution radar imagery, which were interpreted as an S-shaped eddy (Figure 5.48). Dark filamentary structures in the upper part of Figure 5.47 are probably the manifestation of internal waves at the surface recorded in the field of surface roughness.

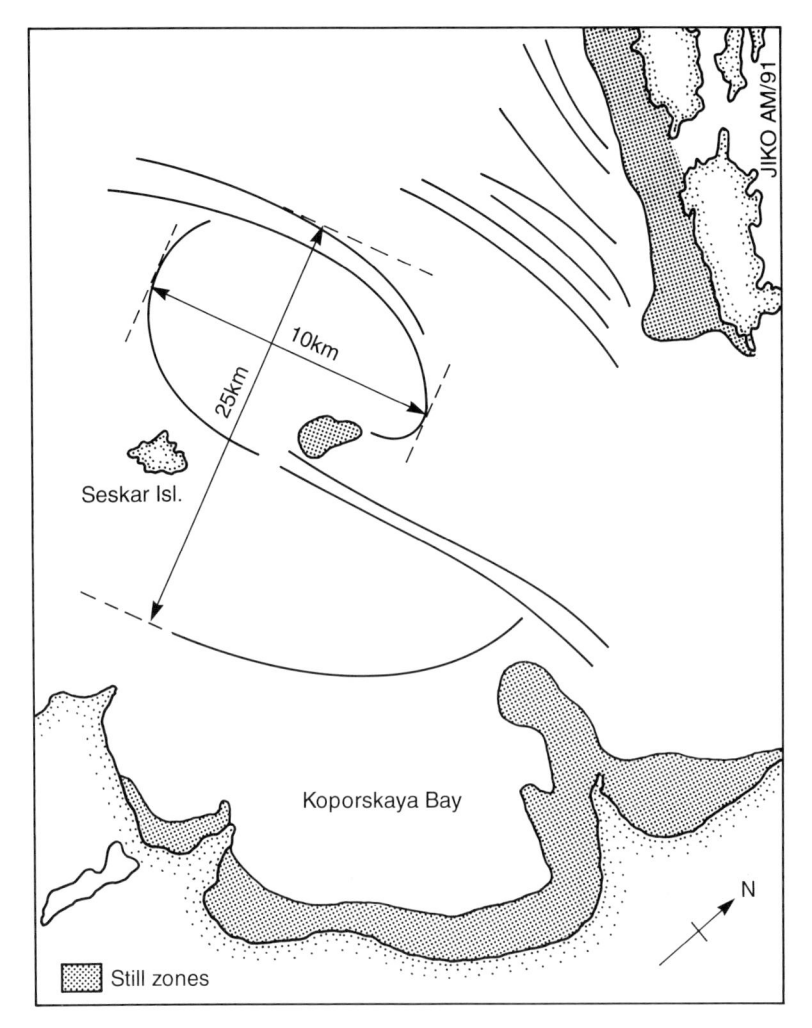

Figure 5.48 Interpretation of SAR image shown in Figure 5.47. Arrow shows north.

5.5 Monitoring of the marine and coastal environments in the Neva Bay

5.5.1 Introduction

Now we pass from the whole-sea scale (Figure 5.49) through the Gulf of Finland scale (Figure 5.50) to the Neva Bay, the easternmost part of the Gulf. In this section I shall try to show how the methods and techniques of regional satellite oceanography work on a local level. Besides being a response to an urgent environmental issue in a 'hot spot', the material to be presented below is also related to two of the 'coastal ocean' problems:

- 'What is the build-up of man-mobilised materials (pollutants and natural) in the coastal zone? How can these materials be used to trace natural processes?
- What are the sedimentation rates in the coastal ocean (shelves, estuaries, continental rises, etc.)?',

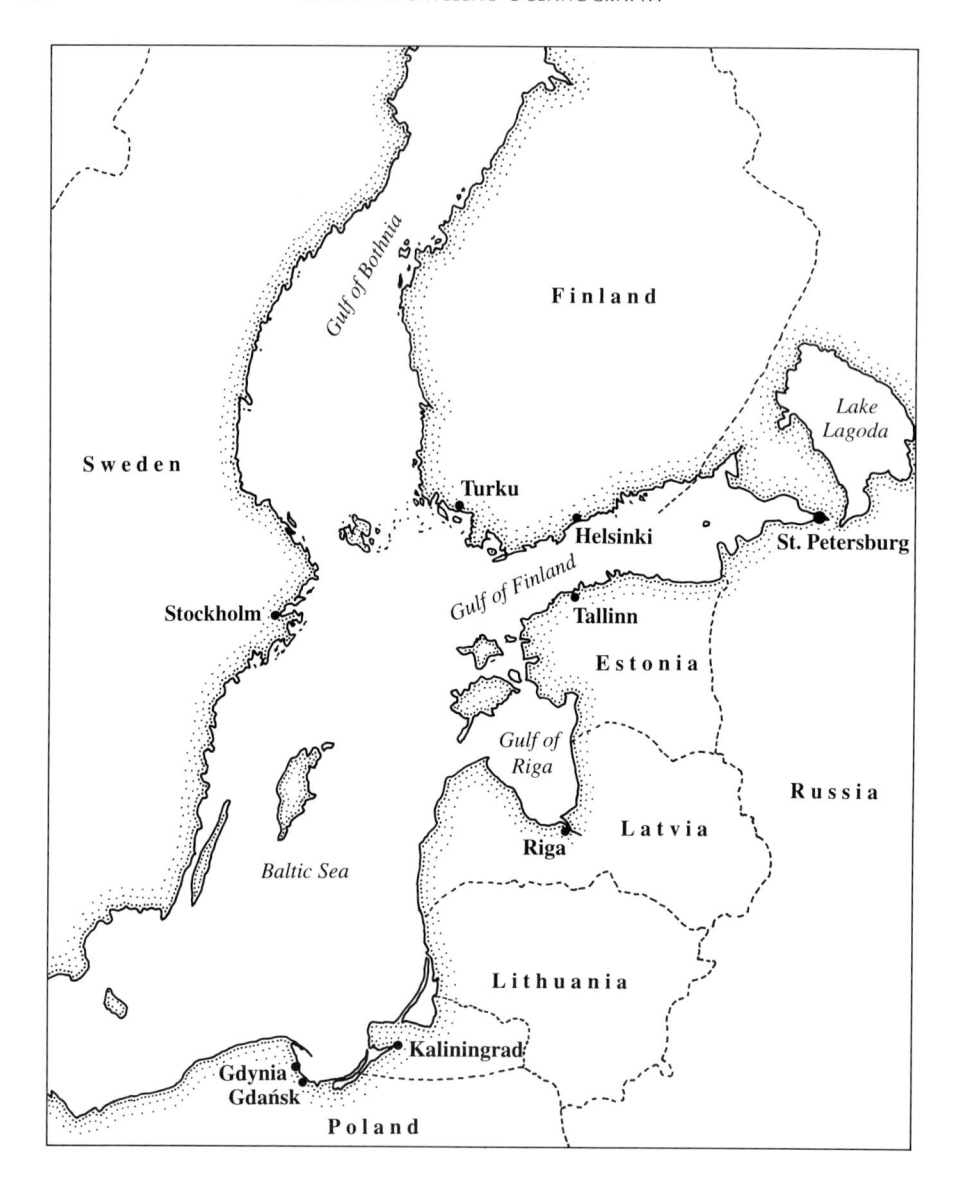

Figure 5.49 The Baltic Sea (geographical map). South-eastern part of the Sea and the Danish Straits are not shown here.

mentioned among other significant specific scientific questions to be addressed in the next decade (Ocean Science for the Year 2000, 1984).

In context of environmental monitoring of coastal waters the multi-year studies of the fields of suspended matter in the Neva Bay will be presented. Of special interest here is the attempt to separate the natural and man-made factors in the patterns recorded in satellite imagery in connection with the construction of a major engineering structure (a term used by the International Association for Bridge and Structural Engineering, see Victorov 1991b) – the St. Petersburg Flood Barrier.

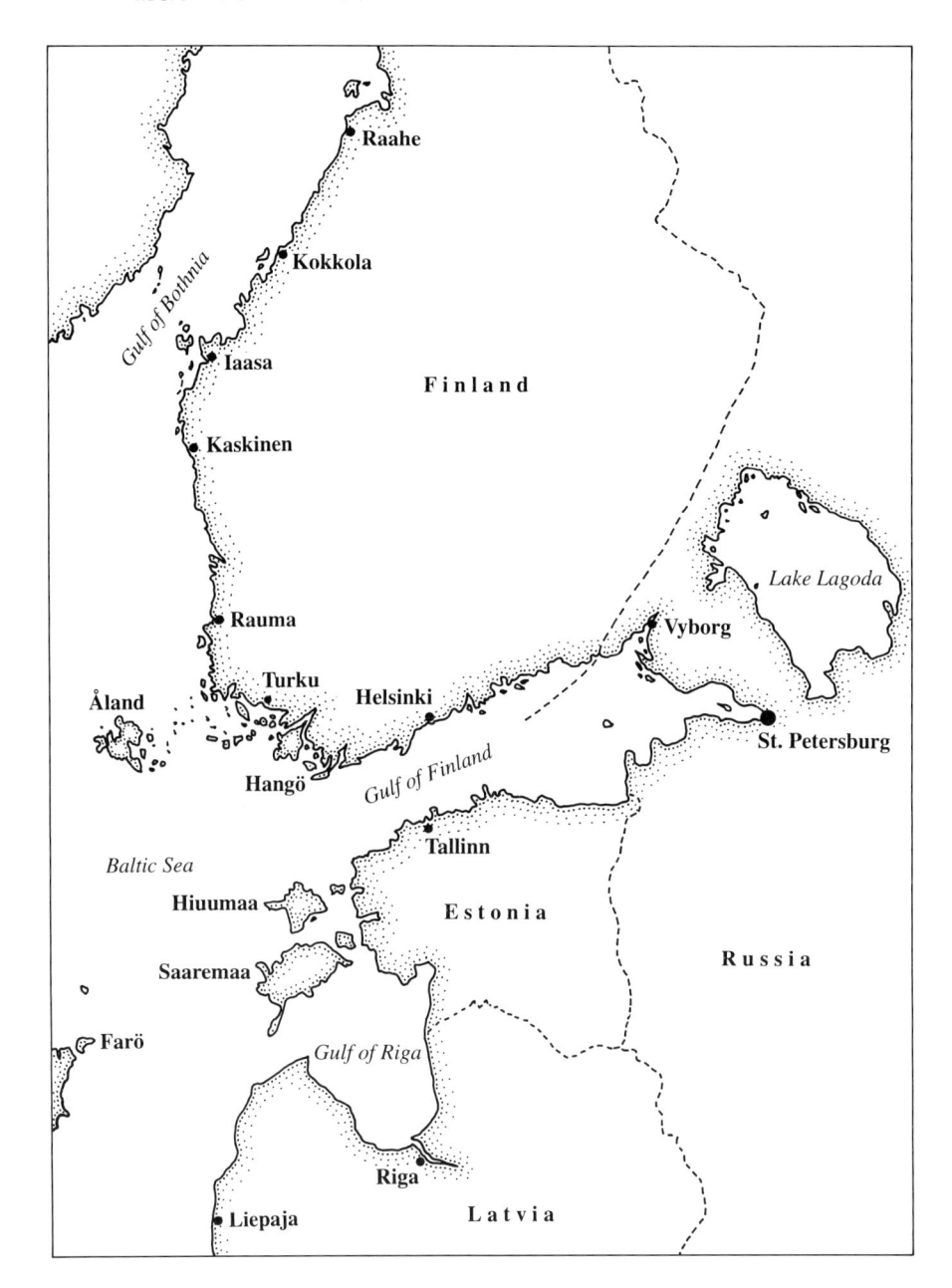

Figure 5.50 The Gulf of Finland (fragment of geographical map).

5.5.2 Neva Bay and the St. Petersburg Flood Barrier

The Neva Bay should be considered as a part of the 'Ladoga Lake–River Neva–Neva Bay–(the eastern part of) the Gulf of Finland' water system (see Figures 5.50 and 5.51). The hydrological characteristics of this system are presented in Figure 5.52.

228 REGIONAL SATELLITE OCEANOGRAPHY

Figure 5.51 LANDSAT image of 20 April 1978. Ladoga Lake, the Neva River and the Neva Bay are shown. Two zones with high concentration of suspended matter can be seen along the northern and the southern coasts of the Neva Bay. (Amateur photo of poster.)

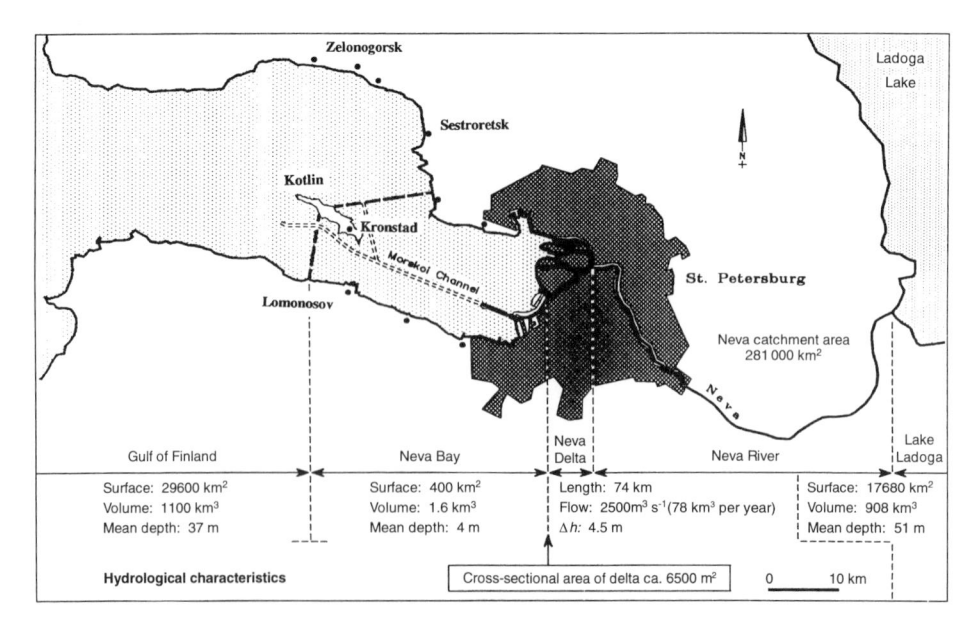

Figure 5.52 Hydrological characteristics of the 'Ladoga Lake–River Neva–Neva Bay–Gulf of Finland' water system.

The River Neva is one of the most full-flowing in Europe. Fresh water surplus from the rivers in the drainage area of the Gulf of Finland are given, along with other characteristics, in Table 5.17.

There are considerable annual variations of discharge from the River Neva, (Figure 5.53), as well as the seasonal variations (e.g., the mean monthly discharge in the winter period is about 1800 m^3 s^{-1}, while in the summer period it is as large as 3000 m^3 s^{-1}).

The Neva Bay is a shallow water body with depths of 0.5–2.0 m within wide bands along the northern and southern coasts. The River Neva flows rather slowly here (see the chart of currents in Figure 5.54) westward through two passages by Kotlin Island (the northern and the southern passages). There are some zones in the Bay with very low flow velocities located along the coasts. The residence time of water in the Bay does not exceed seven days.

In Figures 5.52 and 5.54 and in many satellite images of the Bay one can see the Morskoj (Sea) Channel leading from the Gulf of Finland by Kotlin Island directly into St. Petersburg harbour. The channel is an artificial deep water way enabling cargo ships and passenger ferry boats to reach the harbour. It was dug about a hundred years ago and needs proper maintenance (bottom-deepening dredging operations) to provide the required depth (10–12 m) along the route of ships.

The Neva River and the Gulf of Finland may exchange water masses in different hydrometeorological conditions. Thus the water in the Neva Bay is a mixture of two types of water (Figure 5.55). Figure 5.55 shows the composition of water in the Neva Bay. In normal hydrometeorological conditions there is a very limited exchange

Table 5.17 Some characteristics of rivers in the Gulf of Finland catchment area (Environmental Statistics in Europe and North America 1987)

River	Drainage area (km^2)	Discharge (m^3 s^{-1})
Kymi (Finland)	37235	517
Neva (USSR)	281000	2463
Luga (USSR)	13200	124
Narva (USSR)	56200	437

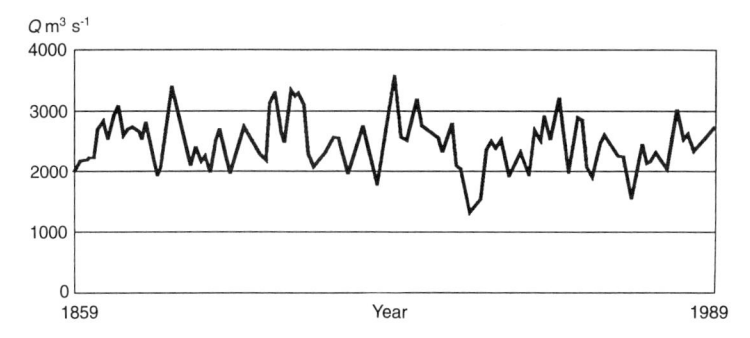

Figure 5.53 Annually averaged discharge of the River Neva.

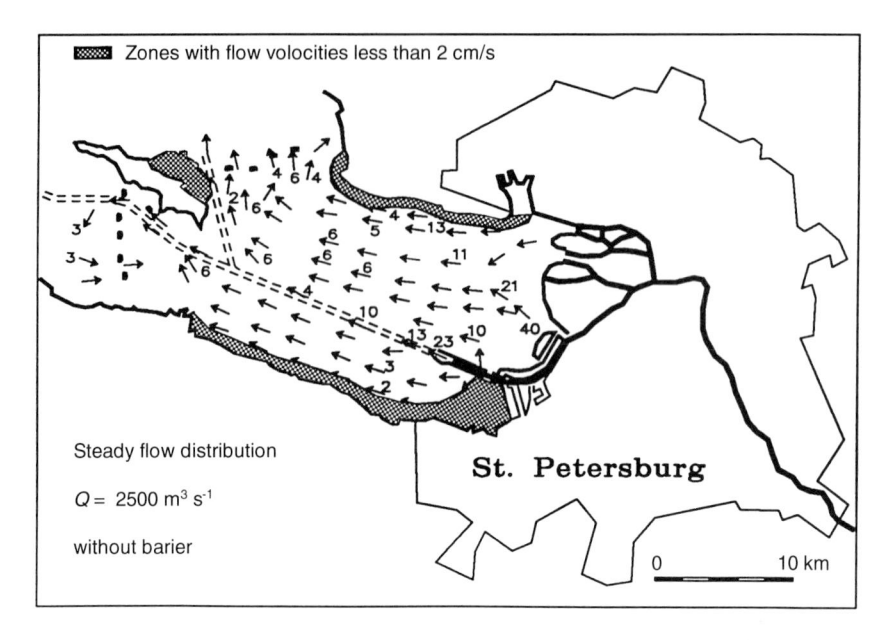

Figure 5.54 Mean current pattern in the Neva Bay for river discharge of 2 500 m^3 s^{-1}.

which is illustrated in Figure 5.56. It shows that only 2% of the water mass from the Gulf can penetrate as far as 5 km inside the Neva Bay; so the water in the Neva Bay is composed mainly of River Neva waters and waste waters from the St. Petersburg metropolitan area.

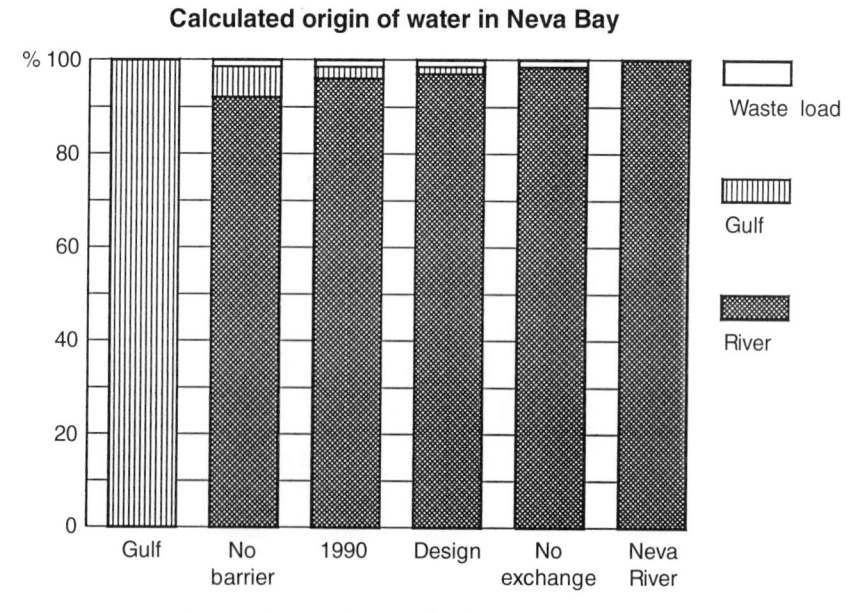

Figure 5.55 Diagram showing the contribution of Gulf water, River Neva water, and waste water to the water mass of the Neva Bay.

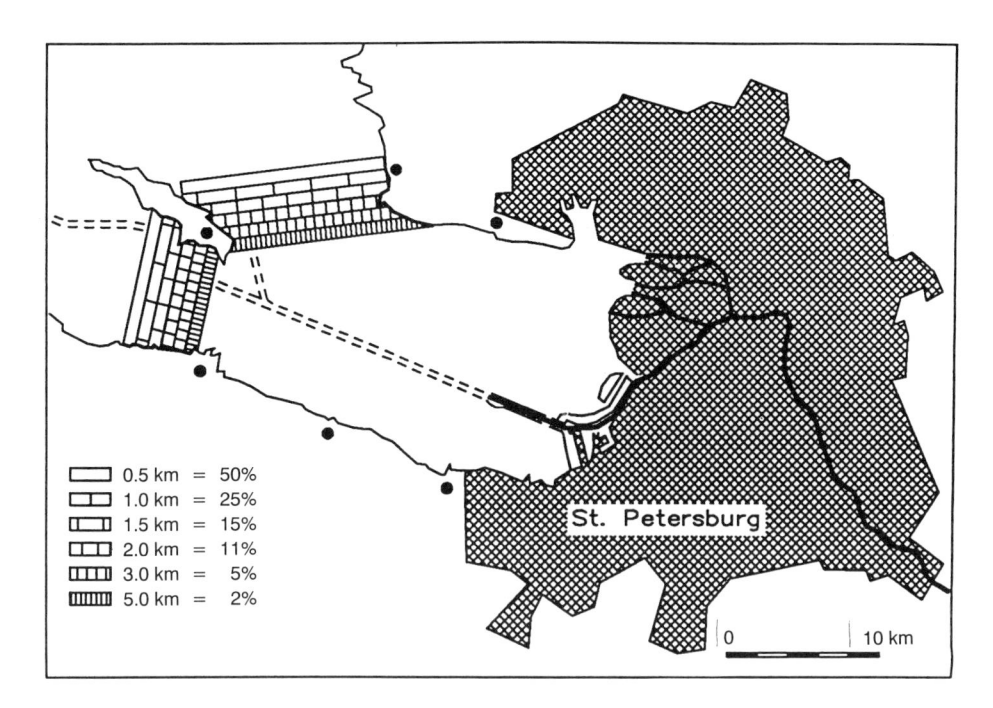

Figure 5.56 Cumulative reach of water packages entering the Neva Bay at a given time.

Now we pass to the environmental situation in the Neva Bay. The figures in Tables 5.17 and 5.18 give some idea of the River Neva (with St. Petersburg) as a source of pollutants.

The city of St. Petersburg (Leningrad in 1924–1992) and its metropolitan area, with its 5.0 million citizens, and a large number of industrial enterprises with inadequate filtering and water treatment systems, is located in the River Neva Delta. The Neva Bay receives about 5 million m³ per day of waste water (1980–1987), of which about 80% is from the St. Petersburg metropolitan area.

Figures 5.57–5.59 show the general situation with treatment of industrial and municipal waters (both are mixed before treatment which leads to additional troubles), installed treatment capacity and location of waste water discharges in the Neva Bay.

Thus the problem of studying the spreading distribution of the River Neva discharges containing pollutants into the Gulf of Finland is obvious. This is not a

Table 5.18 Nutrient emissions by rivers (tonnes per year) to the Gulf of Finland (Environmental Statistics in Europe and North America 1987)

	Total N	Total P
USSR	52500	3460
Finland	1200	620

Treatment of waste water in St. Petersburg (1980-1987)

Figure 5.57 Treatment of waste water in St. Petersburg for the period 1980–1987.

simple scientific problem, and it has not yet been solved adequately. (It is now clear that if scientific knowledge of this and relevant items had been deeper at the beginning of the 1980s, many critical situations and times of social tension could have been avoided.)

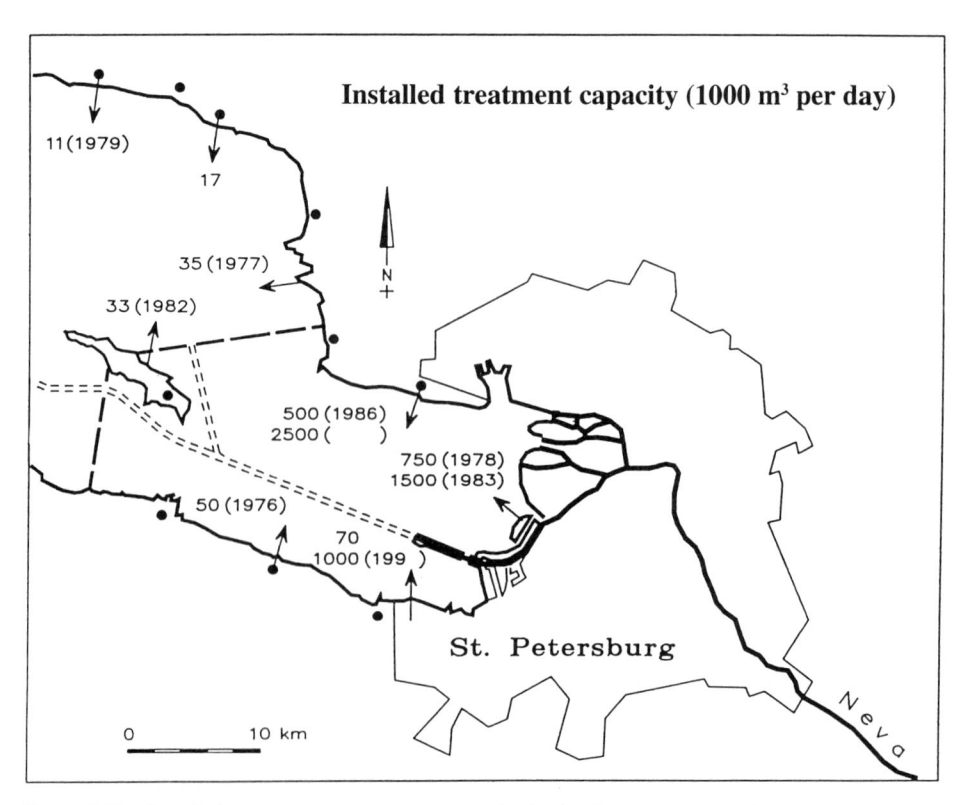

Figure 5.58 Installed waste water treatment capacity in the St. Petersburg region.

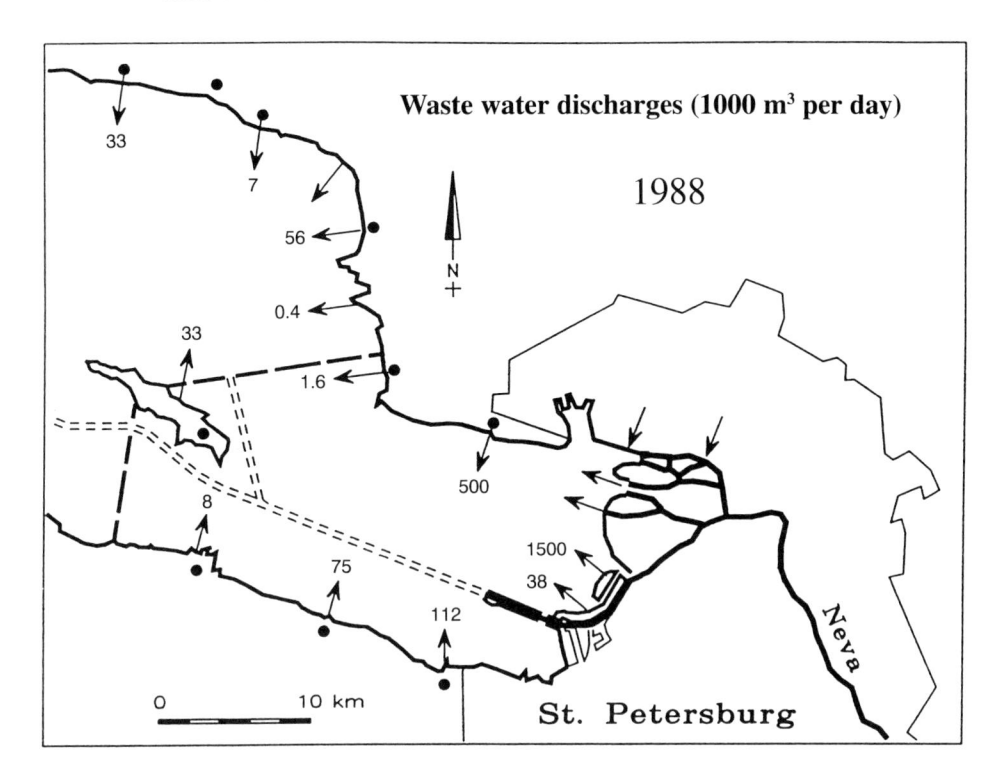

Figure 5.59 Waste water discharges in the St. Petersburg region (1988).

As a result of insufficient waste water treatment the environmental situation in the Bay for a number of years remained critical, especially if the bacterial pollution (the standards for swimming water are exceeded by factors of 10–100) and high concentrations of heavy metals are considered.

This information was made accessible to the public only in recent years as a consequence of general political changes in the former USSR. This period coincided with the construction of the Flood Barrier in the Neva Bay, meant to protect the city against floods. Construction work started in the beginning of the 1980s and was initially scheduled to be finished by 1995, but when the Barrier was nearly built, construction work was practically banned under the pressure of public opinion.

There is no doubt that the city needs protection against floods. Since its foundation in 1703 till 1990 the city has survived 284 floods (a flood event is recorded when the water level rises by more than 160 cm above the zero level). Water oscillations caused by cyclonic activity in the atmosphere over the Baltic Sea, accompanied by strong winds from the west, lead to rapid rises of the water level in the Neva Bay; the flood event lasts for several hours or a day, with a relaxation phase of another several hours. The probability of flood events is shown in Figure 5.60. Without protection the risk of flooding for the city of St. Petersburg is about one per year, which is in sharp contrast with the risk level accepted for the Delta area in The Netherlands (where by law the level is one in ten thousand years!)

Figure 5.61 shows the areas in the city covered by water during the floods which actually occurred in the last three centuries. The city obviously needs to be protected, and the Flood Barrier was designed to do it. Its 25.4 km long body consisting of

234   REGIONAL SATELLITE OCEANOGRAPHY

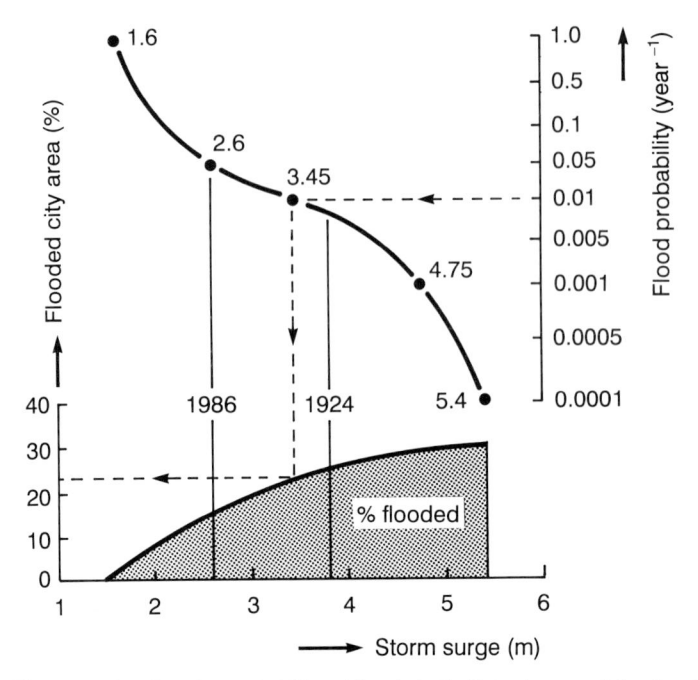

Figure 5.60 Diagrams showing the probability of floods in St. Petersburg and the flooded city areas at various water levels (above the local mean zero level).

Floods in St. Petersburg

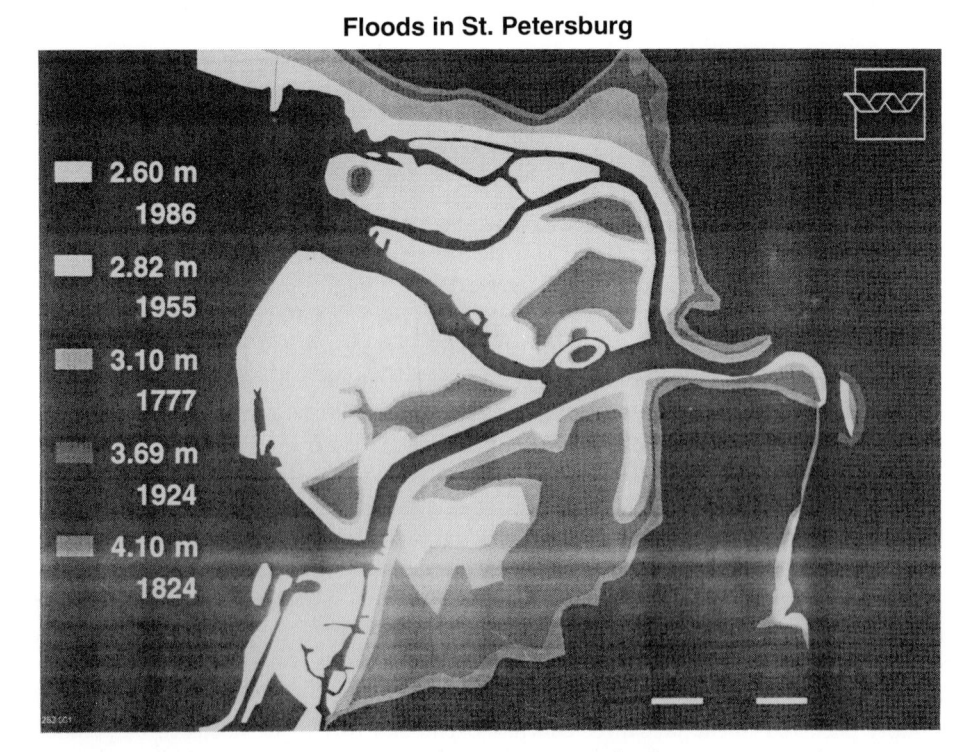

2.60 m 1986
2.82 m 1955
3.10 m 1777
3.69 m 1924
4.10 m 1824

Figure 5.61 City areas covered with water during five historic floods.

ground-filled dams with six water discharge sluices (to allow river outflow) and two navigational passages (for large and small ships) which are permanently open, could become strong protection against waves from the west. At the alarm signal, by a command all the water sluices and navigational passages were supposed to be closed within minutes and protect the city over three days, which exceeds the longest flood period ever observed.

Bearing this in mind, how could it happen that the population of the city meant to be protected still demanded a ban on construction? We recognise the fact that, all over the world, public opinion and public action in the field of environment issues became a powerful driving force, and they should not be neglected. It is also common knowledge that public opinion may be formed by various manipulations with figures, words and images. Two principal questions:

'Has the construction of the Barrier changed the aquatic environment in the Neva Bay and the neighbouring aquatoria?'

and

'Has the Barrier influenced the state of the environment in the Leningrad region?'

were under discussion in the late 1980s and early 1990s. In this section only one aspect of the problem will be considered. Remotely sensed satellite and airborne data have been used to monitor the spatial distribution of suspended matter in this region for more than a decade. Thus we were able to contribute considerably and decisively to the discussions initiated by some non-experts who tried to use satellite images of the Neva Bay as proof that the Barrier was a cork in a bottle preventing the waste waters from the Bay escaping westward, and thus forming a disastrous 'poisoned biological reactor' within the Bay.

5.5.3 Data sources and methods used

Victorov *et al.* (1984), Victorov and Sukhacheva (1984), Victorov and Sazhin (1986), Sukhacheva (1987), and Victorov *et al.* (1989a) discussed the possibilities of using the imagery of low and medium ground resolution from USSR satellites in the studies of optical inhomogeneities cause by suspended matter, along with the reliability of the results obtained.

At the initial stage of the study of the Neva Bay, since 1980, the multispectral images from the METEOR-30 satellite were analysed; later information of better quality from the KOSMOS-1939 satellite became available. Since 1988, data of medium ground resolution of 175 × 200 m from the MSU-SK scanner, and in summer 1989 high-resolution images from the MSU-E device (30 × 45 m), were obtained for the Neva Bay. A limited number of images from photographic satellites were also used in monitoring activities. Data of airborne observations of the Bay carried out by the State Oceanographic Institute, St. Petersburg were also included in the analysis.

The database of satellite and airborne images of the Neva Bay now consists of about 300 items (Usanov *et al.* 1994a). Some examples of satellite imagery of the Bay in a historical context will be presented in Section 5.5.4, all of them being images of the visible and near-IR bands. Radar imagery, which is useful in regional

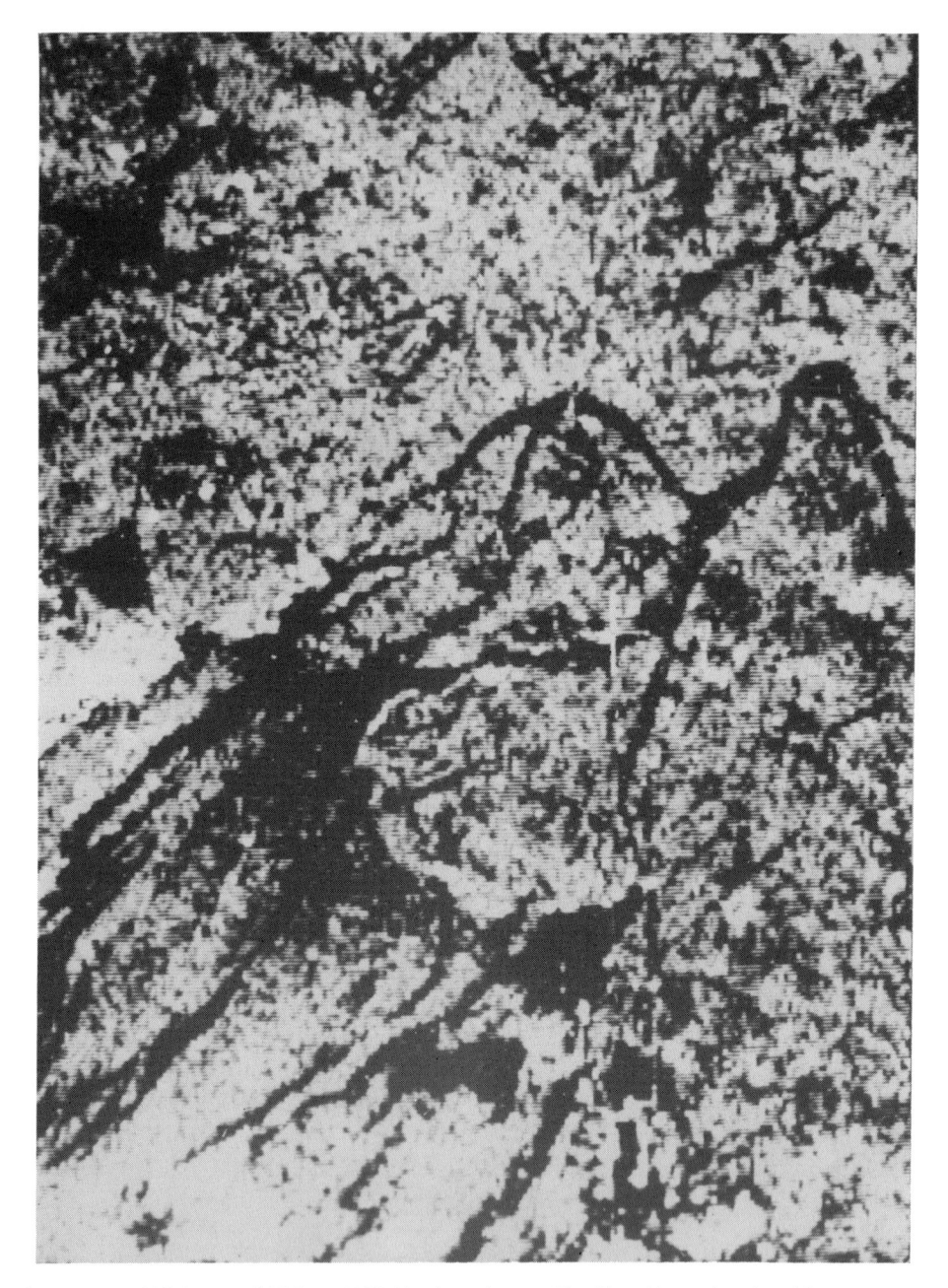

Figure 5.62 SAR image of 23 June 1989. North at the top. The River Neva, the city of St. Petersburg and the Neva Bay are recognisable (Victorov et al. 1990a).

studies of dynamical phenomena (Viter et al. 1989), was also involved, but a few high-resolution images available from the KOSMOS-1870 and ALMAZ satellites (see, e.g., Figure 2.3 in Section 2) appeared not to be very useful in the context of the Neva Bay problems, with the exception of the image of 23 June 1989 (Figure 5.62). The current system in the Bay seemed to be recorded in those images showing some

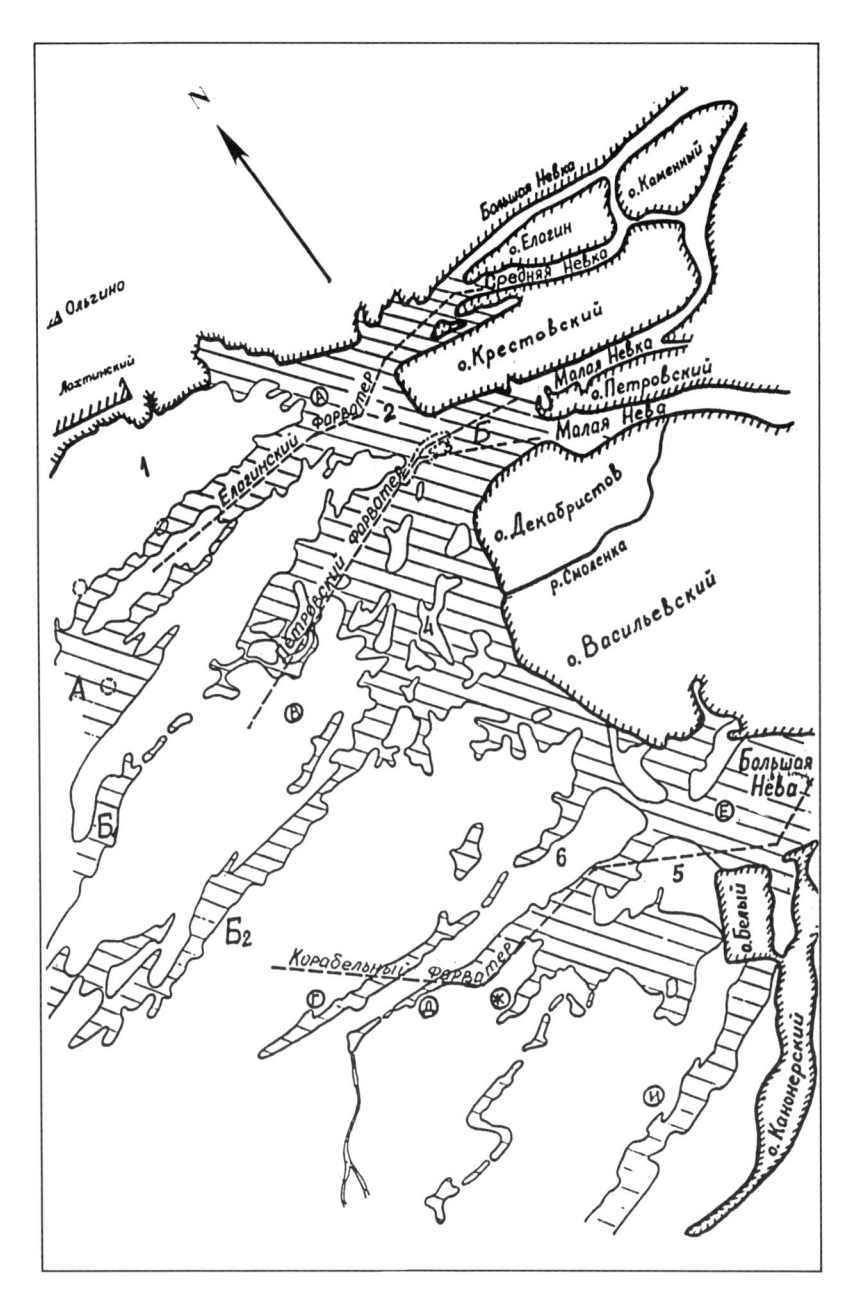

Figure 5.63 Current structure in the Neva Bay as revealed from SAR image of 23 June 1989. Arrow shows North. (The Russian spirit is present in the names of islands and river branches forming the Neva delta.) (Victorov et al. 1990a.)

flows in the Bay, which were probably the continuation of the rivers forming the River Neva Delta (Figure 5.63).

In the analysis of satellite imagery the following additional data were used:

- navigational maps of the Bay;

- data on mean velocities, general structure and schemes of currents;
- characteristics of suspended matter including size distribution of suspended particles and sediments;
- data on seston concentration and the regression characteristics of 'water transparency versus concentration of suspended matter';
- data on the sources of suspended matter;
- historical data on the spatial distribution of suspended matter in various hydrometeorological conditions;
- data on the meteorological situation before, after and at the moment of satellite overpass, including wind, atmospheric pressure and horizontal visibility;
- data on water level in different points in the Bay;
- *in situ* measurements of water transparency, salinity and temperature;
- airborne measurements of sea surface temperature and optical properties of the water upper layer.

The patterns of spreading from the sources and spatial distribution of suspended matter have been recorded in satellite images. Suspended substances could be used

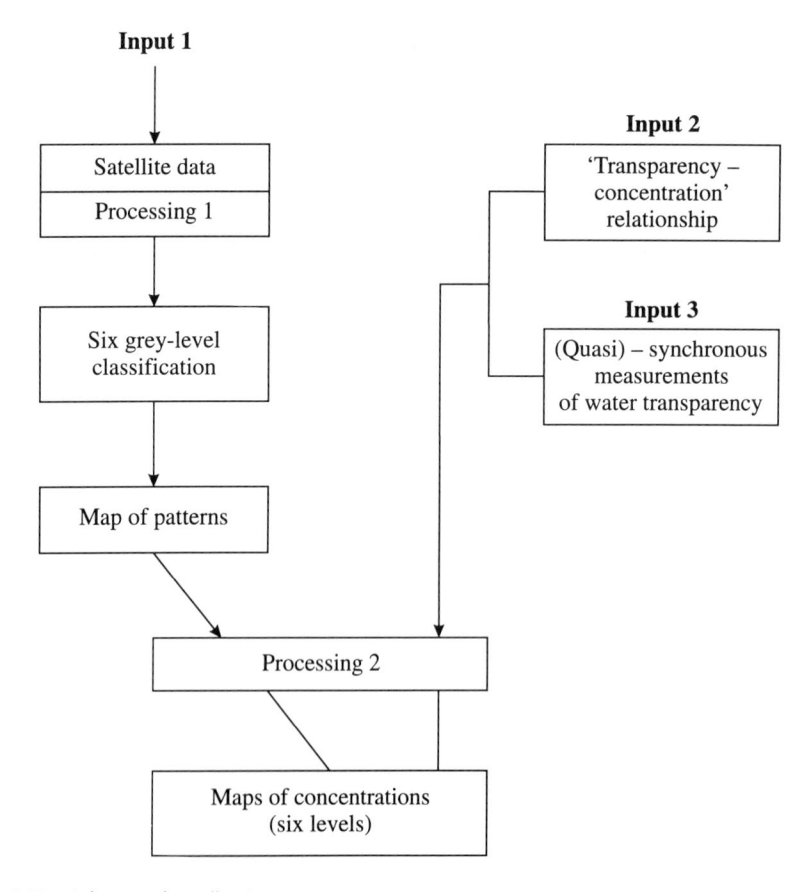

Figure 5.64 Scheme of satellite imagery processing.

as tracers to visualise flows and currents with their fine structure, as well as dynamical phenomena such as eddies and 'mushroom-like' structures. To study the features of fields of suspended matter and the dynamics of water masses in the Neva Bay, satellite images were selected which showed various patterns in various hydrometeorological conditions connected with the winds and water level changes.

In the analysis of satellite images visual and instrumental techniques were used (Sukhacheva 1987). Later, with progress in the quality of satellite imagery a method was developed which enabled us to interpret the images in a quantitative way and determine concentrations of suspended matter in milligrams per litre (mg l^{-1}). To tune the method *in situ* data on water transparency and concentration of suspended substances were used, which were collected during the periods 1959–1960 and 1982–1989. Actually this method could provide quantitative estimates of suspended matter concentration limits at each of the clusters recorded in the satellite images (Victorov *et al.* 1991). Figure 5.64 presents the scheme of data flow in our analysis. This method is based on the reasonable assumption of a horizontally homogeneous atmosphere over this small area, and the assumption of constant sediment properties, confirmed by *in situ* measurements.

Image processing techniques were used to map the spatial distribution of suspended matter in the Neva Bay with six-cluster classification and presentation of results in false colours (Figure 5.65; Plate 6). Victorov *et al.* (1989b) showed that for this water body more than ten levels of concentrations can be determined in high-resolution satellite imagery. But for the sake of consistency we used the six-level scale throughout the study (Victorov 1990a, 1991b; Victorov *et al.* 1990a, 1991; Victorov and Sukhacheva 1992a, 1994; Sukhacheva and Victorov 1994; Sukhacheva and Tronin 1994).

Typical values of transparency and concentration of suspended matter for six zones determined in satellite imagery of the Neva Bay are presented in Table 5.19. Waters with concentrations of less than 10 mg l^{-1} were considered 'clean'; they are usually located in the central part of the Bay. Often extremely high concentrations exceeding 200 mg l^{-1} were measured from the ships.

There are two major databases relevant to the Neva Bay which are to be integrated in a regional 'Neva Bay' GIS (Usanov *et al.* 1994a). The database of conventional hydrological, hydrochemical and hydrobiological data since 1968 consists of about 360 000 measurements of 111 parameters. The database of remotely sensed data consists of some 300 images from satellites, airborne survey charts and synchronous/quasi-synchronous *in situ* measurements relevant to remotely sensed imagery since the end of the 1970s. For a discussion of the integration of both

Table 5.19 Characteristics of zones revealed in remotely sensed data of the Neva Bay

Zone number	Description of the level of turbidity	Transparency (m)	Concentration of suspended matter (mg l^{-1})
1	Clean river water	>1.0	<10
2	Very small	0.7–1.0	10–15
3	Small	0.5–0.7	15–20
4	Medium	0.4–0.5	20–25
5	High	0.2–0.4	25–60
6	Very high	<0.2	>60

databases within the GIS using a mathematical modelling module see Usanov *et al.* (1994a). The 'Neva Bay' GIS is meant to become a consistent part of the Integrated Water Management System in the St. Petersburg region.

5.5.4 Multi-year satellite images of the Neva Bay

In this section some samples from a series of satellite images will be presented with the relevant charts of suspended matter distribution in the Neva Bay. An analysis in terms of 'before Barrier' and 'after Barrier' will be made, and some individual patterns in the satellite images will be discussed. This section is based on the analysis of images from the former USSR satellites, so the development of satellite sensors of the visible band on board those satellites could also be seen (Victorov 1994d,e). (For characteristics of the sensors and satellites see Section 2.3.2.)

Almost all the scenes show the Neva Bay with the city of St. Petersburg in the right part and the island of Kotlin in the centre. The Flood Barrier connects the island with the northern and the southern coasts of the Neva Bay. The bright patches are the areas with high concentrations of suspended matter.

Figure 5.66 shows the Neva Bay as recorded by the MSU-M scanner (ground resolution of about 2000 m) on 12 September 1982. It is a very poor image; only the zones with highest concentration of suspended matter could be seen in imagery of this type (Figure 5.67). It must be noted that the LANDSAT image of the same date (Figure 5.68) was much more informative. But still more informative and more interesting were the images from the Soviet photographic satellites of an even earlier period (see, e.g., Figure 5.69). Alas! They had been classified and not available for the oceanographic community at that time (see Section 2.3.2.8).

Imagery taken by the MSU-S scanner (ground resolution about 200 m) enabled four levels of suspended matter concentration to be determined, with no quantitative analysis (Figure 5.70).

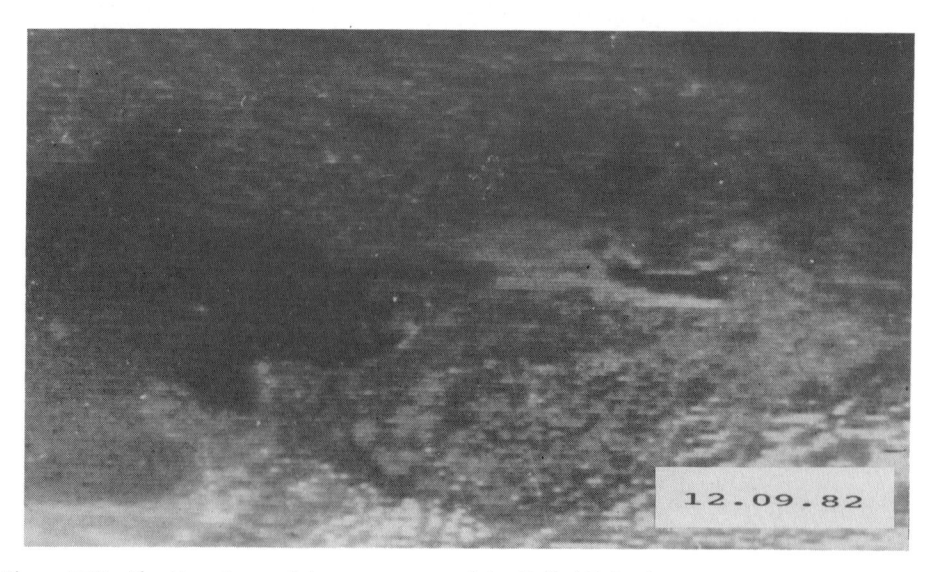

Figure 5.66 The Neva Bay and the eastern part of the Gulf of Finland. MSU-M. 12 September 1982.

REGIONAL SATELLITE OCEANOGRAPHY: CASE STUDY OF THE BALTIC SEA 241

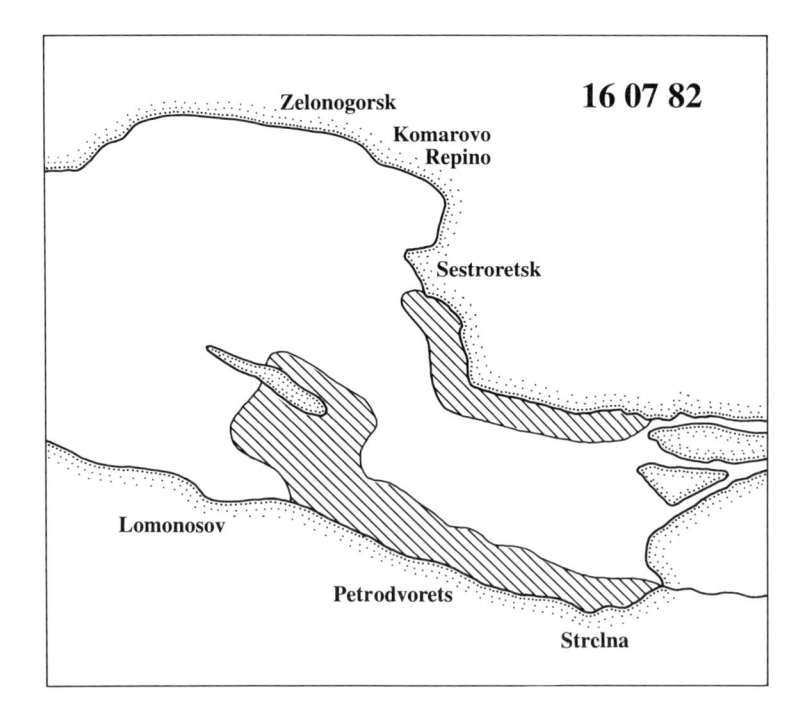

Figure 5.67 Areas with high concentrations of suspended matter in the Neva Bay as revealed from the MSU-M image of 16 July 1982.

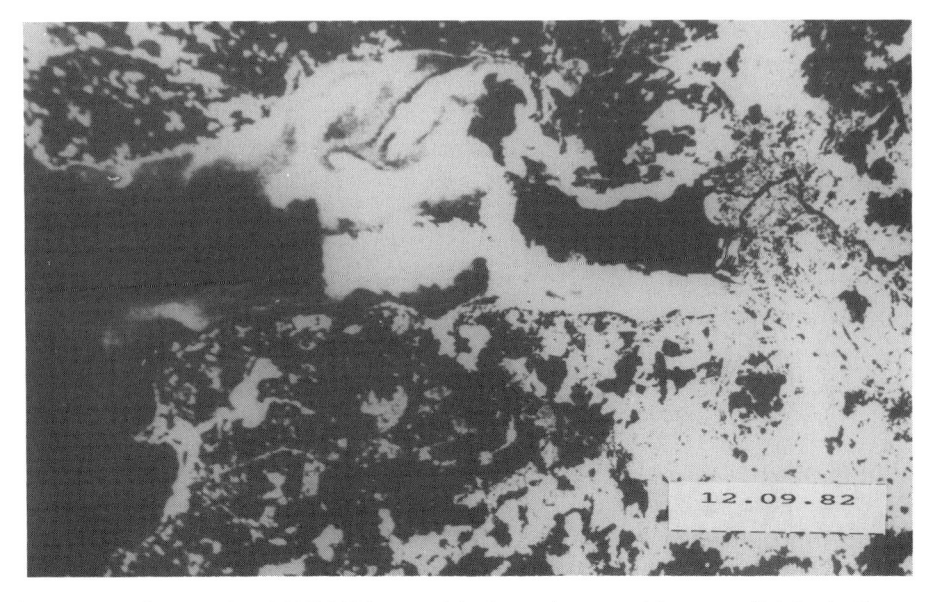

Figure 5.68 The Neva Bay. LANDSAT image of 12 September 1982. (Courtesy of M. Punkari.)

Figure 5.69 The Neva Bay. Image from Soviet photographic satellite of 24 May 1979. Originally black and white film.

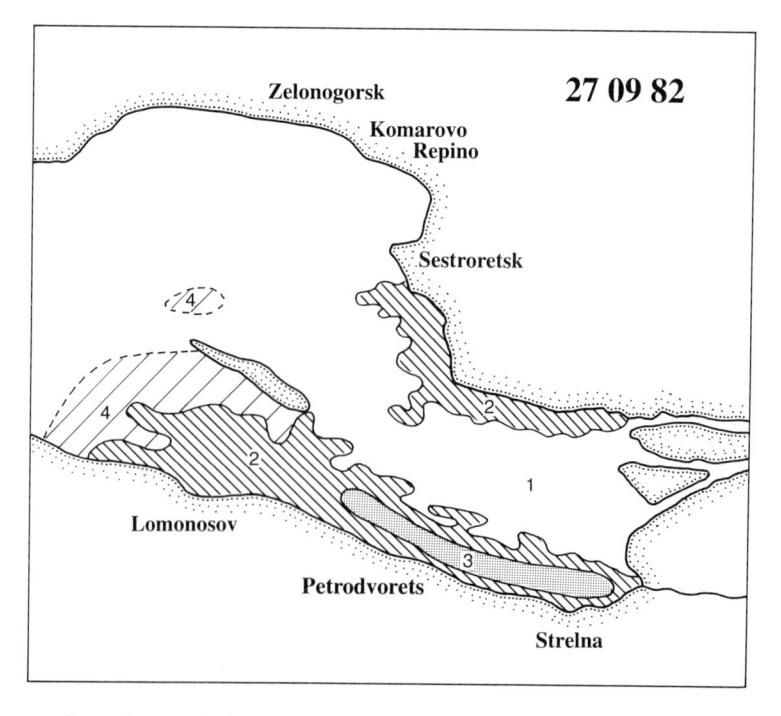

Figure 5.70 Chart of suspended matter distribution in the Neva Bay as revealed from the MSU-S image of 27 September 1982: (1) 'clean' water; (2) high concentration; (3) medium concentration; (4) low concentration.

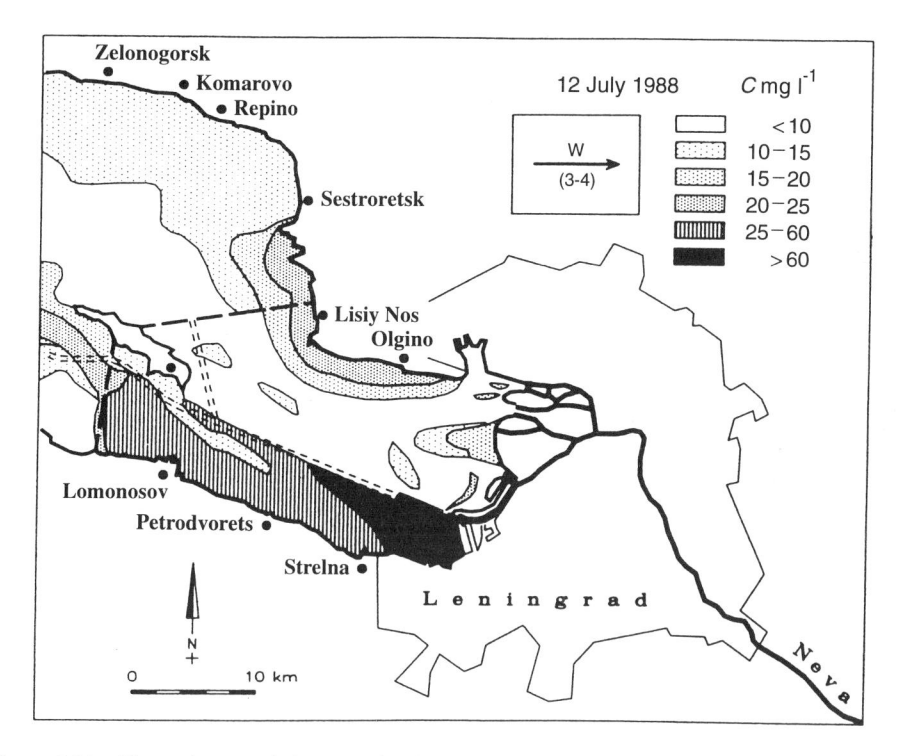

Figure 5.71 Chart of suspended matter distribution in the Neva Bay as revealed from the MSU-SK image of 12 July 1988.

Figure 5.71 shows a chart of the distribution of suspended matter as revealed from the satellite image taken by the MSU-SK scanner on 12 July 1988. Figure 5.72 shows the same water body as recorded by the experimental high-resolution sensor MSU-E on board satellite KOSMOS-1939. It is an example of a good image, and the MSU-E device, in the author's opinion, is the best visible imager ever flown on satellites of the former USSR. And it is imagery of this type (plus a limited amount of *in situ* data) that enabled us to create quantitative value-added informational products: charts of suspended matter distribution in terms of concentration in milligrams per litre and to study fine features in the patterns of these fields (Figures 5.73 and 5.74).

The distribution patterns recorded in Figures 5.71, 5.73 and 5.74 will now be discussed to show how various hydrometeorological factors form the fields of suspended matter in the Neva Bay. With respect to characteristics of water level change at least four types of hydrological situation can be studied:

- smooth change of level (slow decrease);
- level rise;
- sharp level drop;
- period of change of phase (decrease after durable increase).

Figure 5.75 illustrates these four cases. Table 5.20 presents data on the wind situation and water level characteristics, relevant to the four cases, for the particular

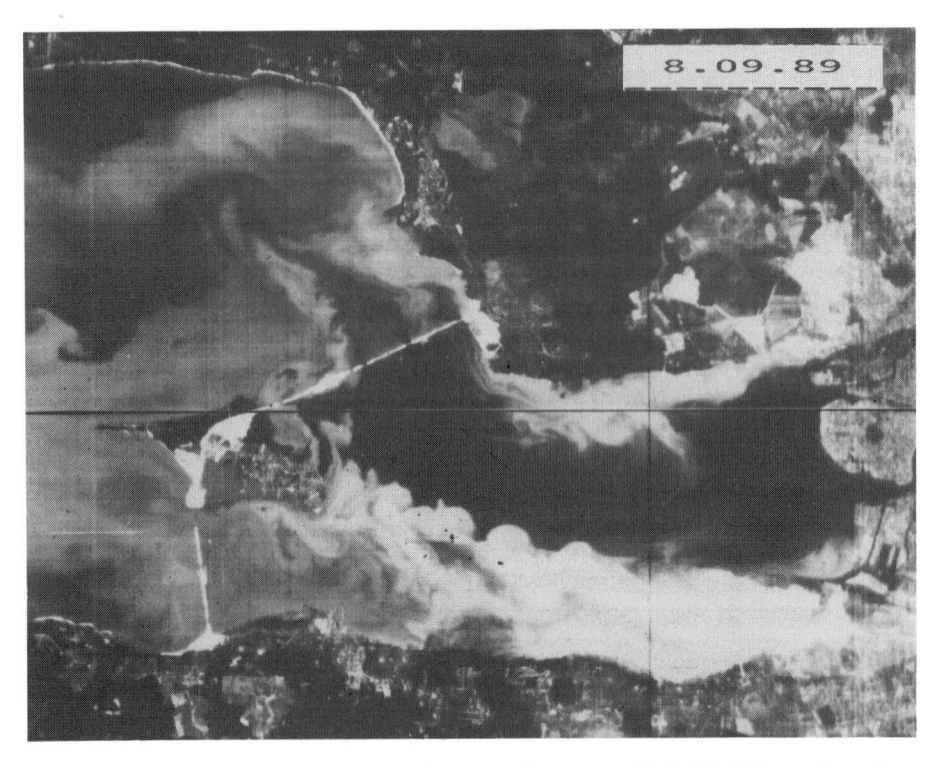

Figure 5.72 The Neva Bay. Satellite image of 8 September 1989. MSU-E, 600–700 nm channel.

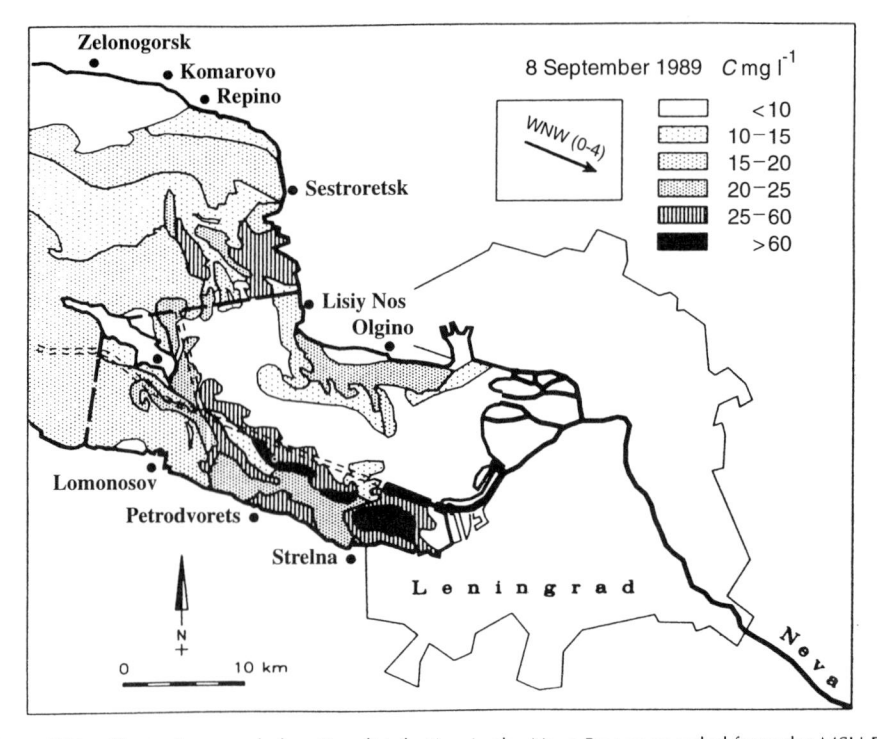

Figure 5.73 Chart of suspended matter distribution in the Neva Bay as revealed from the MSU-E image of 8 September 1989.

REGIONAL SATELLITE OCEANOGRAPHY: CASE STUDY OF THE BALTIC SEA 245

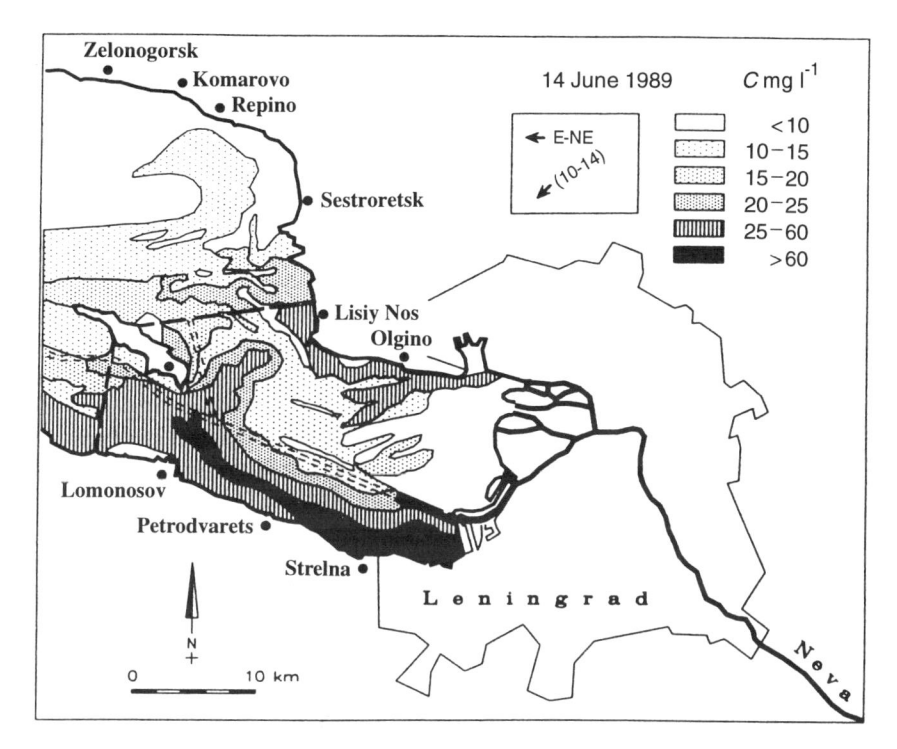

Figure 5.74 Same as in Figure 5.73 for 14 June 1989.

satellite images. Please note that Table 5.20 gives the description of the level change and the wind conditions for the periods *before* the satellite images were taken as well as at the moments of satellite overpasses.

Provided that the location of the 'sources' is frozen (which is not the case), each type of hydrological situation can be related to a specific distribution pattern of suspended matter and the current field as recorded in satellite imagery. (The problem of systematisation and of 'typical cases' arises here. We singled out four 'typical' situations for the Neva Bay, but for the entire eastern part of the Gulf of Finland we decided to single out only three 'typical' cases, see Section 5.5.5.)

Figure 5.71 shows the suspended matter distribution on 12 July 1988 formed at the 'outflow' wind situation (indicated in the figure) at a period of durable smooth change of water level. Within the southern zone of suspended matter one can see a thin stripe of cleaner water, which was identified as the local intrusion of cool saline waters from the Gulf. This was confirmed by *in situ* measurements of salinity on 12–20 July and by airborne measurement of sea surface temperature on 12 and 15 July.

The satellite image of 8 September 1989 (Figure 5.73) was taken at a period of change of phase of water level oscillations. The following features can be traced: outflow currents in the northern passage; a quasi-homogeneous 'mixed' zone at the large area of the Bay to the west of the Barrier; and a peculiar feature along the Sea Channel. The latter was initially interpreted as an eddy chain along the channel; later the *in situ* observations showed that the low stone wall of the channel was damaged in places, and turbid waters from the southern part of the Bay could leak into the clean central zone of the Bay. This was recorded in satellite imagery as semi-circular patches along the channel.

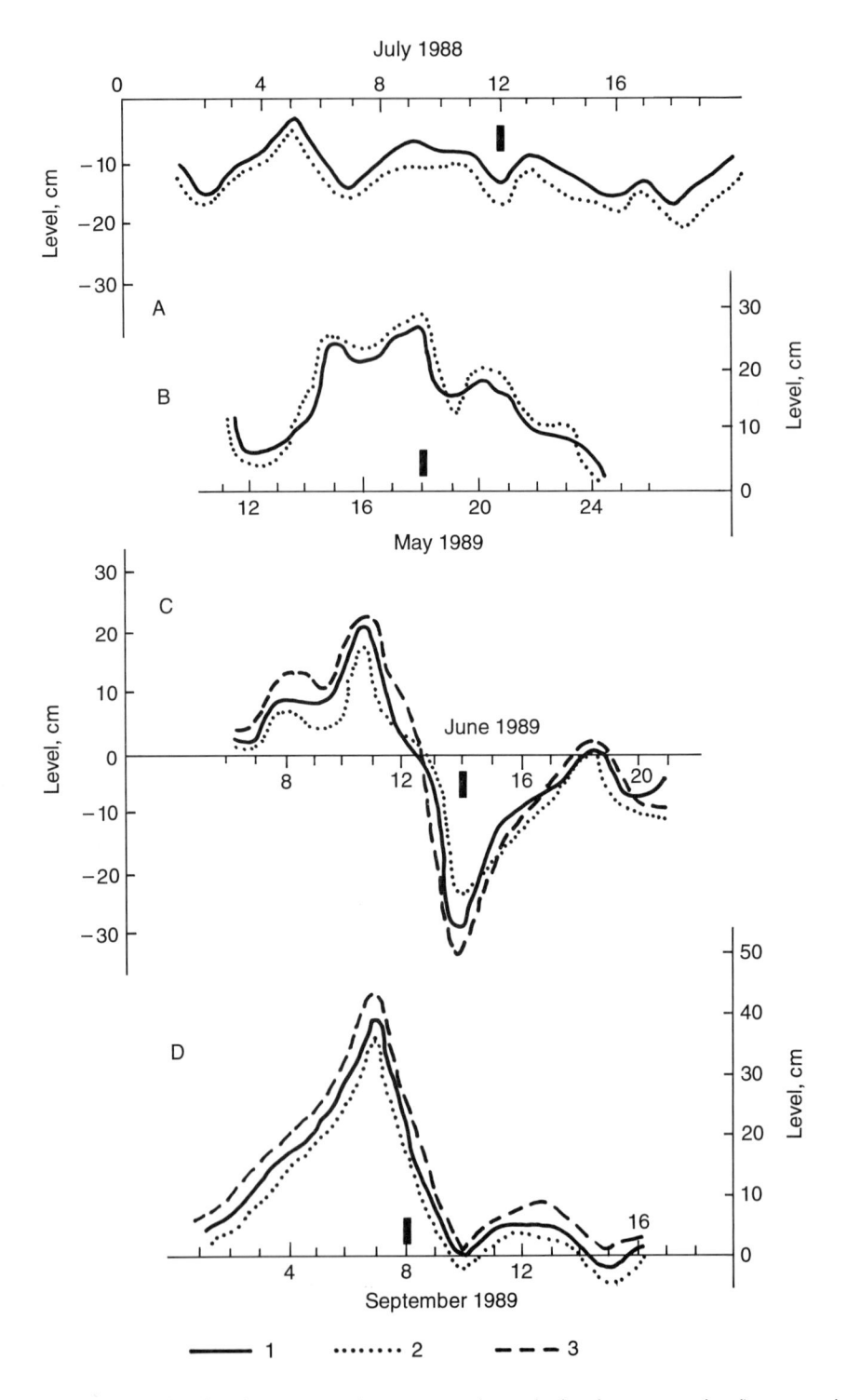

Figure 5.75 Water level in the Neva Bay (centimetres above the local mean zero level) measured at Kronshtadt (1), Lomonosov (2), and Lisij Nos (3). The bold marker indicates the moment of satellite overpass. See also Table 5.20.

Table 5.20 Hydrometeorological situation at the moments of overpass of satellite KOSMOS-1939

No.	Date	Type of sensor, ground resolution (m)	Wind direction and strength	Height of sea surface level (cm)	Level change rate (cm h^{-1})	Characteristics of level change	Note
1	12.07.88	MSU-SK 175 × 200	W 3–4	484	1–2 (slow decrease)	Smooth change	From 5 till 20.07.88 – conservative hydrometeorological situation, weak winds of mainly west directions
2	18.05.89	MSU-SK 175 × 200	W 8–14	529	3–4 (increase)	Increase	From 15 till 18.05.89 – west winds 6–12 m s^{-1}
3	14.06.89	MSU-E 30 × 45	E, NE 10–14	469	7 (sharp decrease)	Decrease	From 13 till 17.06.89 – ENE winds 9–11 m s^{-1}
4	08.09.89	MSU-E 30 × 45	Change from WNW 4 to still	523	3 (decrease after durable increase)	Period of change of phase	From 04.09.89 – wind from westerly directions 4–6 m s^{-1}. On 07.09.89 – maximum value of surface level: 539 cm

The peculiar pattern of suspended matter distribution on 14 June 1989 (Figure 5.74) was recorded at a period of a sharp drop of water level. It corresponds to the situation when an upwelling near the northern coast from Sestroretsk to Repino was registered (*in situ* measurements showed a difference of temperature of 3.5°C and salinity of 1.5 promille here as compared with the adjacent waters) and an anomalous (in sign) position of the water level occurred (a considerable difference of +10 cm of water levels measured at the cross-section at Lomonosov–Kronshtadt–Lisij Nos), generating the current in the opposite direction to the wind direction (see also Figure 5.92).

In similar hydrometeorological situations similar distribution patterns were recorded in satellite imagery irrespective of the existence of the Flood Barrier. Figures 5.69 and 5.76–5.80 are examples of satellite imagery of the Neva Bay in the 'before Barrier' period. In the image dated 30 April 1975 (Figure 5.76), only small areas of suspended matter fields were recorded near both coasts, but a tail of a suspended matter plume can be seen which is spreading through the northern passage. Figure 5.77 shows a complicated pattern with well-pronounced fields of suspended matter, including a source to the south of Kotlin Island. The distance of spreading of suspended matter from this source along the southern coast can be estimated to be as far as the Koporskaya Bay. Contrary to this case, in Figure 5.69 the pattern with the source located at the northern coast was dominant. Those two figures give an idea of the dependence of the pattern on the location of the source.

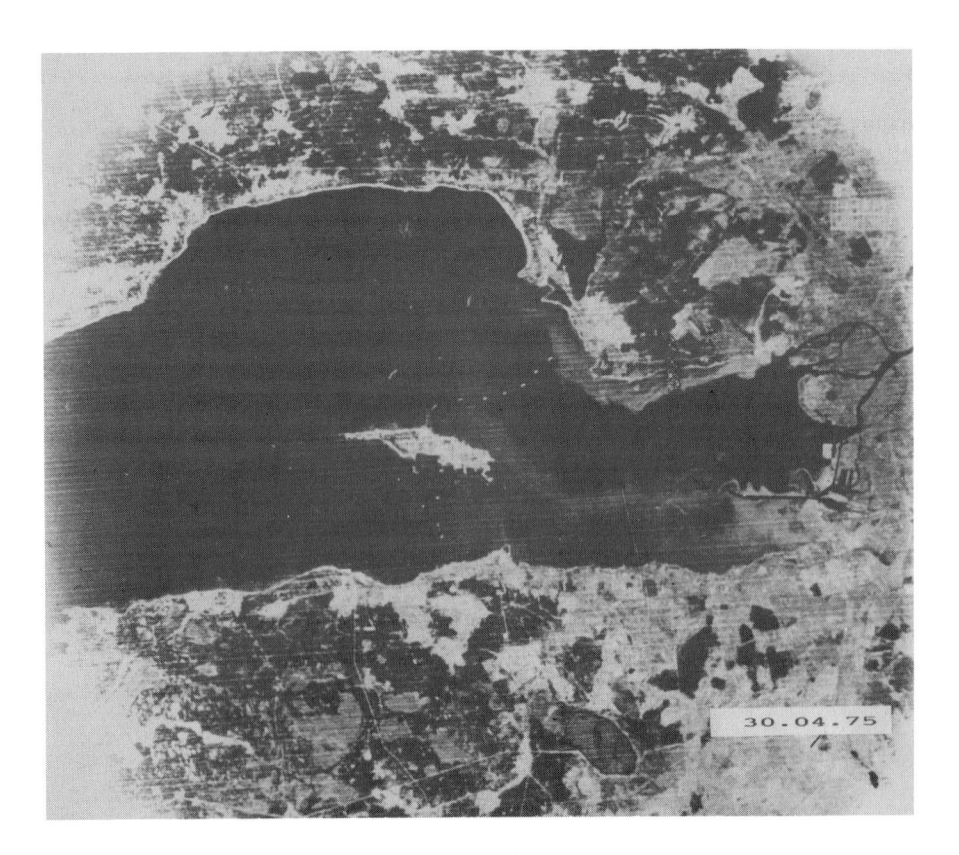

Figure 5.76 The Neva Bay. Satellite image of 30 April 1975.

Figure 5.77 The Neva Bay. Satellite photographic image of 20 September 1976. (The image of the land surface was partly masked and distorted before transfer to scientists.)

The colour photograph in Figure 5.78 (Plate 7) shows fields of suspended substances along both coasts of the Bay. Considerable areas with high concentrations of suspended matter are shown in Figures 5.79 and 5.80. Note that in Figure 5.79 the same dark feature in the southern field of the suspended matter (the surface manifestation of the intrusion of water from the Gulf), as was recorded on 12 July 1988 (Figure 5.71), is visible.

Thus Figures 5.69 and 5.76–5.80 clearly showed that long before the construction of the Flood Barrier started, complicated patterns of fields of suspended substances existed in the Neva Bay and were occasionally recorded in satellite imagery. Depending on the location of the source, one could find dominating areas of suspended matter fields either at the southern or at the northern coasts, or at both. There are four main sources and mechanisms that make water in the Bay turbid:

- wind and wave mixing;
- dredging, bottom-deepening and ground-filling operations, dumping of spoil;
- untreated waste waters;
- dirty ice melting;

of which the first two are dominant in the Neva Bay.

Now we pass to the initial phase of construction of the Flood Barrier. Figures 5.81 and 5.82 (Plate 8a) relate to this period. Figure 5.81 shows clean aquatoria when intensive seasonal dredging, bottom-deepening and ground-filling operations

250   REGIONAL SATELLITE OCEANOGRAPHY

Figure 5.79 The Neva Bay. Satellite photographic image of 5 June 1981. (The image of the land surface was partly masked and distorted before transfer to scientists.)

Figure 5.80 The Neva Bay and the eastern part of the Gulf of Finland. Satellite photographic image of 19 June 1976. Archive. (Courtesy of A. Zubenko.) Special photographic processing to enhance contrasts.

Figure 5.81 The Neva Bay. Satellite photographic image of 6 May 1983.

had not yet started (6 May 1983), while in the colour image of Figure 5.82 a different situation is presented (11 August 1983).

Now we come to the period when the body of the Flood Barrier was constructed. (As soon as the dams were just above the water surface, they became visible from space; in this sense I consider the Barrier to be constructed while analysing satellite imagery of the Neva Bay since 1984.) The Barrier can be easily recognisable as two straight bright lines (or rather linear structures) connecting the island of Kotlin with both coasts. There is still a 940 m gap between the southern part of the Barrier and the island, which is being used now to provide access for ships to St. Petersburg harbour along the Sea Channel. The so-called 'after Barrier' period starts decades in the future.

As in the 'before Barrier' period one can find images with small areas of fields of suspended matter in spring (Figures 5.83 and 5.84). The situation with the dominating southern field of suspended substances was recorded in the satellite image of 3 June 1992 presented in Figure 5.85.

Figures 5.86–5.91 (Figure 5.88 is Plate 8b) are samples of complicated patterns of fields of suspended matter in the Neva Bay as recorded in satellite and airborne imagery. Note that the satellite images of 8 August 1986, 5 July 1989 (colour plate) and 11 May 1990 are very similar, in many features of the suspended substance fields, to the satellite images of the Bay in the 'before Barrier' period (5 June 1981 and 9 July 1981) taken in summer.

All these images show two eddies 10–15 km in diameter generated to the west of the Barrier at the northern coast and in the central part of the Gulf.

Figure 5.91 shows fine structure of the suspended matter field at the northern coast of the Neva Bay near the Barrier, with a remarkable pattern of a 'mushroom-

252   REGIONAL SATELLITE OCEANOGRAPHY

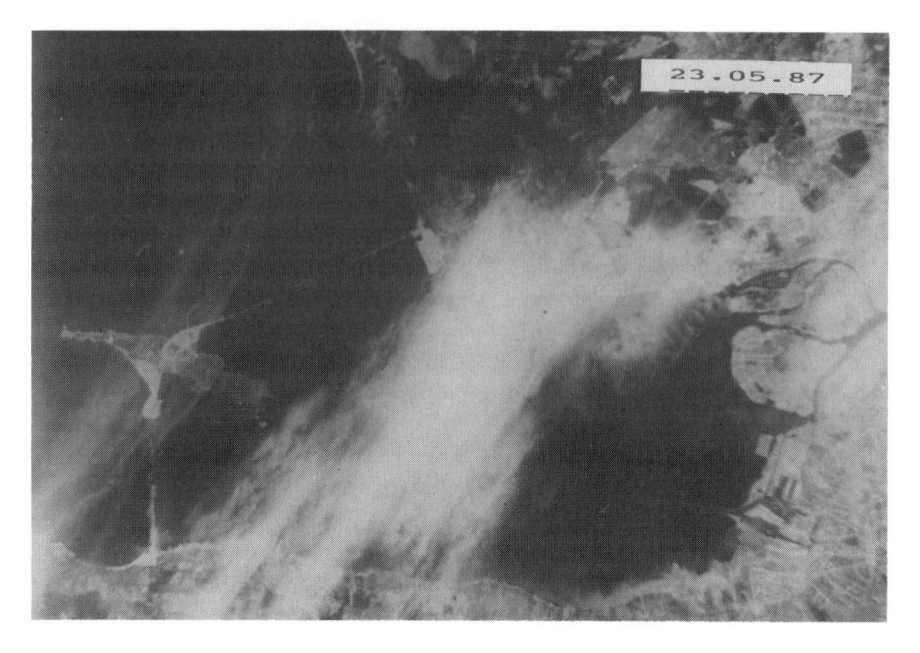

Figure 5.83 The Neva Bay. Satellite photographic image of 23 May 1987. The clean water surface is partly masked with filamentary clouds located across the middle of the Bay and over the island of Kotlin.

like' eddy generated at the sluice of the Barrier, as recorded in the airborne photograph.

Among other remarkable features in remotely sensed images of the Neva Bay and the eastern part of the Gulf of Finland it is worth mentioning the effect of water exchange between two zones with high concentration of suspended matter in the Neva Bay. Figure 5.92 shows the jet from the southern zone approaching the northern zone. This event was caused by the anomalous position of the water surface on 14 June 1989 (see also Figure 5.74 and the relevant discussion above). Next day the 'coupling' of the two water masses was observed from aircraft.

Figure 5.93 shows three schematic charts of the distribution patterns of suspended matter in the eastern part of the Gulf of Finland in the autumn season in winds from the west. The charts were based on satellite images of 12 October 1988, 21 October 1992 and 26 October 1992. The remarkable feature here is the sharp boundary between the turbid water mass running from the Neva Bay and the relatively clean waters of the Gulf, as if the plume had met a solid wall. Note that a similar pattern was also recorded in the LANDSAT image of 12 September 1982 (Figure 5.68). This phenomenon is probably connected with the changing seasonal thermocline structure in this period.

Complex analysis of sets of satellite images of high-resolution and relevant *in situ* measurements enabled us to elucidate a lot of features at the frontal zones separating different water masses in the field of currents and in dynamical structures in the Neva Bay and in the eastern part of the Gulf of Finland. It should be emphasised that they had existed long before the beginning of the construction of the Flood Barrier. It is worth noting that those zones with increased concentration of suspended matter have been traced in satellite imagery even during the periods when

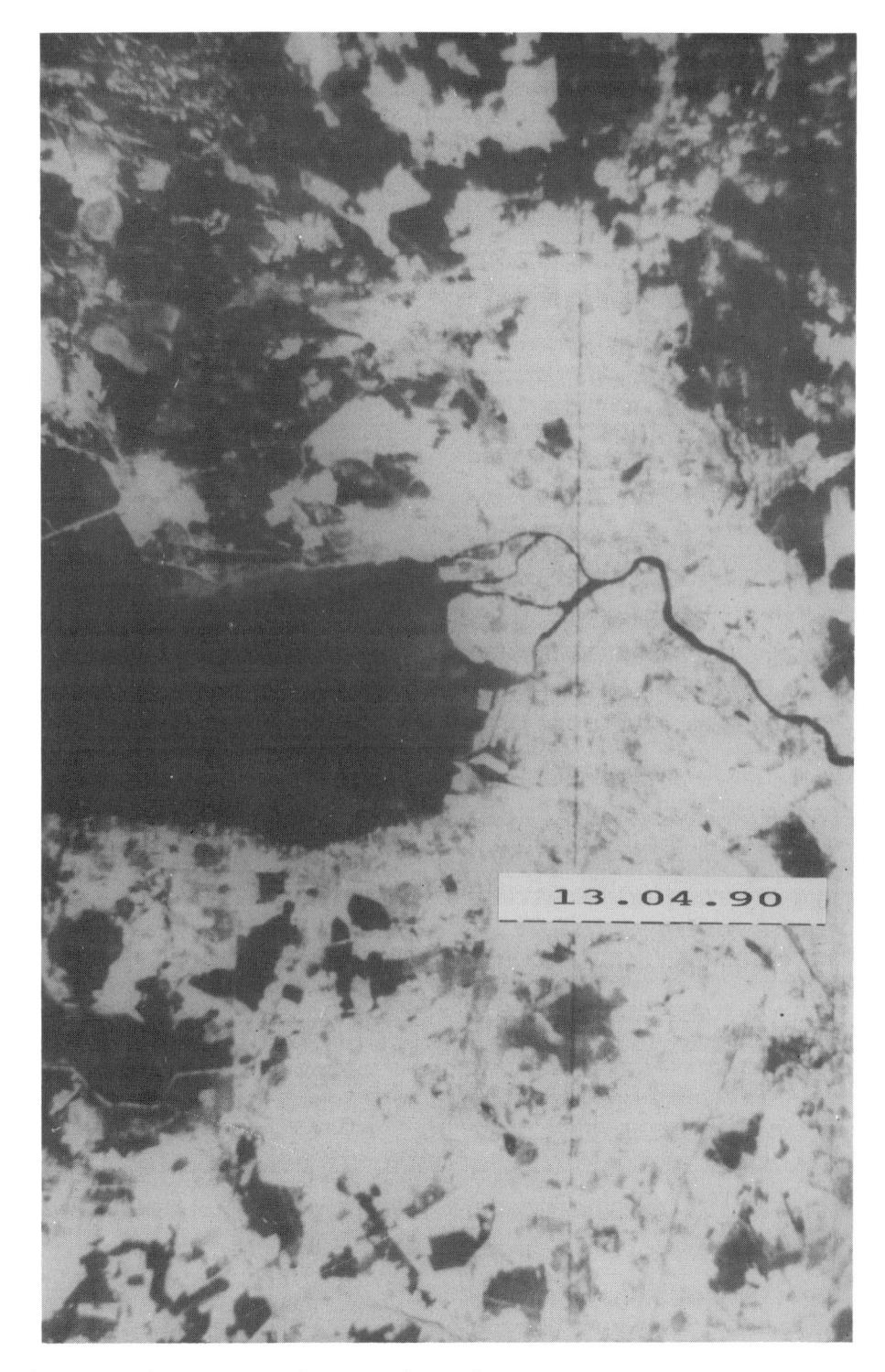

Figure 5.84 The Neva Bay. Satellite image of 13 April 1990.

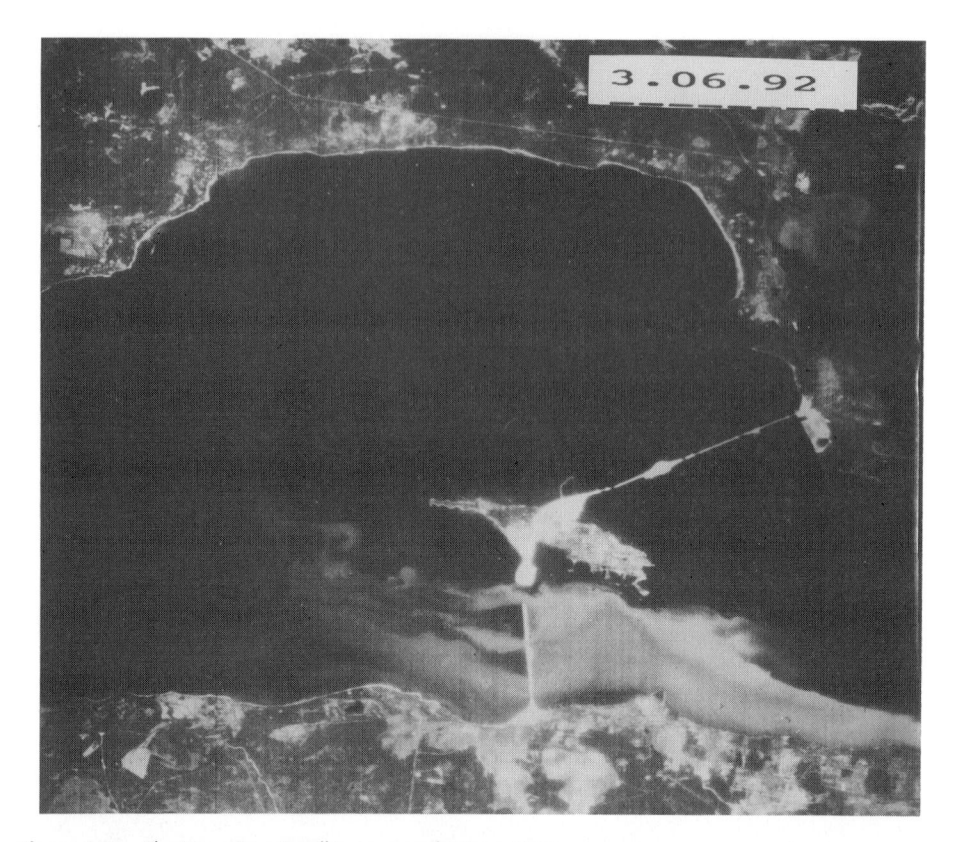

Figure 5.85 The Neva Bay. Satellite image of 3 June 1992.

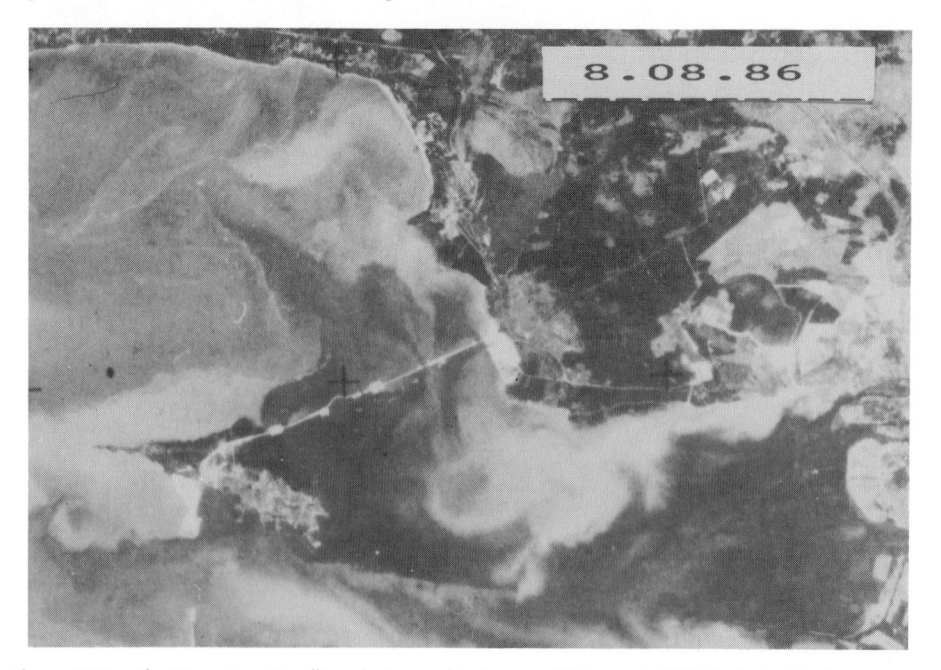

Figure 5.86 The Neva Bay. Satellite photographic image of 8 August 1986. (Label with date in the top right corner.)

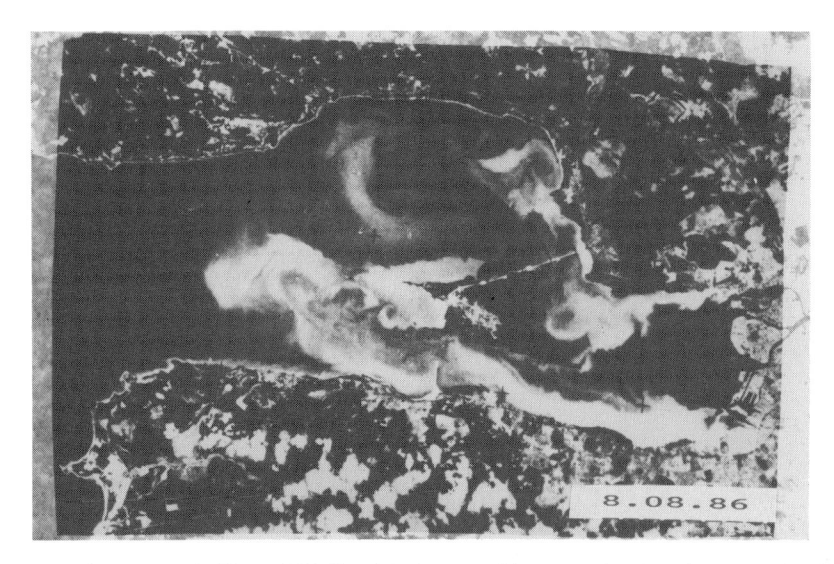

Figure 5.87 The same as in Figure 5.86. Special photographic processing to enhance contrasts. (Label with date in the bottom right corner.)

Figure 5.89 The Neva Bay. Satellite image of 27 July 1989. MSU-E, 500–600 nm channel.

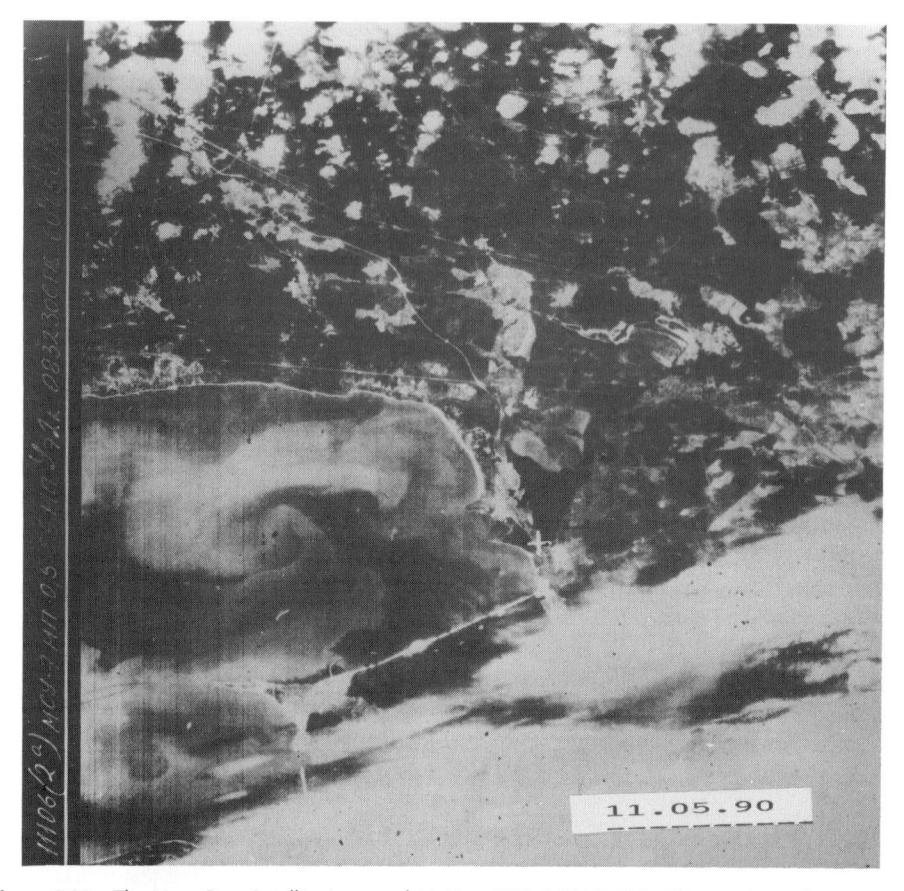

Figure 5.90 The Neva Bay. Satellite image of 11 May 1990. MSU-E, 500–600 nm channel.

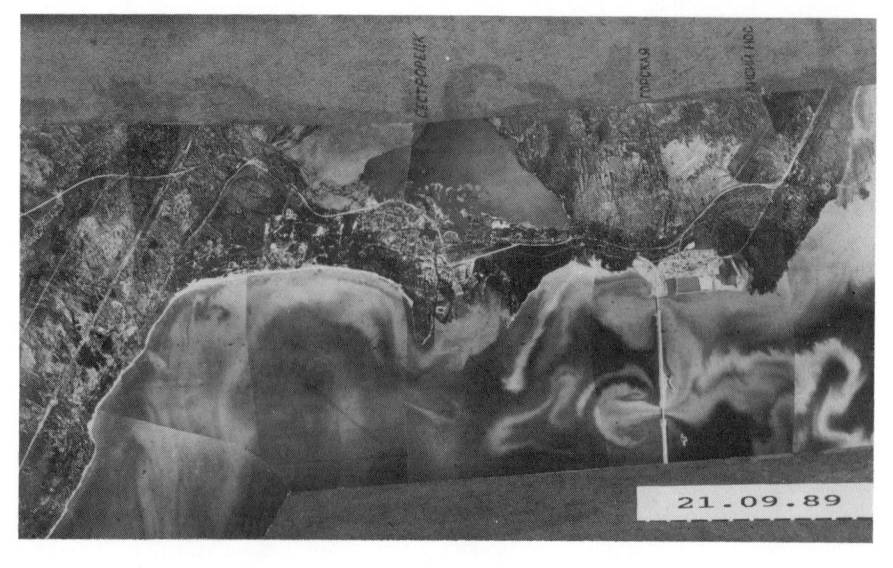

Figure 5.91 Airborne photographic image of the inshore waters near the northern part of the Flood Barrier taken on 21 September 1989.

Figure 5.92 The Neva Bay. Satellite image of 14 June 1989, MSU-E, 500–600 nm channel.

no dredging or bottom-deepening operations were being performed. This fact shows the role of bottom sediment mixing (resuspension) processes in the shallow inshore waters caused by wind- and wave-driven turbulisation.

The analysis of remotely sensed images stored in our database showed that there is a very high variability of pattern characteristics of suspended substance fields on a *synoptic scale* according to the above-mentioned four types of hydrological situation. This means that for a given time interval (say, a month or a particular year)

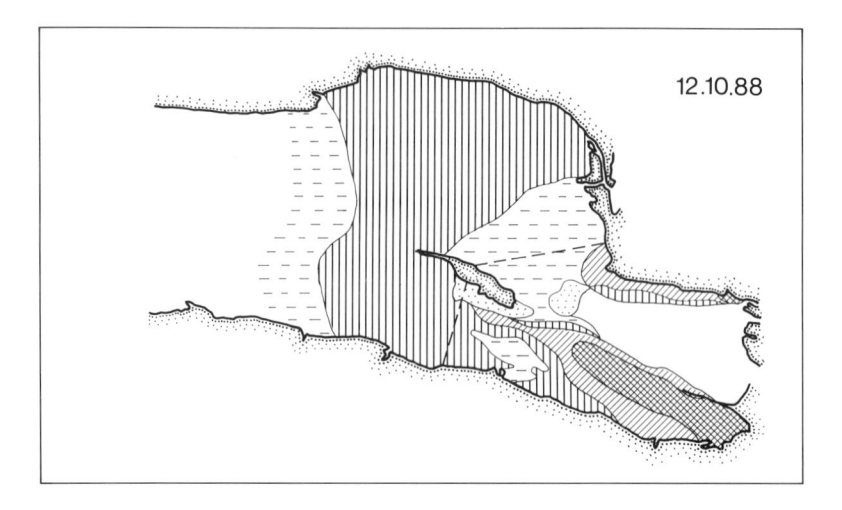

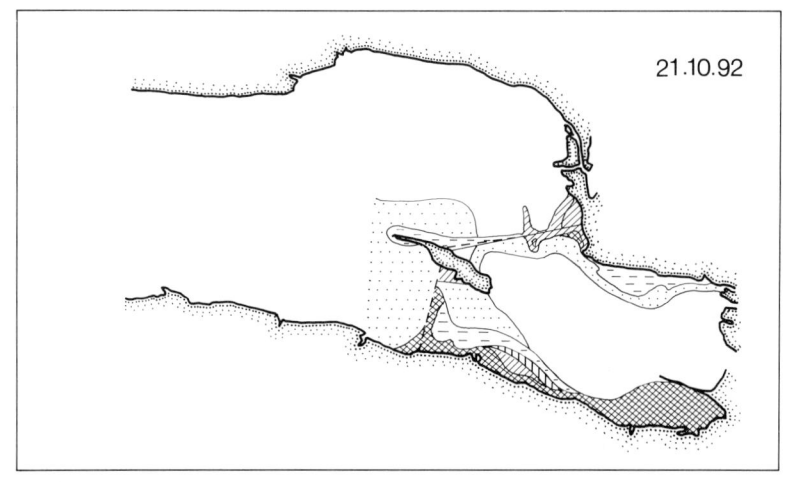

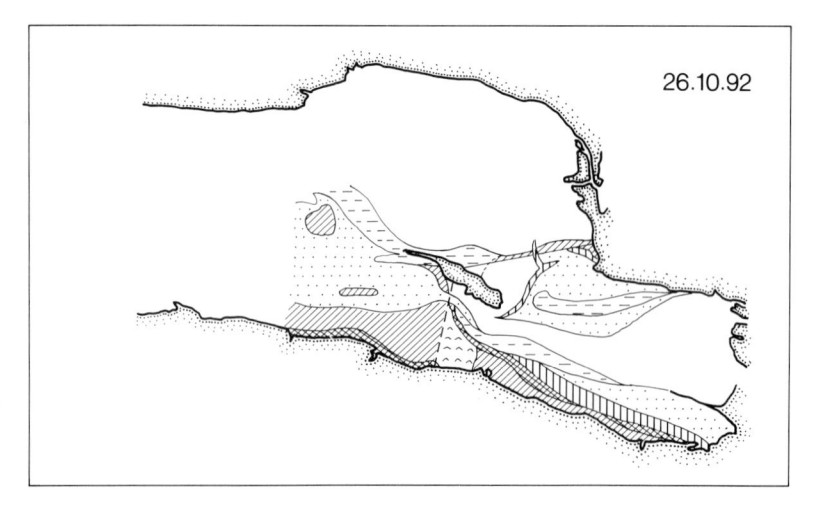

Figure 5.93 The Neva Bay. Schematic charts of suspended matter distribution for three dates in the autumn period. (See text for details).

one can find and select a certain number of images with a 'dirty' Neva Bay (large areas of the Bay are covered with fields of high concentrations of suspended matter), or with a 'clean' Bay (small areas with low concentrations of suspended substances); it all depends on a person's aim. Please, keep this in mind for the discussion to follow!

As for the *seasonal variability*, for a number of years in the 1980s (when routine bottom-deepening and ground-filling operations were being performed on a regular basis), there was a tendency for the total area of turbid water zones to increase from spring to the end of autumn. The reasons might be the increasing activity of dredging and other operations, including the dumping of spoils, in the summer period, and the seasonal growth of wind and wave activity.

It is difficult to show any tendency in the *annual variability*; no reliable remotely sensed data exist which could show any Barrier-dependent changes in general patterns of suspended matter fields before and after the construction of the Flood Barrier.

These were the conclusions we came to after having analysed more than a hundred remotely sensed images from our database (Victorov *et al.* 1990a, 1991; Victorov 1991b). In my opinion, the satellite images shown in this section demonstrate the potential ability of the former Soviet Union satellites – *as such* – to provide the oceanographic research community and relevant regional managers and decision-makers with satellite data useful for regional oceanography and environmental monitoring of the coastal zone. Unfortunately, this potential was not realised completely, due to poor logistics and the weak ground-base segment (see Chapters 2 and 3 for details).

The sore subject of the construction of the Barrier gave rise to some sweeping statements made, as a rule, by non-professional opponents of the Barrier and based on no science data. They agitated public anxiety, appealing not to reason but to emotion. Thus people having no idea of the peculiarities of the hydrological regime of the aquatoria, using a few selected satellite images, tried to present to the public an 'awful' picture of the impact the Barrier would have on the aquatoria. (And I must confess, they were often a success in the general 'anti-Barrier' atmosphere of the late 1980s.) The images they showed were ones taken under different hydrometeorological conditions, at different seasons, and with different numbers and locations of dredging and bottom-deepening machines. More often than not these ignorant 'experts' had a very vague idea of what was recorded in satellite images of different spectral bands; all the optical inhomogeneities in their interpretation were called 'mud'. We shall touch on this matter later, in Section 5.5.6.

5.5.5 Summary of findings

A recent summary of our findings concerning the eastern part of the Gulf of Finland (Sukhacheva and Victorov 1994) is based on the analysis of materials in an extended database which consists of materials mentioned in Section 5.5.3 and SST charts revealed from AVHRR imagery. The bulk of the data were collected during the period 1980–1993, though some satellite and airborne observations were made earlier, since 1975. These findings are related to a somewhat larger area than the Neva Bay.

5.5.5.1 Some general features of spatial and temporal variability of hydrophysical fields in the eastern part of the Gulf of Finland as revealed from remotely sensed data

- *The spatial distribution of the fields of suspended sediments* westward of the northern and southern passages of the Neva Bay depends on the season (through the effects of seasonal thermocline development), on the variability of the wind situation and the sea level, and on the generated surface currents and upwellings. One should note that in the periods of intensive development of phytoplankton the analysis of satellite images of this region in the visible band is rather complicated; in the majority of cases it is still possible to carry out ambiguous interpretation of fields.

- Zones with concentrations of suspended matter exceeding 10 mg l^{-1} can be traced as plumes at a *considerable distance* (60–100 km) from the Neva Bay as a source of them.

- Peculiar features in the spatial distribution of suspended matter in this region were determined for *three meteorological situations*: (1) in durable western winds, (2) in strong north-easterly winds, and (3) in still conditions (regarded as *typical cases*).

- Moreover, under similar meteorological conditions the patterns of suspended sediments may differ considerably in various seasons. Thus the three typical meteorological situations mentioned above should be analysed separately for the spring–summer period (developed thermocline near the surface) and for the autumn period (when the depth of the thermocline is increased, it is being destroyed, and the intensive mixing starts). Extreme cases of seasonal variations were determined.

- A strong *patchiness* in the distribution of hydrophysical characteristics in this region is observed in the summer season (from May till the middle of September), when deep waters may reach the surface, resulting in a temperature difference exceeding 10°C. Often the core of the *upwelling* is localised near Cape Shepelevsky, and during its intensive development the upwelling waters spread eastward to Cape Lebyazhy and northward, almost reaching the opposite coast. In some cases, cold patches with a temperature difference of 3–4°C can be traced westward of the Flood Barrier and in other regions.

- Zones of upwellings are recorded also in satellite imagery of the visible band in the brightness field. One can notice a spatial *correlation* between SST and brightness fields. The appearance frequency of upwellings in this region, and their intensity and duration, differ considerably from year to year, while typical patterns of suspended sediments in the eastern part of the Gulf of Finland were observed during many years *irrespective of the construction of the Flood Barrier* (with the exception of areas closely attached to the barrier, where, in some situations, suspended matter is observed as a tracer of eddy structures).

5.5.5.2 Natural and man-made factors of variability of hydrophysical fields in the eastern part of the Gulf of Finland

- Analysis of a series of remotely sensed data enables us to investigate the multiscale spatial and temporal variability of hydrophysical fields in the eastern part of the Gulf of Finland caused by various natural and man-made factors, namely, annual, seasonal and synoptic variability connected with hydrometeorological conditions, or caused by man-made factors, such as dredging and ground-filling operations, dumping of spoils (including similar operations during the period of Barrier construction), etc.

- We found that ground-filling operations are the main source of suspended materials. The larger the volume of these operations, the more important is the wind and wave mixing factor. Besides, there is a kind of inertia: in autumn even after they had finished work, the zones with high concentrations of suspended matter can still be seen. In spring, when they had not yet started, no zones with high concentrations are observed in practically the whole Neva Bay and the eastern part of the Gulf of Finland.

5.5.5.3 Recent changes in the state of the marine environment in the eastern part of the Gulf of Finland

- Since 1990 the intensity of ground-filling operations has started decreasing and, during the same period, the impact from industrial wastes has decreased. The state of the marine environment seems to be getting better, which was shown with remotely sensed data and with *in situ* measurements.
- We found that the observed decrease in concentrations of heavy metals, bacterial and other types of water pollution was, to a certain degree, connected with a decrease of the concentration of suspended sediments. The available data show a correlation between these two groups of parameters.

5.5.6 The Flood Barrier lessons

The new realities in the former USSR (including access for the general public to environmental data, public discussions on the environmental issues, public hearings on the state of the environment, etc.) encouraged the author to summarise the Flood Barrier lessons which should be learnt (Victorov 1991b).

5.5.6.1 Some lessons in the field of scientific research

- Adequate observation of the aquatoria requires regular satellite monitoring providing high-resolution images once every three or four days and medium-resolution images daily.
- Specific spectrometric remote sensing measurements should be performed along with the existing broad-band satellite images.
- The location of regular *in situ* observation stations must be specified in accordance with the current system and spatial distribution of suspended matter (as recorded in remotely sensed data).
- Prior to and during construction of Major Engineering Structures (MES), the storage of a homogeneous time series of remote sensing and relevant *in situ* data is necessary.
- These data should be accumulated in integrated Geographical Information Systems (GIS) of the type suggested in Section 5.2.

- Further analysis of satellite imagery demands the efforts of specialists in hydrodynamical modelling who could use satellite data to obtain the new, or to specify the already known, initial and boundary conditions and some parameters in numerical models. These models for the Neva Bay should account for two-dimensional sources of pollution.
- The 'zero-solution approach' suggested for the Great Belt Link (Denmark) is a good example to follow in future MES projects.

5.5.6.2 Some lessons in the field of public relations

- For MES that could possibly affect a great number of citizens or a large aquatoria, or cause trans-generational effects, international examination of the project is recommended.
- Scientists should not ignore public discussions. Sometimes scientists simply ignore non-specialists' opinions. Frequently, a large number of non-professional opinions expressed in letters, published in newspapers or appearing in TV programmes quickly become so well established in the public mind that it becomes almost impossible to convince the public of the true state of affairs.
- Try to avoid manipulating with terms: pollution, mud, clean, dirty: all these terms should be defined strictly before starting discussion with laymen.
- It should be made clear that two environmental events occurring at the same time interval or one after another might not necessarily be caused by one another. Even this very simple thought has to be explained to many people over and over again.
- Experts in public relations in the field of the environment should have a good knowledge of modern sociology, psychology, mass-media, communicative theory, etc.
- First ideas of ecology and environment must come to people at a very early age, perhaps in their childhood (see my concept of the ECOLOGIUM in Section 5.2).

5.5.7 Further development

Now some remarks on how the situation with the Flood Barrier developed.

> In the Autumn of 1987 concerned citizens appealed to Secretary General Gorbatchev. They felt that the pollution of the bay was aggravated by the construction of the barrier and they protested against its completion. In response several sessions on this subject were held in the Academy of Sciences of the USSR. In short succession three commissions of Soviet experts were nominated and set to work. They submitted their reports early 1989. (Final Report of the International Commission of Experts 1990)

As the three national commissions came to contradicting conclusions on the Barrier and its role in the environmental situation in the Neva Bay and the region, on a governmental level Delft Hydraulics (The Netherlands) was requested in June 1990 to submit a proposal to set up an international commission, to co-ordinate its work and to provide support. After consultation with Soviet authorities, Ir. H. Engel was invited to be chairman for the Commission. A former director of the Dutch 'Delta

Directorate', 'he was responsible for the flood protection and the protection of the environment of the estuaries of the Rhine, Meuse and Scheldt rivers, which together form the Dutch Delta area' (Rodenhuis 1992). Another 10 members, representing six nationalities, 'were selected for their specialisation, their involvement in similar projects and for their knowledge of conditions in the area' (Rodenhuis 1992).

Members of the International Commission and its Work Unit had a number of extensive discussions with local experts. They studied thoroughly the materials submitted by local scientists. Remotely sensed data on flow patterns and sedimentation patterns were submitted by S. Victorov and L. Sukhacheva. These materials, along with other original data, became part of the Final Report of the International Commission. Of major importance was its conclusion that the ecology of the 'Ladoga Lake–River Neva–Neva Bay–(the eastern part of the) Gulf of Finland' system

> ... has already, over a long period of time, been adversely impacted by man's activities, including principally: the discharge of untreated or inadequately treated waste water, the destruction of wetlands for urban expansion, dredging and dumping of spoil.
>
> During the 10 years the Barrier was under construction, its effects on the water quality of Neva Bay have been negligible compared to the above impacts. (Final Report 1990)

The International Commission of experts formulated its findings and detailed recommendations on the key issues of the complex regional environmental problem: (1) the Ladoga–Neva system; (2) water and environmental quality; (3) Flood Protection Barrier; and (4) organisation (Final Report 1990; for a digest version see Rodenhuis 1992). Following Dr. G. Rodenhuis, I have not given references to individual reports or datasets on which the figures presented in Section 5.5.2 were based. All of them were presented in this section as they appeared in Rodenhuis (1992); here the Final Report (1990) is given as the sole reference.

Possible economic developments in the St. Petersburg region could be:

- new harbour developments outside the old city, but protected from flooding and well connected to road and railway transportation;
- reconstruction and reallocation of industrial activities outside the old city;
- planning and execution of an intensified programme for upgrading of the sewerage system and municipal waste water treatment;
- construction of an orbital highway, including the western route over the Flood Barrier and the island of Kotlin (EBRD 1994).

Taking these projects into account, the regional policy should include the following issues of primary importance:

- The St. Petersburg water system is *one* system and should be studied and managed as *one* in all its physical, chemical and biological aspects.
- As scientific evidence and international expert opinion show that the completion of the Flood Barrier would have no negative effect on the water quality of the Neva Bay, the Flood Protection Barrier should be completed to protect the city, the infrastructure and future investments in developments.
- The environmental conditions of the waters surrounding St. Petersburg should be improved, and water quality standards should be developed in the framework of a masterplan for Integrated Water Management (EBRD 1994).

It is doubtless that remotely sensed data will be widely used in monitoring the marine and coastal environments of this region as a consistent part of modern economic development.

5.6 Concluding remarks

Looking back some 15 years, and comparing the general level of RSO activities in the Baltic Sea 'now' and 'then', one can see considerable progress. In the beginning of the 1980s using METEOR data, oceanographers in the 'Eastern' Baltic states could study only very generalised features of the oceanographic fields; their 'Western' colleagues had access to satellite data of higher quality, but did not use the data on a regular basis. As a result of the development of various sides of RSO, 30 m spatial resolution satellite images have been widely used by the Baltic oceanographers, time series of satellite data could be made available as autonomous data acquisition stations have been installed in some regional institutions, the feasibility and importance of satellite imagery use have been proved during this period, valuable practical experience of data processing and analysis has been gained, and a good number of scientifically sound results have been obtained. Moreover, younger scientists now get some basics of satellite oceanography in the universities, which makes it easier to use satellite data in their current research and operational activities.

The 'traditional' oceanographers seem to get used to satellite data and nowadays they consider remotely sensed data as (at least) an interesting source of oceanographic information (while even in 1985 some of them refused to believe that the patchy structure of the SST field, as recorded in satellite imagery, was a reality and not just 'noise').

In this chapter the wide scope of regional studies of dynamical oceanographic phenomena, as well as long-term satellite monitoring of some local sites in the Baltic sea, have been presented, thus reviewing the activities in nearly all the major issues of RSO conducted in the Laboratory for Satellite Oceanography, in St. Petersburg. It was a fruitful period though the technological level of our data processing facilities was much lower than the world level, and we had no access to the satellite data from satellites operated by the US agencies (except for the AVHRR/NOAA data in APT mode and on a non-operational basis).

It is worth noting that a large proportion of those results had been obtained by the middle of the 1980s. By that time

'(1) we were able to receive, process, and analyse satellite data of various bands and started to use these data in the studies of some features of the Baltic Sea ecosystem;

(2) we could share our experience with other users; but

(3) we were not in a position to conduct these activities on a regular basis "on-line"' (Internal Report 1984).

As we tried to think about non-scientific users and their problems, we stated that two bottlenecks had to be considered in this connection, namely:

(a) satellite data from centralised sources were not regular, they were not properly calibrated, and their formats were not suitable for users;

(b) the potential users had no up-to-date facilities for processing the satellite imagery and no trained personnel (Internal Report 1984).

For purely scientific applications of satellite imagery, the 1980s could still be referred to as an encouraging period in RSO development. Since then there has been very slow progress in the field of RSO in the former USSR. If it was a period of 'stagnation', then it is difficult indeed to find other words than 'hard times' to characterise the situation in Russian RSO in the mid-1990s (Victorov 1994b).

Elsewhere in the Baltic region the situation nowadays is much more favourable. In Sweden, Germany and Finland the AVHRR/NOAA data in AVHRR mode are being regularly received, processed, stored and analysed. SPOT and ERS-1 data are being widely used by many operational agencies and scientific groups in these countries as well as in Denmark. Many institutions are networked with satellite acquisition stations of the highest level located in Norway and Italy.

Interesting new tendencies can now be traced in Baltic Sea oceanography in the European context. The Institute for Remote Sensing Applications of the Joint Centre of the European Communities (Ispra, Italy) has launched 'collaborative programmes linking the remote sensing activities' of this institute to two other institutes – the Institute of Oceanology of the Polish Academy of Sciences (Sopot, Poland) and the Institute for Baltic Research (Rostock–Warnemunde, Germany). '... the motivations of the European Commission to promote the use of satellite data for coastal management ...' are given by Hoepffner *et al.* (1994). In March 1994 scientists from Ispra participated in a cruise to the Bornholm Basin on board the research vessel *Alexander von Humboldt* (Warnemunde) (Zuelicke and Schlittenhardt 1994). Numerical studies of fronts between the Baltic subbasins were started by Elken (Estonian Marine Institute, Tallinn) at the Institute for Marine Research, Kiel, Germany, using their computational facilities and continued at the Institute for Baltic Research (Warnemunde) (Elken 1994).

A three-dimensional hydrodynamic model of Gdansk Bay was based on the TRISULA software package (Delft Hydraulics, The Netherlands) (Robakiewicz and Karelse 1994). The same package and a mathematical model of water quality, DELWAQ, also developed in Delft, were used to study the eutrophication of the Gdansk Bay as a joint Polish–Dutch project (van der Vat 1994). Urbanski (1994) demonstrated the use of a set of SST AVHRR NOAA images for planning the position of monitoring stations in the Gdansk Bay.

The Nordic Council of Ministers financed several shipborne expeditions of Estonian scientists in 1993–1994 who explored the Suur Strait – one of the straits connecting the Gulf of Riga with the Baltic Proper (Astok *et al.* 1994; Suursaar and Kullas, 1994).

The EC-funded Project ODER (Oder Discharge Environmental Response) which will form the basis for a comprehensive mapping of the area in the GIS context, is a joint effort of the Department of Oceanography, Academy of Agriculture (Szczecin, Poland) and the Department of Geology and Geophysics, University of Edinburgh (UK) (Radziejewska and Shimmield 1994).

> The enlarged cooperation with the science community outside the Baltic area and the integration of problems of the Baltic Sea into larger regional concepts find expression also in the ongoing activities within the framework of the European Community. The aim is to expand existing EC programmes in marine science by interdisciplinary Baltic Sea ecosystem studies and to suggest new initiatives. (Brosin *et al.* 1994)

In this context a new body – The European Committee on Ocean and Polar Sciences – which was established in 1990 by Directorate General XII of the EC and the European Science Foundation (see Hempel 1991) is expected to play a decisive role in formulating the basics of the science policy in the Baltic region.

I would dare to make a forecast of future development of RSO in the Baltic region for this decade. (While speaking about a certain country I will mean only the oceanographic community and even only its part involved in RSO activities.) Denmark and Germany (the latter playing the leading role) together with the new (since 1995) member states of Sweden and Finland, with gradual step-by-step involvement of Poland, and later Lithuania, Latvia and Estonia, will possibly create a scientific and technological pool providing access to new satellite data, modern processing facilities and informational technologies, and value-added informational products. Joint field exercises and cruises are likely to be performed. Some kind of Human Resources Mobility Programme is likely to be launched for these states.

In this context the 'European Russia–Europe' collaboration in RSO seems to be less certain in this decade. At the same time there must be no obstacles for the development of co-operation in RSO based on personal involvement of scientists from Russia or on a laboratory-to-laboratory basis or within the framework of professional European organisations, say, the EARSeL (European Association of Remote Sensing Laboratories) through the mechanism of the Special Interest Groups (Allewijn 1994).

Conclusion

In summer 1983 in San Francisco, USA, two naval officers and one civilian addressed the civilian OCEANS '83 Conference with the following optimistic words:

> The array of satellites flying and planned for launch ... show that oceanography from space is healthy, growing, and focused on Navy target parameters. The Navy's program is a key element in this national effort in space. (Malay *et al.* 1983)

Now, in 1995, I would agree partly with this statement in that satellite oceanography is not in bad health and may soon become even more healthy. The year 1995 may become a remarkable year indeed for regional satellite oceanography as two new sensors, the SeaWiFS on SeaStar and SAR on Radarsat, are expected to be launched.

As for another part of the quotation, on the *key* and *focus*, we are living in the post-cold-war era, aren't we? And, if *yes*, my message is:

Let us *focus* on regional satellite oceanography, and in particular on coastal zone monitoring and management.

Let us make the synergetic approach the *key* element in our space-based studies of marine and coastal environments.

The general perspectives for regional satellite oceanography are favourable. Mick *et al.* (1994) even think that 'these are exciting times for commercial satellite remote sensing'. Along with the well-known programmes (see Section 2.3), some others seem to merge.

> With the close of the Cold War, the Federal Government appears willing to consider the licensing of National Security satellite technology for civil–commercial use. At the same time, the Land Remote Sensing Commercialization Act was amended in 1992 to allow different treatment for truly commercial systems. These two government policies coincide with significant growth in GIS markets for remote sensing data worldwide and a proliferation of products that can take advantage of these policy shifts.
>
> As 1993 comes to a close, one application for a commercial remote sensing system has been granted and another has progressed toward approval. Either or both of the World View and Lockheed systems would dramatically change the character of space-derived data and the applications of such data to public or commercial uses. World

View proposes three-meter spatial resolution and Lockheed's Commercial Remote Sensing System proposes stereo spatial resolution at the one-meter level. (Mick *et al.* 1994)

From the other side of the former iron curtain Zaitsev (1994) published a list of satellites launched in 1993 with an indication of photo-reconnaissance satellites, while Lukashevich (1994) reported on the details of these dual-purpose satellites. So the number of satellite sensors seems to grow, and data from dual-purpose satellites may soon be available also for civilian users.

All this relates to the satellite payload. Before turning to data processing, image transfer and GIS-related problems, at least in the Conclusion to this book, I cannot but respond to the eternal challenging question of remote sensing – 'satellite or aircraft?' Nowadays this issue is given much attention, which is partly caused by the development of airborne imaging spectrometers (for a survey of these instruments see, for example, Vaughan 1994); and the First International Airborne Remote Sensing Conference and Exhibition held in Strasbourg, France on 12–15 September 1994, demonstrated a high technological and methodological level achieved by some companies in both land and marine applications. And it is a remarkable fact that the WMO/IOC Working Group on Oceanic Satellites and Remote Sensing (OSRS) is going to make an assessment of up-to-date aircraft and sensors, and their actual information capabilities of providing the oceanographic community with the required data on a regional scale, using the same analytical approach as for the satellite systems (Sherman 1994; Ocean Remote Sensing Group Meets 1994; Report on Satellite Systems and Capabilities 1995).

Broadly speaking, the best way is to combine the satellite and airborne data. In the short term the economic reasons will govern the situation. Generally, a single short-term aerial campaign seems to be expensive, and it could well happen that it will be difficult to run such campaigns very often. If the money is available *now*, it is reasonable to collect airborne data. In the long term, retrospective satellite data are usually available from archives/databases, and one can purchase them *later*, when the money is available. For monitoring purposes the aircraft should be regarded as a self-consistent tool; they can also be used for 'additional observations' when satellite-based information alone brings unambiguous results.

Returning to satellite oceanography, we will touch on the data processing and transfer issues. There are dozens of commercial image processing facilities comprising hardware and software of different levels (and prices), and it is a problem for the user to make the proper choice. In the present market situation the user can also choose the satellite product, as there is an overlap of various products in their basic characteristics (see Section 2.3 for details), and strong competition between operating agencies exists. The inability to provide information services on a modern level leads to loss of customers and rapid decline in incomes. As was shown in Chapters 1 and 2, the ground-based segment was always the bottleneck of the Earth observing system of the former USSR, while the satellite sensors of the visible band, as such, could provide good-quality raw data.

Insufficient acquisition capacity, the 'low technologies' used for data tackling, poor communication lines, the lack of up-to-date services (browsers, CD-ROM catalogues, user-oriented value-added products on sale, flexible prices), which is the fact in the Russia of today, is no longer admissible in the present world market situation. The Swedish Space Corporation (SSC) will build a kind of by-pass to avoid using the Russian ground-based segment, thus showing a new tendency in international technological co-operation in this area.

> During the spring of 1995 a new reception and processing chain, at SSC's facilities at Esrange in northern Sweden, will be ready for the Resurs data ... At SSC Satellitbild in Kiruna, the images will be turned into user-specified products/information and delivered through data networks, CD-ROMs, paper prints or any other media asked for by the user. (Stern 1995)

The reader can easily see that the SSC will actually complement the raw Russian 'Resurs' satellite data with modern services providing end-users with imagery from MSU-E and MSU-SK sensors worldwide. Not the best way for the Russian Space Agency to do business in general, but perhaps the best option nowadays (?).

Satellite data processing is the core of the *operationalisation* problem. The special International Symposium *Operationalisation of Remote Sensing* held in Enschede, The Netherlands in 1993, showed how different are the levels of operationalisation in different countries and in different institutions. An attempt to use the existing operational hydrometeorological network in the utilisation of satellite-derived oceanographic information is being made in some regions (see, for example, Victorov *et al.* 1993b) following the example of colleagues in Scandinavia (see Section 4.1). The satellite data transfer problem becomes a bottleneck in these activities, so long as one is becoming involved in truly *operational* services. An approach based on cellular telephony may help in some cases.

Satellite data processing and utilisation is closely connected with the implementation of GIS technologies meant to provide multilayer data storage, fusion and analysis. The effective use of high-resolution satellite imagery in the GIS structure needs adequate georeferencing of images, geocoding of *in situ* data, and the harmonisation of the data spatial resolution and the scales of basic maps used. This is indeed the problem which will be faced by the marine community in the lesser developed countries, as well as in the remote regions for which no adequate maps are available.

As we have touched upon the maps–communications–GIS technologies issue, we inevitably should touch upon the problem which can be referred to as the *North–South* problem, the well-known socio-political issue. It is common knowledge that 'there is a global inequality of consumption between the north and south' (Kondratyev *et al.* 1994), and this inequality can be easily traced even in the 'consumption' of satellite images for regional marine and coastal monitoring activities. Within the framework of regional satellite oceanography activities it is possible to compare the efforts (manpower and financial resources) put into the studies of various regional seas and coastal areas. For example, an impressive and even exciting difference can be easily seen if one compares the regional/local activities in satellite data application for the marine and coastal environments in the *Gulf of Mexico* (comprehensive eight-year time series stored in a digital database of 850 CZCS images and related bio-optical information products run on a set of modern computers, and supported by super computers for data archiving, as described by Oriol *et al.* (1994); the database of over 200 AVHRR scenes for a four-year period with relevant *in situ* datasets, as presented by Stumpf *et al.* (1994), to mention just two recently reported works), the eastern part of the *Gulf of Finland* (old-fashioned archive of about 300 satellite and airborne images and charts for a 20-year period, with inadequate facilities for image processing, as described in Section 5.5), and the

Redang Island archipelago (with only several satellite images during a five-year project, as presented by Mohamed 1994). To be honest, this striking difference in the number of satellite images used may not actually be a matter of money alone, but of the coverage of various sea areas with satellite imagery, if it was not ordered in advance. Money comes first as we consider current problems of practical applicability of remote sensing and integrated GIS technologies in developing countries.

The many years of experience of British universities, which educated and trained a large number of students and post-graduate students from the developing countries, should be acknowledged. I had an opportunity to be acquainted with the remote sensing educational programmes currently run at the universities of Dundee, Sheffield and Edinburgh, and also with the numerous theses submitted 'in part fulfilment of the requirements for the degree of Master of Science in Remote Sensing, Image Processing and Applications' at the Department of Applied Physics and Electronics and Manufacturing Engineering, University of Dundee (many of them were devoted to water dynamics, transport of sediments and coastal zone monitoring using visible, IR and radar imagery). I am sure that those who got their MSc degrees can start practical work in the remote sensing area. Coming back home, they bring experience, methodologies and the general culture of satellite remote sensing applications.

Another problem arises – the efficiency of the work of these experts. During the celebration in May 1994, in Toulouse, of the 20th anniversary of GDTA (Groupement pour le Développement de la Télédétection Aérospatiale), the French institution that did a lot to promote remote sensing in many regions worldwide (and particularly in French-speaking developing countries), through a system of training, there were interesting discussions on training in remote sensing and GIS applications. One of the most crucial issues was the next steps of a trained person after his/her return back home. Two issues need to be discussed:

- the student will bring knowledge and expertise, but will he/she bring the sophisticated hardware and software she/he got used to?
- does a proper national/regional/local infrastructure exist that could accommodate the young expert?

Considering that these questions are very important in the context of regional satellite oceanography, I planned to discuss the whole problem here. Fortunately, an article on this topic by Perera and Tateishi (1995) has just appeared, and I would like to present here some of their findings based on a thorough analysis.

Perera and Tateishi (1995) brought evidence that the application of new technologies 'is showing large differences in developed and developing countries.' They identified a number of barriers and common problems of acquiring, applying and developing remote sensing and GIS in developing countries, based on their personal experience gained by studying, teaching and research conducting in Sri Lanka, Japan and the USA, and also on the results of an international survey – a questionnaire was posted to 36 geography departments of universities in 26 randomly selected countries, with responses from only 12 universities (Perera and Tateishi 1995). Among the common barriers they mention:

- *low economic strength* (for example, in Sri Lanka the total annual income of seven universities in 1985 was about US$12 million, as compared to the annual income (1984–85) of the University of Sheffield of about US$61 million (Craig 1987). GIS hardware at a minimum level requires about US$10 000 (Yapa 1991), while the geography departments of two leading universities, in Sri Lanka and Ethiopia, have reported an annual income as less than US$2000. With the price of a personal computer with graphic facilities in Sri Lanka in 1993 of US$2000, 'practically it is economically difficult to implement a GIS laboratory in most of the developing countries without foreign aid' (Perera and Tateishi 1995).);

- *educational barriers* (the scarcity of experts in remote sensing and GIS; weakness of the decision-makers, who stick to their own academic methodologies and are pessimistic about application of new technologies; institutional weakness);
- *social factors* (lack of communication among intellectuals; language barriers, as all the manuals and research materials are written, and all international conferences are conducted, in languages other than the native languages in developing countries);
- *poor information flow* (for example 'private construction companies rarely consult or exchange ideas with universities, and governments are very poor in ascertaining public opinion about research and development projects');
- *political factors* (the lack of political stability may cause critical changes in development plans with a change of government, the influence of political powers on the master plans, etc. As an example, Perera and Tateishi (1995) wrote that a 'compromise between Sri Lanka and India might influence the reception of satellite data from Indian resource observation satellites to Sri Lanka');
- *weakness in international co-operation* (inadequate involvement of university research communities in relevant programmes).

To promote remote sensing and GIS in developing countries through structural developments Perera and Tateishi (1995) suggested an approach based on 'a cycle of solutions'. With this realistic and pessimistic background, with the large difference in satellite data availability, computer hardware and application software modules used in the developed and developing countries, we must still say that an increasing number of up-to-date studies of regional seas and coastal zones carried out in many regions are merging (and the case study, being part of the article by Perera and Tateishi (1995), also confirms this statement).

We started this Conclusion with some optimistic forecasts on the further development of satellite oceanography. I would say that this is particularly related to *regional* satellite oceanography. 'Bearing in mind the limited amount of trained manpower and other resources, it seems more natural for national and regional authorities to invest in regional environmental monitoring (in the broad meaning of the terms 'monitoring' and 'environment'), than to participate in global programmes. After all, collecting, storing and analysing a time series of information on the state of the regional environment (including the marine and coastal zones) is a natural task of each country or geographical cluster of countries, and it is the main task of regional/local authorities to improve the quality of life, of which the state of the environment is an important constituent part. The data collected in the course of these efforts may also be used for global climate research, but for each country the highest priority will always be these data for regional planning and development of industry and agriculture' (Victorov 1994e). In this context the climate for the development of regional satellite oceanography is favourable and its prospects seem to be encouraging, particularly in the rather broad problem area covered by the term 'coastal zone management'. There is a practical socially motivated interest in the studies and monitoring of natural and antropogenic impacts on coastal waters, shores and coasts: the transitional zone between the land and the sea. (We will probably need stricter definitions of terms widely used in this problem area in the near future.) The growing interest of users in this subject can be indirectly illustrated by the data presented by Troost et al. (1994). They analysed the response to their famous computer-based lessons and obtained the following figures: 44% of users of module 4 of the training package expressed their interest in 'coastal zone management', while only 15% were interested in 'marine geology', 26% in 'marine pollution', 28% in 'marine biology' and 33% in 'physical oceanography' (Troost et al. 1994). The rapidly growing interest in the implementation of regional satellite oceanography techniques in the coastal zone monitoring and management area was taken into account, when I planned the content of the review Chapter 4 of this book; as a result the coastal zone-related issues are dominant in this chapter.

We have just mentioned the response from the users; it is high time to turn again to the topic of 'the users' in the context of regional satellite oceanography, though this issue was already given much attention in Chapter 1. Some 10–15 years ago there were psychological problems with those users for whom it was very difficult to get used to new information from satellites. Recently Stjernholm and Windolf (1994) reported that this problem still exists. While discussing the role of centralised and regional (local county authorities) factors in the infrastructure of environmental monitoring, under Danish conditions, they wrote:

> At present there are two major constraints for implementing the application of remote sensing in environmental monitoring in Denmark. The first constraint is related to the fact that the existing system is well-developed, and it is difficult to convince the decision-makers of the potentials of new methodologies. The second constraint is related to the fact that the relatively 'cheap' satellite based data has either a too coarse spatial resolution and/or an inadequate spectral design. (Stjernholm and Windolf 1994)

So the well-developed and established system of environmental monitoring can be an obstacle for the implementation of new technologies. From this point of view the implementation of remote sensing techniques and other marine high technologies was much easier in the regions where, by the beginning of the satellite era, the 'well-developed' network of oceanographic observations had not been established. Let me give just one characteristic example. The SeaWatch/Thailand marine environmental monitoring and forecasting system located in the Gulf of Thailand is currently operated by a group of Thai organisations. It is based on seven moored data buoys, TOBIS, designed and manufactured by OCEANOR, the Oceanographic Company of Norway, and uses the ARGOS satellite communication system (Singhasaneh 1994). The attempt to install a similar system in the Gulf of Finland appeared to be much more difficult, and the project has been suspended for several years.

In spite of the overall optimistic forecasts on the development of satellite oceanography, the problem of *the user* remains. Dealing with regional users, one should bear in mind one more aspect of satellite data implementation in the present situation in the remote sensing products international market. Of great importance is the preparatory phase, the period between the moment when the new satellite/sensor is first advertised and the moment of launch of this sensor. Adequate information on the sensor, its potential advantages as compared to existing sensors, broad international airborne campaigns with a sensor prototype, dissemination of simulation products (images) from the new sensor, and workshops with invitations to potential users, are commonplace nowadays. Excellent lessons in this respect have been given in recent years by ERS-1, SeaWiFS, Radarsat teams. Total negligence/ignorance of those important items have been demonstrated by the famous KOSMOS-1500 side-looking radar team. There is another aspect of activities during the preparatory phase. Long before the launch of a new satellite/sensor a thorough feasibility study of the new information products must be carried out at a national level (and, for larger countries, at a regional level). For example the following detailed analytical notes have been produced and the relevant recommendations have been worked out in The Netherlands before the ERS-1 launch, by Wijsmuller Engineering B.V., Delft Hydraulics, and National Aerospace Laboratory NLR. A special study to

- 'analyse the information requirements in various marine domains;
- analyse how remote sensing can contribute to these information needs;
- analyse the costs and savings' (Drenth *et al.* 1990)

was supported by the Netherlands Remote Sensing Board (BCRS) and resulted in a set of documents (Drenth *et al.* 1990, with appendices; de Valk *et al.* 1990). It is remarkable that the report by de Valk *et al.* (1990) on the extraction of marketable information from ERS-1 data addressed two different groups – the remote sensing community and the people involved in marine activities – potential users interested in what ERS-1 data may add to proven information sources they currently used and had got accustomed to.

The problem of the *user* remains crucial for the true commercialisation of satellite remote sensing. This is how Davis *et al.* (1994) assess the situation with the use of remote sensing technology in the USA:

> The lack of understanding concerning the value of remote sensing data to US industry is widespread. Even as sales of remote sensing data are increasing on a yearly basis the largest market for these data is still the US government (Asker 1992) .... The use of remote sensing technology by the US industry has progressed too slowly ... Whatever the reason, it is apparent that barriers exist to industry's use of remote sensing technology as a viable source of information. These barriers range from the prohibitive costs (at least for small to medium sized businesses) of data and data analysis to a complete lack of knowledge concerning access to remote sensing data and its potential applications. (Davis *et al.* 1994)

As one of the attempts to improve the situation, the Visiting Investigator Program (VIP) was launched at Stennis Space Center, Mississippi, which provides a non-risk opportunity for potential users to train their personnel, to utilise the specialised resources and to obtain hands-on experience in using remotely sensed data. VIP is a unique venture offered to US industry 'at no cost, 90 day basis and, as such, serves as a catalyst for future work in other NASA commercial program opportunities' (Davis *et al.* 1994). It would be interesting to see whether this programme is a success, what are the current results, and what lessons can be learned from this VIP initiative in the international context. Hands-on training is an important constituent part of VIP. It is good if the professional training in a particular application area of remote sensing and GIS is based on previously obtained general knowledge of basics of remote sensing.

In this respect, of great interest is an attempt in the USA to give the basics of remote sensing to schoolchildren, thus, figuratively speaking, starting the process of breeding the future users of satellite data from childhood. Wagner and Czurylo (1994) presented the Space Technology Education Program (STEP) which 'provides support services to the educational community with the goal of using this technology to motivate student interest in learning marine sciences and coastal studies, among other important science content areas'. In partnership with NASA, the Environmental Research Institute of Michigan (ERIM) implemented the STEP in 1989 to promote science learning in K-12 schools through uses of space science and technology. The STEP was motivated by 'the fact that the US students are consistently below the science achievements of many European and East Asian students' which is 'symptomatic of our failure to innovate and to develop effective methods for motivating science learning in our classrooms. This failure is evident in the area of marine science and coastal studies, where most students today receive only cursory exposure, if any at all' (Wagner and Czurylo 1994). STEP promotes the applications of low-cost satellite data acquisition stations which are being installed in the classrooms and receive images from geostationary and polar orbiting environmental satellites. In the framework of STEP several hundred educators have been trained in installing and operating these stations. STEP also provides information services, training materials and technical support to teachers. Currently National Science Education Standards are under development (the initiative put forward in 1989 by President Bush and the National Governors Association, led by then Governor Bill Clinton). In the context of regional satellite oceanography I was pleased to know that 'there is no question that the standards will contain concepts important to the understanding of marine science and coastal processes ... Those schools located in coastal areas will have opportunities to meet the standards by teaching a broad range of studies applicable to that environment' (Wagner and Czurylo 1994). STEP involves children in the process of 'live' satellite imagery acquisition, provides hands-on experience, from just entertainment and games at the kindergarten level, to simple analysis and comparison of satellite-derived information with *in situ* observations (say, carried out in the neighbouring bay), at the senior high school level. Some real examples of high school students doing marine science, illustrating STEP in action, were presented by Wagner and Czurylo (1994).

I could have finished my Conclusion with these optimistic pictures of children mastering remote sensing technologies in the kindergarten, thus providing future prospects of this science ... but as the book is on regional satellite oceanography, I feel that I must finish with something *regional*. As the reader will have understood from the text of the Preface, the author is not a supporter of an overcentralised system in any field of human activity, so I will not use the term 'Centre' below. I will say a few words on the *International Focal Points* – the possible methodological and training institutions dealing with development of the techniques relevant to regional satellite oceanography. I propose the establishment of a few test areas in the seas with the most advanced *in situ* measuring tools installed, with the most sophisticated facilities for satellite data processing running, with up-to-date communication links available, and where the satellite databases are already integrated into GIS. New ideas in satellite oceanography could be tested here, new techniques implemented, and the new knowledge could spread from here. I know one area which can be regarded as a prototype; this area is nowadays almost what the marine and remote sensing communities dream of (please see Section 4.2.1). This site could be an excellent Focal Point for the Americas, and it practically already exists. The second Focal Point in regional satellite oceanography/coastal zone monitoring and management could act somewhere in the Asian–Australian region. I am sure it will be set up. Will a united Europe run the third one?

References

AITSAM, A., ELKEN, J. (1982). Synoptic scale variability of hydrographical fields in the Baltic Proper on the basis of CTD measurements, in *Hydrodynamics of Semi-enclosed Seas* (ed. J. C. J. Nihoul), Amsterdam, Elsevier Scientific Publishing Company, pp. 433–468.

ALLABERT, A. V., VICTOROV, S. V., DRABKIN, V. V., FLORENSKAYA, M. L. (1981). Study of representability of marine test sites as part of the problem of satellite payload calibration (in Russian), in Abstracts of papers presented at the Fourth All-Union Seminar on *Remote Methods and Techniques for Measurement of Oceanographic Parameters* (Odessa, 1981).

(1982). On airborne studies of sea surface temperature fields at experimental subsatellite test areas (in Russian), in Abstracts of papers presented at the First All-Union Conference on the *Biosphere and Climate Studies Based on Spaceborne Observations* (Baku, 29 November–3 December 1982), Baku, ELM, pp. 320–322.

ALLAN, T. D. (1979). Monitoring the sea surface, in *Proceedings of Technical Conference on Use of data from Meteorological Satellites* (Lannion, France, October 1979), Document ESA SP-143, pp. 205–215.

ALLEWIJN, R. (1994). EARSeL special interest group on 'Water Applications', in Proceedings of the EARSeL workshop *Remote Sensing and GIS for Coastal Zone Management* (Delft, The Netherlands, 24–26 October 1994), pp. 2–6.

ALTSHULER, V. M., SHUMAKHER, D. A. (1985). On the influence of blow off winds on the hydrological regime of the Neva Bay (in Russian), in Proceedings of the Leningrad Hydrometeorological Centre, No. 2, 15.

ANDERSON, J. R., HARDY, E. E., ROACH, J. T., WITMER, R. E. (1976). *A Land Use and Land Cover Classification System for Use with Remote Sensor Data*, US Geological Survey Professional Paper 964, Washington, DC, 28 pp.

APEL, J. R. (1977). *Past, Present and Future Capabilities of Satellites Relative to the Needs of Ocean Science*. Report to IOC, UNESCO, Paris, 42 pp.

A prototype of GIS 'Environment and Population Health' for St. Petersburg Metropolitan Area (1992):

Part 1. SHIHUKIN, G., VICTOROV, S., MIKHAILOV, N., FEDOROV, A., VOROZHKO, N. General Structure and Module 'Atmosphere'.

Part 2. VICTOROV, S., SMOLYANITSKY, V., SUKHACHEVA, L. Module 'Satellite Monitoring of Neva Mouth'. Paper presented at the International Seminar *Ecological Problems of Metropolitan Areas: Key Indicators, Modelling and Decision Making*. Centre INENCO, St. Petersburg, Russia, 17–21 February 1992.

ARMSTRONG, R. A. (1994). Ocean fronts in the northeastern Caribbean revealed by satellite ocean colour imaging, *Int. Journal Remote Sensing*, **15**, No. 6, pp. 1169–1171.
ASKER, J. (1992). Remote sensing sales grow with expanding data needs, *Aviation Week and Space Technology*, July 13, pp. 46–51.
ASTOK, A., HANNUS, M., KULLAS, T., NOMM, A., OSTMANN, M., SUURSAASR, U. (1994). Suur Strait – big enough? in abstracts of papers presented at the Nineteenth Conference of Baltic Oceanographers (Sopot, Poland, 29 August–1 September), p. 103.
ASTOK, V., NOMM, A., TAMSALU, R. (1986). Some ideas on the optimization of the Baltic Sea monitoring, in Baltic Sea Monitoring Symposium, 1986, Baltic Sea Environment Proceedings, No. 19, Helsinki, pp. 297–306.
AUSTIN, R. W., PETZOLD, T. J. (1981). The determination of the diffuse attenuation coefficient of sea water using the Coastal Zone Color Scanner, in *Oceanography from Space*, ed. J. F. R. Gower, Plenum Press, New York, p. 239.
AVANESOV, G. A. (1994). Local space monitoring of the Earth, *Space Bulletin*, **1**, 4, pp. 8–11.
BALL, D., BABBAGE, R. (1990). *Geographic Information Systems. Defence Applications*, Pergamon Press, 220 pp.
BALTIC MARINE ENVIRONMENT PROTECTION COMMISSION – HELSINKI COMMISSION (1986). First Periodic Assessment of the state of the marine environment of the Baltic Sea area, 1980–1985 background document. Baltic Sea Environment Proceedings, No. 17 B. Helsinki, 351 pp.
BALTIC SEA DECLARATION (1990). Final document of the Ministerial meeting (Ronneby, Sweden, 2–3 September 1990).
BALTIC SEA MONITORING SYMPOSIUM (1986). Baltic Sea Environment Proceedings No. 19, Helsinki, 553 pp.
BARALE, V., SCHLITTENHARDT, P. (1994). Remote sensing and coastal zone monitoring in the European seas: activities of the JRC EC, in Proceedings of EARSeL workshop on Remote Sensing and GIS for Coastal Zone Management, Rijkswaterstaat Survey Department (24–26 October 1994, Delft, The Netherlands), pp. 47–56.
BASHARINOV, A. E., GURVICH, L. S., EGOROV, S. T. (1974). *Radioemission of the Earth as a Planet* (in Russian), Moscow, Nauka, 188 pp.
BELYAEV, M. M., DRABKIN, V. V., LAZARENKO, N. N., KOURPACHEVSKIJ, V. G. (1977). Experimental oceanographic test sites (in Russian), in First Congress of Soviet Oceanologists (abstracts), Moscow, part 1, pp. 173–174.
BERESTOVSKIJ, I. F., BROSIN, H.-J., VICTOROV, S. V. (1983). USSR–GDR oceanographical subsatellite experiment (in Russian), *Earth Studies from Space*, **1**, pp. 121–122.
(1984). Internationales Subsatelliten-Experiment in der Ostsee, *Beitrage zur Meereskunde*, Heft. **50**, s. 5–7.
BERESTOVSKIJ, I. F., VICTOROV, S. V. (1981). Role of perspective spaceborne systems of observation in carrying out oceanographic section of world climate research program (in Russian), *Meteorology and Hydrology*, **10**, pp. 113–119.
(1982). Spaceborne methods in a system of oceanographic experiments of world climate research program (in Russian), in abstracts of papers presented at the First All-Union Conference *Biosphere and Climate Studies Based on Spaceborne Observations* (Baku, 29 November–3 December 1982), Baku, ELM, pp. 5–8.
BIVINS, L. E., PALMER, H. D. (1989). 'Smart Maps'. A new look at ocean resources: geographic information systems will present a whole new way for ocean data management, analysis, display, *Sea Technol.*, **30**, No. 11, pp. 49–51, 53, 55–56.
BLATOV, A. S., IVANOV, V. A., KOSAREV, A. N., et al. (1983). Medium-scale eddies in the world ocean and their geographical distribution (in Russian), Vestnik MGU (Moscow State University News), *Geography*, **4**, pp. 28–36.
BLUMBERG, A. F., SIGNELL, R. P., JENTER, H. L. (1993). Modelling transport processes in the coastal ocean. *J. Marine Env. Eng.*, **1**, No. 1, pp. 31–52.

BOLSHEV, L. N., SMIRNOV, N. V. (1983). *Tables of Mathematical Statistics* (in Russian), Moscow, Nauka, 416 pp.

BONDARENKO, V. D., MAKOVENKO, E. Z., SVETLISTKIJ, A. M. (1986). Method of synthesis of test polygon network in a selected region of the ocean (in Russian), in *Remote Methods and Techniques for Measurement of Oceanographic Parameters*, Proceedings of the Fifth All-Union Seminar (Moscow, 20–23 September 1983), Moscow, Gidrometeoizdat, pp. 8–11.

BOS, W. G., KONINGS, H., PELLEMANS, A. H. J. M., JANSSEN, L. L. F., VAN SWOL, R. W. (1994). The use of spaceborne SAR imagery for oil slick detection at the North Sea, in Proceedings of EARSeL Workshop on Remote Sensing and GIS for Coastal Zone Management. Rijkswaterstaat Survey Department (24–26 October 1994, Delft, The Netherlands), pp. 57–66.

BOWDEN, K. (1988). *Physical Oceanography of Coastal Waters* (in Russian), Moscow, Mir Publishers, 324 pp.

BOYARINOV, P. M. (1985). Structure and physical peculiarities of large lake fronts (in Russian), in *Rotation of Mass and Energy in Water Basins*, Irkutsk, pp. 57–59.

BROSIN, H.-J., GOHS, L., SEIFERT, T., SIEGEL, H., BYCHKOVA, I., VICTOROV, S., DEMINA, M., LOBANOV, V., LOSINSKIJ, V., SMOLJANITSKIJ, V. (1986). Some investigations on mesoscale phenomena in the southeastern part of the Baltic Sea in the frame of ground-truth experiments, in Proceedings of the Fifteenth Conference of Baltic Oceanographers (Copenhagen, Denmark, October 1986), pp. 109–118.

BROSIN, H.-J., GOHS, L., SEIFERT, T., SIEGEL, H., BYCHKOVA, I. A., VICTOROV, S. V., DEMINA, M. D., LOBANOV, V. YU., LOSINSKIJ, V. N., SMOLJANITSKIJ, V. M. (1988). Mesoskale strukturen in der sudöstlichen Ostsee im Mai 1985. *Beitrage zur Meereskunde*, Heft **58**, s. 9–18.

BROSIN, H.-J., KULLENBERG, G., VOIGT, K. (1994). Cooperation efforts with respect to marine scientific research in the Baltic sea, in R. Platzoder and P. Verlaan, eds, *The Baltic Sea: New Developments in National Policies and International Cooperation*, vol. 2 in series: European workshops on the Law of the Sea, Stiftung Wissenschaft und Politik, Ebenhausen, June 1994, pp. 296–301.

BROSIN, H.-J., VICTOROV, S. V. (1984). Joint complex oceanographic subsatellite experiments of the USSR and GDR on the Baltic Sea, in Proceedings of the Fourteenth Conference of Baltic Oceanographers (Gdynya, Poland, September 1984), **1**, pp. 95–103.

BULATOV, A. S., TUZHILIN, V. S. (1980). Hydrological structure and energetics of medium-scale eddies in the World Ocean (in Russian), in *Problems of Geography*, No. 125, Moscow, Misl Publishers.

BUSHUEV, A. V. (1984). Ice observations in the seas and oceans, studies of sea ice distribution and dynamics (in Russian), in *The Earth's Nature from Space*, compiled by A. Tishenko and S. Victorov, edited by N. Kozlov. Leningrad, Gidrometeoizdat, pp. 107–110.

BYCHKOVA, I. A. (1994). Utilisation d'une méthode d'étalonnage basée sur l'utilisation de surfaces naturelles pour la correction des effets atmosphériques sur les mesures de température de surface de la Mer Baltique, Actes Sixième Symposium International *Mesures Physiques et Signatures en Télédétection* (Val d'Isère, 17–24 January 1994), 773–776.

BYCHKOVA, I., SUKHACHEVA, L., VICTOROV, S. (1994). Multilevel data assimilation in monitoring of the Gulf of Finland, in abstracts of papers presented at the Nineteenth Conference of Baltic Oceanographers (Sopot, Poland, 29 August–1 September), p. 128.

BYCHKOVA, I. A., VICTOROV, S. V. (1985). Studies of the allowed non-synchronicity of observations of various levels in Complex Oceanographic Subsatellite Experiments (in Russian), *Earth Studies from Space*, **6**, pp. 93–100.

 (1987). Elucidation and systematization of upwelling zones in the Baltic Sea based on satellite data (in Russian), *Oceanology*, **XXVII**, 2, pp. 218–223.

(1988). Parameters of eddy Structures and 'mushroom-like' currents in the Baltic Sea as revealed from satellite imagery (in Russian), *Earth Studies from Space*, **2**, pp. 29–35.

BYCHKOVA, I. A., VICTOROV, S. V., DEMINA, M. D. (1989a). Using regular satellite data of infra-red band in studies of thermobar and upwelling phenomena (in Russian), *Oceanology*, **XXIX**, 5, pp. 759–766.

(1989b). Using the method of satellite sensors calibration at natural polygons in regional oceanography (in Russian), in *Oceanography from Space*, USSR Academy of Sciences, Leningrad, Express-Information, 4–89, pp. 3–4.

BYCHKOVA, I. A., VIKTOROV, S. V., DEMINA, M. D., LOBANOV, V. YU., LOSINSKIJ, V. N., SMIRNOV, V. G., SMOLYANITSKIJ, V. M., SUKHACHEVA, L. L., BROSIN, H.-J. (1990a). Experimental satellite monitoring of the Baltic Sea in the 1980s, ICES Paper C.M. 1990 (Hydrography Committee, Session P).

BYCHKOVA, I. A., VICTOROV, S. V., EVDOKIMOV, S. N. (1981). On the use of satellite information of IR band in studies of temperature regime of the Caspian Sea (in Russian), in abstracts of papers presented at the Fourth All-Union Seminar *Remote Methods and Technique for Measurement of Oceanographic Parameters* (Odessa 1981).

(1982a). On the use of satellite information of IR band in studies of hydrological regime of the Seas (in Russian), in abstracts of papers presented at the First All-Union Conference *Biosphere and Climate Studies Based on Spaceborne Observations* (Baku, 29 November–3 December 1982), Baku, ELM, pp. 10–11.

BYCHKOVA, I. A., VICTOROV, S. V., LOSINSKIJ, V. N. (1987). Structure of the coastal fronts of the Baltic Sea as revealed from satellite data of infra-red band (in Russian), in abstracts of papers presented at the Third Congress of Soviet Oceanologists (Leningrad, 14–19 December 1987). Section *Physics and Chemistry of Ocean and Atmosphere, Spaceborne Oceanology*. Leningrad, Gidrometeoizdat, 1988, pp. 64–65.

BYCHKOVA, I. A., VICTOROV, S. V., SHUMAKHER, D. A. (1988a). On a relationship between the large-scale atmospheric circulation and the origin of coastal upwellings in the Baltic Sea (in Russian), *Meteorology and Hydrology*, **10**, pp. 91–98.

BYCHKOVA, I. A., VICTOROV, S. V., SMIRNOV, V. G. (1986a). Some peculiarities of using satellite data of infra-red band in studies of the seas in cloud situations (in Russian), *Earth Studies from Space*, **3**, pp. 60–66.

BYCHKOVA, I. A., VICTOROV, S. V., SMIRNOV, V. G., DUBRA, J. J., BROSIN, H.-J. (1990b). Studies of the river discharge spreading in the south-eastern part of the Baltic Sea on the basis of satellite imagery, in Proceedings of the Seventeenth Conference of the Baltic Oceanographers (Norrkoping, Sweden, September 1990).

BYCHKOVA, I. A., VICTOROV, S. V., VINOGRADOV, V. V. (1985a). Using satellite data for studies of upwelling and frontogenesis in the Baltic Sea (in Russian), *Earth Studies from Space*, **2**, pp. 12–19.

(1988b). *Remote Sensing of Sea Surface Temperature* (in Russian), Leningrad, Gidrometeoizdat, 224 pp.

BYCHKOVA, I. A., VICTOROV, S. V., VINOGRADOV, V. V., LOSINSKIJ, V. N., BROSIN, H.-J. (1985b). Airborne and spaceborne observations of advective eddies in the central part of the Baltic Sea (in Russian), *Earth Studies from Space*, **1**, pp. 118–122.

BYCHKOVA, I. A., VICTOROV, S. V., VINOGRADOV, V. V., SMIRNOV, V. G., VISHNEVSKIJ, A. E., POPOV, S. S., TARASOV, V. S., RODIONOV, D. D., FILIMONOV, V. I. (1986b). Technical facilities of experimental autonomous satellite data acquisition station and technology of digital processing of satellite data of infra-red band for studies of the Baltic Sea (in Russian), in *Remote Methods and Techniques for Measurement of Oceanographic Parameters*, Proceedings of the Fifth All-Union Seminar (Moscow, 20–23 September 1983), Moscow, Gidrometeoizdat, pp. 219–223.

BYCHKOVA, I. A., VINOGRADOV, V. V., SMIRNOV, V. G. (1982b). Use of satellite IR-information in studies of thermal processes at the boundary 'ocean–atmosphere' (in Russian), in Abstracts of papers presented at the First All-Union Conference *Biosphere and Climate Studies Based on Spaceborne Observations* (Baku, 29 November–3 December 1982), Baku, ELM, pp. 8–10.

BYRNE, H. M. (1983). Review summary of the URI satellite sea surface temperature workshop, September 1982, in Proceedings of the OCEANS '83 Conference (San Francisco, USA, 29 August–1 September 1983), **1**, pp. 338–339.

CALKOEN, C. J., KOOI, M. W. A. VAN DER, HESSELMANS, G. H. F. M., WENSINK, G. J. (1993). The imaging of sea bottom topography with polarimetric P-, L-, and C-band SAR. Report BCRS project 2.1/AO-02. Netherlands Remote Sensing Board. Delft.

CARLSTROM, A., ULANDER, L. M. H., HAKANSSON, B. G. (1994). Model for estimating surface roughness of level and ridged sea ice using ERS-1 SAR in proceedings of EARSeL Symposium (Goteborg, Sweden, 1994).

CARSTENS, T., MCCLIMANS, T. A., NILSEN, J. H. (1986). Satellite imagery of boundary currents, in *Remote Sensing of Shelf Sea Hydrodynamics* (ed. J. C. J. Nihoul), Elsevier Oceanogr. Series, 38. Amsterdam.

CARTER, D. T. J., CHALLENOR, P. G., SROKOSZ, M. A. (1988). Satellite remote sensing and wave studies into the 1990s, *Int. J. Remote Sensing*, **9**, No. 10/11, pp. 1835–1846.

CAVALIERI, D. J., BURNS, B. A., ONSTOTT, R. G. (1990). Investigation of the effects of summer melt on the calculation of sea ice concentration using active and passive microwave data, *J. Geophys. Res.*, **95**, No. C4, pp. 5359–5369.

CAYULA, J.-F. (1988). Edge detection for SST images, MS thesis, Department of Electrical Engineering, University of Rhode Island, 91 pp.

CAYULA, J.-F., CORNILLON, P. (1992). Edge detection algorithm for SST images, *Journal of Atmospheric and Ocean Technology*, **9**, pp. 67–80.

CHAMPAGNE, P. M., GUEVEL, D., FROUIN, R. (1982a). Etude du front de Malte: à partir de données de télédétection et de mesures in situ, *Ann. Hydrogr*, **10**, No. 757, pp. 65–98.

CHAMPAGNE, P. M., HARANG, L., VOURCH, J. (1982b). La télédétection des fronts thermiques, *Bull. Soc. Franc. Photogramm. et Télédétect.*, **86**, pp. 5–23.

CHELOMEI, V. N., EFREMOV, G. A., et al. (1990). Sea surface imaging by the high-resolution KOSMOS-1870 radar (in Russian), *Earth Research from Space*, **1**, pp. 64–78.

COAKLEY, J. A. JR. (1983). Properties of multilayer cloud systems from satellite imagery, *J. Geoph. Res.*, **88**, No. 15, pp. 10818–10828.

COAKLEY, J. A., BRETHERTON, F. P. (1982). Cloud cover high-resolution scanner data: detecting and allowing for partially filled fields of view, *J. Geoph. Res.*, **82**, No. C7, pp. 4917–4932.

COASTAL OCEAN PROGRAM COASTWATCH (1994). Leaflet distributed at the John C. Stennis Space Center, Mississippi in January 1994.

CORINE DATA BASE MANUAL (1989). CEC, 15 June 1989.

COSKUN, H. G., ORMECI, C. (1994). Water quality monitoring in Halic (Golden Horn) using satellite images, in Proceedings of EARSeL Workshop on Remote Sensing and GIS for Coastal Zone Management. Rijkswaterstaat Survey Department (24–26 October 1994, Delft, The Netherlands), pp. 84–93.

COUNTWELL, B. J. (1984). Regular motions in turbulent flows (in Russian), in *Vikhri i Volni (Eddies and Waves)*, Moscow, Mir Publishers, pp. 9–79.

CRACKNELL, A. P., HAYES, L. W. B. (1991). *Introduction to Remote Sensing*, Taylor & Francis, London, 293 pp.

CRAIG, T. (1987). *Commonwealth University Year Book*, London: Association of Commonwealth Universities.

CREPON, M., RICHEZ, C., CHARTIER, M. (1984). Effects of coastline geometry on upwelling, *J. Phys. Oceanogr.*, **14**, 8.

CROUT, R. L., ORIOL, R. A., ARNONE, R. A. (1994). An environmental database for the United States West Coast, in Proceedings of the Second Thematic Conference *Remote*

Sensing for Marine and Coastal Environments (New Orleans, USA, 31 January–2 February 1994), **1**, pp. 335–345.

CSANADY, G. T. (1971). On the equilibrium shape of the thermocline on a shore zone, *J. Phys. Oceanogr.*, **1**, pp. 263–270.

DARNELL, W. L., HARRIS, R. C. (1983). Satellite sensing capability for surface temperature and meteorological parameters over the ocean, *Int. J. Remote Sens.*, **1**, pp. 65–92.

DAVIDAN, I. N. (1981). Scientific basis for studies of the Baltic Sea, in Proceedings of the Twelfth Conference of the Baltic Oceanographers and the Seventh Workshop of Experts on Water Balance of the Baltic Sea (Leningrad, USSR, 14–19 April 1980), Leningrad, 1981, pp. 9–12.

DAVIS, B. A., CARR JR., H. V., SCHMIDT, N., HICKERSON, L. P. E. (1994). Opportunities in the commercial use of remote sensing and GIS technologies: an overview of NASA's Visiting Investigator Program at Stennis Space Center. Material distributed at Stennis Space Center in January 1994, 7 p.

DE LISLE, D. A., DRAPEAU, G., LAROUCHE, P., BJERKELUND, C. (1994). Coastal evolution monitoring using remote sensing, in Proceedings of the WMO/IOC Technical Conference on Space-Based Ocean Observations (September 1993, Bergen, Norway), pp. 119–124.

DE VALK, C. F., WENSINK, G. J., VAN SWOL, R. W., VENEMA, J. C. (1990). Extraction of marketable information from ERS-1 data. BCRS Report No. 90-03, June 1990.

DOBLAR, R. A., CHENEY, R. E. (1977). Observed formation of a Gulf Stream cold core ring, *J. Geophys. Res.*, **7**, No. 6, pp. 944-946.

DOBSON, J. E., BRIGHT, E. A. (1991). CoastWatch — detecting change in coastal wetlands, *Geo. Info. Systems*, January, pp. 36–40.

DOBSON, J. E., FERGUSON, R. L., FIELD, D. W., WOOD, L. L., HADDAD, K. D., IREDALE III, H., KLEMAS, V. V., ORTH, R. J., THOMAS, J. P. (1993). *NOAA CoastWatch Change Analysis Project Guidance for Regional Implementation*, CoastWatch Change Analysis Project, Coastal Ocean Program, NOAA, US Department of Commerce.

DOTSENKO, S. V. (1980). Calibration of remote sensors using polygon measurements (in Russian), in *Satellite Hydrophysics*, Sevastopol, 1980, pp. 86–96.

DOTSENKO, S. V., NELEPO, B. A., SALIVON, L. G. (1981). Optimal calibration in the remote sensing of the ocean (in Russian), in *Remote Methods and Techniques for Measurement of Oceanographic Parameters*, Proceedings of the Third All-Union Seminar, Moscow, Gidrometeoizdat, 1981, pp. 101–104.

DRABKIN, V. V., ALLABERT, A. V. (1982). On calculation of economic benefit of the use of airborne and satellite information at the test areas (in Russian), in *Satellite and Aerial Oceanography* (Proceedings of the State Oceanographic Institute), **166**, Leningrad, Gidrometeoizdat, pp. 37–40.

DRABKIN, V. V., ALLABERT, A. V., FLORENSKAYA, M. L. (1982). Objectives and general principles of setting up of oceanographic subsatellite test areas (in Russian), in *Satellite and Aerial Oceanography* (Proceedings of the State Oceanographic Institute), **166**, Leningrad, Gidrometeoizdat, pp. 30–36.

DRAPEAU, G. (1980). Shoreline evolution at the northern end of Iles-de-la-Madelene, in Proceedings of the Canadian Coastal Conference '80, NRC, pp. 294-308.

DRENTH, F., WENSINK, G. J., BOER, S. (1990). Remote sensing for marine activities, BCRS Report No. 89-41, February 1990, 239 pp. plus Appendices (separate volume).

DUBRA, JU. (1978). Vadens balansas (in Lithuanian), in *Kursin marios*, Vilnius, Mokslas, Part II, pp. 50–70.

DUTRIEUX, E., DENIS, J. (1992). Baseline study of the Tambora–Tatan gas field (Mahakam delta, East Kalimantan, Indonesia), Total-Indonesia, 114 pp.

DYADIUNOV, V. N., KRIULKOV, V. A., PRISTAVKO, G. V. (1986). Optimization of network of hydrochemical observation stations at the marine test area (in Russian), in

Remote Methods and Techniques for Measurement of Oceanographic Parameters, Proceedings of the Fifth All-Union Seminar (Moscow, 20–23 September 1983), Moscow, Gidrometeoizdat, pp. 16–20.

DYBERN, B. I., HANSEN, H. P. (eds) (1989). Baltic Sea Patchiness Experiment – PEX 86. International Council for the Exploration of the Sea, Cooperative Research reports No. 163, pp. 100–156.

EARSeL NEWSLETTER (1994). No. 17, pp. 14–16.

EBRD (1994). Materials of the Annual Meeting of the Board of Governors of the European Bank of Reconstruction and Development (St. Petersburg, Russia, 18-19 April 1994).

ELACHI, C. (1987). *Introduction to the Physics and Techniques of Remote Sensing* (Wiley series in remote sensing), Wiley Interscience, John Wiley, NY, 413 pp.

ELKEN, J. (1994). Numerical study of fronts between the Baltic sub-basins, in abstracts of papers presented at the Nineteenth Conference of Baltic Oceanographers (Sopot, Poland, 29 August–1 September), p. 123.

ELKEN, J., KAHRU, M., HANSEN, H.-P., AITSAM, A. (1984). An eddy found east from the Bornholm, in Proceedings of the Fourteenth Conference of Baltic Oceanographers (Gdynya, Poland, September 1984), Paper P-28.

ENVIRONMENTAL MANAGEMENT IN THE BALTIC REGION (1992). Collection of papers presented at International Workshop (Leningrad, USSR, 21–24 November 1989). INENCO Proceedings No. 1, St. Petersburg, 1992, 263 pp.

ENVIRONMENTAL STATISTICS IN EUROPE AND NORTH AMERICA (1987). An Experimental Compendium. UN, New York.

FEDOROV, K. N. (1977). Remote sensing methods of ocean studies (a review) (in Russian), *Progress in Science and Technology, Oceanology*, **4**, pp. 132–165.

 (1980). Expectations and realities of space-based oceanology (in Russian), *Earth Studies from Space*, **1**, pp. 64–78.

FERGUSON, R. L., WOOD, L. L. (1990). Mapping submerged aquatic vegetation in North Carolina with conventional aerial photography, in *Federal Coastal Wetland Mapping Programs*, eds S. J. Kiraly, F. A. Cross, J. D. Buffungton, US Fish and Wildlife Service Biological Report, **90**, 18, pp. 125–133.

FETTERER, F., GINERIS, D., JOHNSON, C. (1993). Remote sensing aids in sea-ice analysis, *EOS, Transactions, American Geophysical Union*, **74**, No. 24, 15 June 1993, pp. 265–268.

FINAL REPORT OF THE INTERNATIONAL COMMISSION OF EXPERTS (1990). Document produced by the Commission set up to work out recommendations on the state of the environment in the Neva Bay in connection with the Flood Barrier construction (in English and Russian).

FU, L. L., HOLT, B. (1982). *Seasat views oceans and sea ice with synthetic aperture radar*. JPL publication 81-120, Pasadena, California, USA.

GARELIK, N. S., GRIN, A. M., TSVETKOV, D. G. (1977). Aerospace test sites, their objectives and ground-based observations (in Russian), in *Space-based Research of the Earth's Resources*, Moscow, pp. 333–347.

GERSON, D. L., GABORSKI, P. (1977). Determination of ocean front positions using digital satellite data, *EOS Trans. Amer. Geophys. Union*, **58**, No. 9, p. 891.

GIDHAGEN, L. (1984). Coastal upwelling in the Baltic – a presentation of satellite and *in situ* measurements of sea surface temperatures indicating coastal upwelling, SMHI Reports Hydrology and Oceanography RHO 37, part I and II, 35, resp., 59 pp.

GIDHAGEN, L., HAKANSSON, B. (1987). A synergetic approach to analyse remotely sensed data within the 1986 Patchiness Experiment (PEX), Paper presented at the Workshop on PEX, April 1987.

GINZBURG, A. I., FEDOROV, K. N. (1984a). The development of mushroom-like currents in the ocean (in Russian), *Doklady Akademii nauk SSSR*, **276**, 2, pp. 481–484.

 (1984b). Mushroom-like currents in the ocean (based on the analysis of satellite images) (in Russian), *Earth Studies from Space*, **3**, pp. 18–26.

GONZALEZ, F. I., BEAL, R. C., BROWN, W. E., et al. (1979). Seasat SAR; ocean wave detection capabilities, *Science*, **204**, No. 4400, pp. 1418–1421.

GOSSMANN, H. (1982). Können Satellitendaten Thermal-befliegungen ersetzen?, *Beitr. Akad. Raumforsch. und Landesplan*, **62**, pp. 69–96.

GOULD, R. W., ARNONE, R. A., MUELLER, J. L. (1994). Optical relationships on Case II waters and their application to coastal processes and remote sensing algorithms, in Proceedings of the Second Thematic Conference *Remote Sensing for Marine and Coastal Environments* (New Orleans, USA, 31 January–2 February 1994), **2**, pp. 283–294.

GREMILLION, T. (1993). MRL Satellite Image Processing System User's Guide (MSIPS), Sverdup Technology, Inc.

GUDDAL, J., STROM, G. D. (1993). Ocean models and needs for ocean satellite data, in Proceedings of the Commission for Marine Meteorology Technical Conference on Ocean Remote Sensing (26 April 1993, Lisbon, Portugal). Report No 28. WMO Technical Document No. 604, pp. 41–52.

GUIDE TO SATELLITE REMOTE SENSING OF THE MARINE ENVIRONMENT (1992). Intergovernmentnal Oceanographic Commission. Manuals and Guides 24, Document SC-92/WS/30, 178 p.

GUIDELINES FOR THE BALTIC MONITORING PROGRAMME FOR THE SECOND STAGE (1984). Baltic Sea Environmental Proceedings No. 12, Helsinki, 251 pp.

GULF OF MEXICO REGIONAL NODE (1994). CoastWatch leaflet distributed at the John C. Stennis Space Center, Mississippi in January 1994.

GUNDLACH, E. R., MAYER, M. O. (1978). Classification of coastal environment in terms of potential vulnerability to oil spill impact, *Marine Technol. Soc. J.*, **12**, pp. 18–27.

GUPTILL, S. C. (1989). Evaluating geographic information systems technology, *Photogramm. Eng. and Remote Sens.*, **55**, No. 11, pp. 1583–1587.

GUREVICH, I. YA., SHIFRIN, K. C. (1982). Physical principles of optical remote sensing methods of detection of oil slicks at the sea surface (in Russian), in *Satellite and Aerial Oceanography* (Proceedings of the State Oceanographic Institute), **166**, Leningrad, Gidrometeoizdat, pp. 96–112.

HAKANSSON, B. (1989). Remote sensing of total suspended matter from the Glomma River in the Skagerrak (in Swedish), *Vatten*, **45**, 4, pp. 271–277.

HAKANSSON, B., MOBERG, M., THOMPSON, T. (1995). Real-time use of ERS-1 imagery for ice service and icebreaking operations in the Baltic Sea. Submitted to *International Journal of Remote Sensing*.

HAKANSSON, B. G., MOBERG, M., THOMPSON, T., BACKMAN, A. (1994). Evaluation of real-time use of ERS-1 SAR imagery for icebreaking operations in the Baltic, in Proceedings of the First ERS-1 Pilot Project Workshop (Toledo, Spain, 22–24 June 1994), Document ESA SP-365, October 1994, pp. 153-156.

HARI, J. (1984). Correlation between the area of increased temperature around a thermal power plant and the ice conditions of the Gulf of Finland, in Proceedings of the Fourteenth Conference of Baltic Oceanographers (Gdynya, Poland, September 1984), Paper P-22.

HEMPEL, G. (1991). Introduction – The European Committee on Ocean and Polar Sciences (ECOPS). European Science Foundation, The Ocean and the Poles. European cooperation in Ocean and Polar Research, pp. 2–4.

HESSELMANS, G. H. F. M. (1990). Bathymetry in Indonesian coastal waters: a first pilot project, Delft Hydraulics Report H 964, November 1990.

HESSELMANS, G. H. F. M., WENSINK, G. J. (1994). The use of optical satellite observations to optimize ship-based sand inventories in coastal areas, in Proceedings of EARSeL Workshop on Remote Sensing and GIS for Coastal Zone Management. Rijkswaterstaat Survey Department (24–26 October 1994, Delft, The Netherlands), pp. 139–146.

HESSELMANS, G. H. F. M., WENSINK, G. J., CALKOEN, C. J. (1994a). The use of optical and SAR observations to assess bathymetric information in coastal areas, in Proceedings of the Second Thematic Conference *Remote Sensing for Marine and Coastal Environments* (New Orleans, USA, 31 January–2 February 1994), **1**, pp. 215–224.

(1994b). The use of ERS-1 SAR data to support bathymetric surveys, in Proceedings of EARSeL Workshop on Remote Sensing and GIS for Coastal Zone Management. Rijkswaterstaat Survey Department (24–26 October 1994, Delft, The Netherlands), pp. 147–154.

HILLEN, R., DE HAAN, TJ. (1993). Development and implementation of the coastal defence policy for the Netherlands, in Hillen and Verhagen (eds), *Coastlines of the Southern North Sea*, ASCE, New York.

HOEPFFNER, N., FERRARI, G. M., SCHLITTENHARDT, P. (1994). Study of the marine processes in the Baltic: a joint effort between IRSA/NE and other institutes for the use of satellite data, in Abstracts of papers presented at the Nineteenth Conference of Baltic Oceanographers (Sopot, Poland, 29 August–1 September), p. 92.

HOGG, J., GAHEGAN, M. (1986). Regional analysis using geographic information systems based on linear quadtrees, in *Mapping from Modern Imagery*, Proc. of Intern. Symp. Comis. IV Intern. Soc. Photogram. Remote Sensing, 1986, pp. 113–122.

HORSTMANN, U. (1983). Distribution patterns of temperature and water color in the Baltic Sea as recorded in satellite images: indicators for phytoplankton growth, *Berichte aus dem Christian Albrecht Universität Kiel*, 106, **1**, 147 pp.

(1986). The use of satellite data for the monitoring of the Upper Pelagial in the Baltic Sea, abstract of paper presented at the Conference on Baltic Sea Monitoring System, Tallinn, USSR, 10–15 March 1986, in *Baltic Sea Environment Proceedings*, No. 19. Helsinki, HELCOM, p. 547.

(1988). Satellite remote sensing for estimating coastal offshore transports, in *Lecture Notes on Coastal and Estuarine Studies*, **22**, B.-O. Jansson (ed.), *Coastal-Offshore Ecosystems Interactions*, Springer Verlag, Berlin, Heidelberg, pp. 50–66.

HORSTMANN, U., HARDTKE, P. G. (1981). Transport processes of suspended matter, including phytoplankton, studied from Landsat images of the southwestern Baltic Sea, in *Oceanography from Space* (ed. J. F. R. Gower), Plenum Press Corp., pp. 429–438.

HORSTMANN, U., ULBRICHT, K. A., SCHMIDT, D. (1978). Detection of eutrophication processes from air and space, in Proceedings of the Twelfth International Symposium on Remote Sensing of Environment, Manila, Philippines, pp. 1379–1389.

HORSTMANN, U., VAN DER PIEPEN, H., BARROT, K. W. (1986). The influence of river water on the southeastern Baltic Sea as observed by Nimbus 7/CZCS, *AMBIO*, **15**, 5, pp. 286–289.

HUGHES, B. A., GASPAROVICH, R. F. (eds) (1988). Georgia Strait and SAR internal wave signature experiments, in *J. Geophys. Res.* Special Section, pp. 12217–12345.

HUPFER, P. (1982). *Baltic – Small Sea, Big Problems* (in Russian), Leningrad, Gidrometeoizdat, 136 pp.

HYDROGRAPHICAL OBSERVATIONS ON SWEDISH LIGHTSHIPS IN 1951–1959, 1953–1961, Elanders Boktr. Akt. Goteborg.

IKEDA, M., EMERY, W. J. (1984). Satellite observations and modelling of meanders in the California Current System off Oregon and Northern California, *J. Phys. Oceanogr.*, **14**, 9, pp. 1434–1450.

IKEDA, M., MYSAK, L. A., EMERY, W. J. (1984). Observation and modelling of satellite-sensed meanders and eddies off Vancouver Island, *J. Phys. Oceanogr.*, **14**, 1, pp. 3–21.

ILYIN, YU. P., LEMESHKO, E. M. (1994). Quantitative description of Black Sea coastal zone dynamical processes by satellite multi-time images, in Proceedings of EARSeL Workshop on Remote Sensing and GIS for Coastal Zone Management. Rijkswaterstaat Survey Department (24–26 October 1994, Delft, The Netherlands), pp. 181–190.

INTERNAL REPORT (1984). On the regime and some features of the Baltic Sea ecosystem as revealed from satellite imagery, State Oceanographic Institute, Leningrad, 119 pp.

Irwin, P. A., Manley, T. O. (1994). Volumetric visualisation: an effective use of GIS technology in the field of oceanography, in Proceedings of Conference *Oceanology International '94* (8–11 March 1994, Brighton, UK), vol 2.

Ismailov, T. K. (1980). Development of methods and equipment for subsatellite observations (in Russian), *Earth Studies from Space*, **1**, pp. 35–39.

Ivanov, M. F., Gerbek, E. E., Kazanskij, A. V. (1983). Software and equipment for producing charts of surface temperature of the ocean using satellite data (in Russian), in Proceedings of the Seventh Conference on Cosmonautics, *Studies of the Earth's Natural Resources from Space*, Moscow, IHST, USSR Academy of Sciences, pp. 185–189.

Ivanov, M. F., Kazanskij, A. V., Victorov, S. V. (1986). Shipborne systems for acquisition and processing of satellite remote sensing measurements (in Russian), in *Science and Technology Review, VINITI, Atmosphere, Ocean, Space*, Program 'Razrezi', Moscow, **7**, pp. 189–199.

Jendro, L. M., Bernstein, R. L. (1994). The US Coast Guard's initial use of a modern environmental satellite receiving system on polar icebreakers, in Proceedings of the Second Thematic Conference *Remote Sensing for Marine and Coastal Environments* (New Orleans, USA, 31 January–2 February 1994), **2**, pp. 201–212.

Jensen, J. R., et al. (1990). Environmental sensitivity index (ESI) mapping for oil spills using remote sensing and geographic information system technology, *Int. J. Geog. Inf. Syst.*, **4**, No. 2, pp. 181–201.

Johnston, J. B., Summers, K., Bourgeois, P. E., Sclafani, V. (1994). The use of GIS technology to support environmental monitoring and assesssment programs: estuaries activities, in Proceedings of the Second Thematic Conference *Remote Sensing for Marine and Coastal Environments* (New Orleans, USA, 31 January–2 February 1994), **1**, pp. 126–130.

Kagan, B. A. (1978). *Tracers in the World Ocean* (in Russian), Leningrad, Gidrometeoizdat, 58 pp.

Kahru, M. (1981a). Variability of 3-dimensional structure of chlorophyll field in the Baltic Proper (in Russian), *Oceanology*, **21**, 4, pp. 685–690.

(1981b). Spatial variability of chlorophyll concentration in the coastal waters of the Baltic Sea (in Russian), *Oceanology*, **21**, 5, pp. 879–881.

(1986). Vertical inhomogeniety and the feasibility of remote sensing measurements of chlorophyll in the Baltic Sea (in Russian), *Oceanology*, **XXVI**, 4, pp. 667–672.

Kahru, M., Hakansson, B., Rud, O. (1995). Distributions of the sea-surface temperature fronts in the Baltic Sea as derived from satellite imagery, *Continental Shelf Research*, **15**, No. 6, pp. 663–679.

Kahru, M., Leppanen, J.-M., Rud, O. (1993). Cyanobacterial blooms cause heating of the sea surface, *Marine Ecology Progress Series*, **101**, pp. 1–7.

Kalmykov, A. I., et al. (1984). Side-looking radar onboard satellite 'KOSMOS-1500' (in Russian), *Earth Research from Space*, **5**, pp. 84–93.

Karpov, A. (1991). *Soviet Geostationary Operational Satellite GOMS: current status and perspectives for wind data extraction*, in Proceedings of the EUMETSAT/NOAA/WMO Workshop on wind extraction from operational meteorological satellite data, Washington, DC, 1991, p. 39.

Kazmin, A. S., Sklyarov, V. E. (1981). Use of video-information from 'Meteor' satellites in the studies of oceanic phenomena (in Russian), *Earth Studies from Space*, No. 6, pp. 48–57.

Kielmann, J. (1982). Grundlage und Anwendung eines numerischen Modells der geschichteten Ostsee, *Berichte Institut fur Meereskunde Universität Kiel*, No. 87 a, b. 115, resp., 158 pp.

Klemas, V. V., Dobson, J. E., Ferguson, R. L., Haddad, K. D. (1994). A coastal land cover classification system for use with remote sensors, in Proceedings of the Second Thematic Conference *Remote Sensing for Marine and Coastal Environments* (New Orleans, USA, 31 January–2 February 1994), **1**, pp. 131–141.

KLEMAS, V. V., THOMAS, J. P., ZAITZEFF (eds) (1987). *Remote Sensing of Estuaries, Proceedings of a Workshop*, US Department of Commerce, NOAA and US Government Printing Office, Washington, DC.

KOCHAROV, G. E., VICTOROV, S. V., KOVALEV, V. P., et al. (1975). Variations of lunar surface chemical composition in the contact zone 'mare-highland' (Report to XVII session of COSPAR, 1974), in *Space Research*, Akademie Verlag, Berlin, XV, p. 587.

KOCHAROV, G. E., VICTOROV, S. V., VOROPAEV, O. M., et al. (1972). Investigation of the chemical composition of lunar surface along the route of 'Lunokhod-1' (Report to XIV session of COSPAR, 1971), in *Space Research*, Akademie Verlag, Berlin, XII, p. 13.

KONDRATYEV, K. YA. (1983). *Satellite-borne Climatology* (in Russian), Leningrad, Gidrometeoizdat, 263 pp.

(1987). International Geosphere–Biosphere Program (IGBP): the role of spaceborne observations (in Russian), *USSR Academy of Sciences Express-Information*, Leningrad, **4**, pp. 3–25.

KONDRATYEV, K. YA., BOBYLEV, L. P. (1994). The state of the problem of ecological monitoring of the city of St. Petersburg and its region, in *Remote Sensing and Global Climate Change* (ed. A. Vaughan, A. P. Cracknell). Proceedings of the NATO Advanced Study Institute Summer School on Remote Sensing and Global Climate Change (Dundee, Scotland, July 19–August 8, 1992). NATO ASI Series I, **24**. Springer-Verlag, pp. 445–458.

KONDRATYEV, K. YA., CRACKNELL, A. P., VAUGHAN, R. A. (1994). Politics and climate change, in *Remote Sensing and Global Climate Change* (ed. A. Vaughan, A. P. Cracknell). Proceedings of the NATO Advanced Study Institute Summer School on Remote Sensing and Global Climate Change (Dundee, Scotland, July 19–August 8, 1992). NATO ASI Series I, **24**. Springer-Verlag, pp. 465–473.

KONDRATYEV, K. YA., POKROVSKY, O. M. (1978). Planning of multipurpose experiments on remote indication of parameters of the environment and the natural resources (in Russian), *USSR Academy of Sciences Izvestiya, Geography*, **3**, pp. 83–89.

KONDRATYEV, K. YA., VICTOROV, S. V. (1989). Some problems of the environmental monitoring in the 'Baltic Europe', Paper presented at the International Workshop *Environmental Management in the Baltic Region* (Leningrad, USSR, November 1989).

KOZLOV, V. F., MAKAROV, V. G. (1985). A hydrodynamical model of the development of mushroom-like currents in the ocean (in Russian), *Doklady Akademii nauk SSSR*, **281**, pp. 1213–1215.

KRAMER, H. J. (1994). *Observation of the Earth and Its Environment. Survey of Missions and Sensors*, second edition, Springer-Verlag.

KULLENBERG, G. (ed.) (1984). Overall report on the Baltic Open Sea Experiment 1977 (BOSEX). International Council for the Exploration of the Sea. Cooperative Research Reports No. 127, 82 pp.

KUMARI, B., SOLANKI, H. U., RAMAN, M., NARAIN, A. (1994). Role of remote sensing in effective utilisation of oceanographic features for long term fisheries forecast: a case study in the northwestern waters of India, in: Proceedings of the Second Thematic Conference *Remote Sensing for Marine and Coastal Environments* (New Orleans, USA, 31 January–2 February 1994), **1**, pp. 105–111.

LANDMARK, F., HAMNES, H. (1993). Near real time services for marine applications: commercial aspects, in Proceedings of International Symposium *Operationalization of Remote Sensing* (19–23 April 1993, Enschede, The Netherlands), **7**, pp. 111–116.

Landscape Ecology and Geographical Information Systems, 1993, ed. Young, R. H., Green, D., Cousins, S. Taylor & Francis, 296 pp.

LANGRAN, G. (ed.) (1992). *Time in Geographic Information Systems*, Taylor & Francis, 180 pp.

LAURITSON, L., NELSON, G. J., PORTO, F. W. (1979). Data extraction and calibration of Tiros-N/NOAA Radiometers. NESS-107. Technical Memorandum. Washington, DC, 58 pp.

LEBEDINTSEV, A. A. (1910). *Hydrological and Hydrochemical Studies in the Eastern Part of the Baltic Sea in August–September 1908* (in Russian), St. Petersburg, M. P. Frolov Publisher.
LEE, T. F., ATWATER, S., SAMUELS, C. (1993). Sea ice-edge enchancement using polar-orbiting environmental satellite data, *Weather and Forecasting*, September 1993.
LEGECKIS, R. (1977). Detection and interpretation of oceanic fronts from satellite sea surface temperature measurements, *EOS Trans. Amer. Geophys. Union*, **58**, No. 9, p. 890.
  (1978). A survey of worldwide sea surface temperature fronts detected by environmental satellites, *J. Geophys. Res.*, **83**, No. C9, p. 4501.
LEGECKIS, R., BANE, J. M. (1983). Comparison of the Tiros-N satellite and aircraft measurement of Gulf Stream surface temperatures, *J. Geoph. Res.*, **83**, No. C8, pp. 4569–4577.
LOTZ-IWEN, H.-J. (1994). Online access to remote sensing data and ancillary information for global environmental monitoring, in Proceedings of the Second Thematic Conference *Remote Sensing for Marine and Coastal Environments* (New Orleans, USA, 31 January–2 February 1994), **2**, pp. 403–410.
LUKASHEVICH, E. L. (1994). The space system 'RESURS-F' for the photographic survey of the Earth, *Space Bulletin*, **1**, 4, pp. 2–4.
LYON, K. J., WILLARD, M. R. (1993). SeaStar TM: a private sector operational remote sensing satellite system, in Proceedings of the International Symposium *Operationalization of Remote Sensing* (19–23 April 1993, Enschede, The Netherlands), **7**, *Operationalization of Remote Sensing for Coastal and Marine Applications*, pp. 129–140.
MAKTAV, D., KAPDASLI, S. (1994). An investigation of the capability of remote sensing technology for the coastal areas, in Proceedings of EARSeL Workshop on Remote Sensing and GIS for Coastal Zone Management. Rijkswaterstaat Survey Department (24–26 October 1994, Delft, The Netherlands), pp. 223–229.
MALAY, J. T., BROWN, D. N., MCCANDLESS, JR, S. W. (1983). Space-based ocean remote sensing — capabilities and deficiences in the 1980s, in Proceedings of the Conference OCEANS '83 (San Francisco, USA, 29 August–1 September 1983), **1**, pp. 326–330.
MANUALS ON COMPLEX USE OF SATELLITE INFORMATION FOR STUDIES OF THE SEAS (1987). (In Russian), Leningrad, Gidrometeoizdat, 144 pp.
MARBLE, D. F. (ed.) (1980). *Computer Software for Spatial Data Handling*, Ottawa, **1**, pp. 1–272; **2**, pp. 273–576; **3**, pp. 577–1042.
MATTHAUS, W. (1987). The history of the Conference of Baltic Oceanographers, *Beitrage zur Meereskunde*, Berlin, **57**, pp. 11–25.
MAUL, G. A., BAIG, S. R. (1977). Fluctuations of the Gulf Stream front as inferred from infrared satellite measurements, *EOS Trans. Amer. Geophys. Union*, **58**, No. 9, p. 899.
MCCLAIN, E. P. (1985). Overview of satellite data applications in the Climate and Earth Science Laboratory of NOAA's NESDIS, in Proceedings of US–India Symposium cum Workshop on Remote Sensing Fundamentals and Applications, Library: Space Applications Centre, Ahmedabad, India, p. 247.
MCFARLAND, B. A., HALL, A. J. (1994). Observer training: an integral part of remote sensing at oil spills, in Proceedings of the Second Thematic Conference *Remote Sensing for Marine and Coastal Environments* (New Orleans, USA, 31 January–2 February 1994), **2**, pp. 393–402.
METALNIKOV, A. P., TOLKACHEV, A. YA. (1985). Program 'Razrezi' (Energy-Active Zones of the Ocean and Climate Change) (in Russian), in *Complex Global Monitoring of the World Ocean*, Proceedings of the First International Symposium (Tallinn, 2–10 December 1983), Leningrad, Gidrometeoizdat, pp. 132–140.
MICK, M., ALEXANDER, T. M., WOOLEY, S. (1994). Mississippi high accuracy reference network and coastal remote sensing, in Proceedings of the Second Thematic Conference *Remote Sensing for Marine and Coastal Environments* (New Orleans, USA, 31 January–2 February 1994), **1**, pp. 237–247.

MILLOT, C. (1982). Analysis of upwelling in the Gulf of Lions, in *Hydrodynamics of Semi-Enclosed Seas*, Proceedings of the Thirteenth International Liege Colloquium on Ocean Hydrodynamics, pp. 143–153.

MOBERG, M., HAKANSSON, B. G. (1993). Monitoring of remotely sensed blue–green phytoplankton in the Baltic Proper during Summer 1991, *Nordisk Romvirksomhet*, No. 5, pp. 14–16.

MOHAMED, M. I. HJ. (1994). Remote sensing applications to environmental monitoring of the coastal zone, in *Remote Sensing and Global Climate Change* (ed. A. Vaughan, A. P. Cracknell), Proceedings of the NATO Advanced Study Institute Summer School on Remote Sensing and Global Climate Change (Dundee, Scotland, 19 July–8 August 1992). NATO ASI Series I, **24**, Springer-Verlag, pp. 429–438.

MORGAN, C. W., BISHOP, J. M. (1977). An example of Gulf Stream eddy-induced water exchange in the Mid-Atlantic Bight, *J. Phys. Oceanogr.*, **7**, No. 3, pp. 427–429.

MULDER, N. J. (1994). Progress in the integration of 4-dimensional GIS and RS image analysis, applied to coastal zone monitoring, in Proceedings of EARSeL Workshop on Remote Sensing and GIS for Coastal Zone Management. Rijkswaterstaat Survey Department (24–26 October 1994, Delft, The Netherlands), pp. 241–249.

NAGLER, R. G., MCCANDLESS, S. W., JR. (1975). *Operational Oceanographic Satellites. Potentials for Oceanography, Coastal Processes and Ice*, JPL/NASA, 13 pp.

NARAYANA, A. (1992). Normalisation of multidate digital remote sensing data using scene statistics, *Photonirvachak, Journal of the Indian Society of Remote Sensing*, **20**, Nos 2 & 3, pp. 165–172.

NAULT, J. M. (1994). Image compression technique for NOAA operational satellite image browse capability, in Proceedings of the Second Thematic Conference *Remote Sensing for Marine and Coastal Environments* (New Orleans, USA, 31 January–2 February 1994), **2**, pp. 423–429.

NAZIROV, M. (1982). *Ices and Suspended Substances as Hydrodynamical Tracers* (in Russian), Leningrad, Gidrometeoizdat, 159 pp.

NEEDHAM, B. H. (1983). The effective use of NOAA satellite data by the marine community, in Proceedings of the Conference OCEANS '83 (San Francisco, USA, 29 August–1 September 1983), **1**, pp. 138–141.

NELEPO, B. A. (1979). Space-based oceanography: problems and perspectives (in Russian), in *Problems of Exploration and Mastery of the World Ocean*, Leningrad, pp. 111–133.

—— (1980). Current problems of satellite oceanology (in Russian), *Earth Studies from Space*, **1**, pp. 55–63.

NELEPO, B. A., ARMAND, N. A., KHMYROV, B. E., et al. (1982). Experiment 'Ocean' onboard satellites 'KOSMOS-1076' and 'KOSMOS-1151' (in Russian), *Earth Research from Space*, **3**, pp. 5–12.

NELEPO, B. A., BULGAKOV, N. P., BLATOV, A. S., et al. (1984). Systematization of synoptic eddy structures in the world ocean (in Russian), Preprint, MGI AN Ukraine SSR, Sevastopol, 40 pp.

NELEPO, B. A., DOTSENKO, S. V., POPLAVSKAYA, M. G. (1976). Optimal reconstruction of the field using remote sensing data (in Russian), *Hydrophysical Research*, **2**, pp. 46–55.

NELEPO, B. A., GRISHIN, G. A., et al. (1986). *Optical Methods of Satellite Hydrophysics* (in Russian), Part 1, Kiev, Naukova Dumka, 157 pp., Part 2, Kiev, Naukova Dumka, 143 pp.

NELEPO, B. A., KOROTAEV, G. K. (1985). Satellite monitoring of ocean climate (in Russian), in *Complex Global Monitoring of the World Ocean*, Proceedings of the First International Symposium (Tallinn, 2–10 December 1983), Leningrad, Gidrometeoizdat, pp. 163–171.

NELEPO, B. A., TEREKHIN, YU. V., et al. (1983). *Satellite Hydrophysics* (in Russian), Moscow, Nauka, 253 pp.

NIELSEN, A., HANSEN, P. (1983). Plankton distribution as shown by satellite pictures,

Paper presented at the Workshop on the Patchiness Experiment in the Baltic Sea, (Tallin, March 1983).

NIERENBERG, W. A. (1980). Oceanography from space: looking back twenty years – looking forward ten. COSPAR/SCOR/IUCRN Symposium *Oceanography from Space*, Venice, Italy.

NIILER, P. P. (1977). *Le Gulf Stream La Recherche*, **79**, pp. 517–526.

NIKANOROV, V. A., SVETLITSKIJ, A. M. (1986). The ways to manage subsatellite measurements to obtain standard hydrophysical fields at test polygon (in Russian), in *Remote Methods and Techniques for Measurement of Oceanograhic Parameters*, Proceedings of the Fifth All-Union Seminar (Moscow, 20–23 September 1983), Moscow, Gidrometeoizdat, pp. 4–7.

NIKITIN, P. A. (1991). The sea ice satellite monitoring, Proceedings of the Fifth AVHRR Data Users Meeting, Tromso, 25–28 June 1991.

NOAA's COASTAL ASSESSMENT FRAMEWORK (1994). Leaflet distributed at the John C. Stennis Space Center, Mississippi in January 1994.

NOAA COASTWATCH (1993). Informational flyer, December 1993.

NOVOGRUDSKIJ, B. V., SKLYAROV, V. E., FEDOROV, K. N., SHIFRIN, K. S. (1978). *Studies of the Ocean from Space (A Review)* (in Russian), Leningrad, Gidrometeoizdat, 54 pp.

OCEAN REMOTE SENSING GROUP MEETS (1994). *International Marine Science Newsletter*, No. 72, fourth quarter 1994, p. 7, UNESCO.

OCEAN SCIENCE FOR THE YEAR 2000 (1984). IOC, UNESCO, Paris, 95 pp.

OGURTSOV, A. P. (1988). *Disciplinary Structure of Science* (in Russian), Moscow, Nauka, 256 pp.

ORIOL, R. A., MARTINOLICH, P. M., ARNONE, R. A. (1994). Development of an 800 meter (m) bio-optical database from satellite ocean color for applications in coastal processes, in Proceedings of the Second Thematic Conference *Remote Sensing for Marine and Coastal Environments* (New Orleans, USA, 31 January–2 February 1994), **2**, pp. 271–280.

OSADCHIJ, V. YU. (1983). Experimental Optical Study of Oil Slicks at the Sea Surface (in Russian), Ph.D. Thesis, USSR Academy of Sciences P. P. Shirshov Institute of Oceanology, Moscow.

OSADCHIJ, V. YU., SHIFRIN, K. S. (1979). Laboratory device for study of the brightness coefficient of the pure sea water and the surface covered with oil slicks (in Russian), in *Optical Methods of Study of Oceans and Inland Waters*, Novosibirk, Nauka, pp. 199–204.

OSTROM, B. (1976). Fertilization of the Baltic Sea by nitrogen-fixation in the blue-green alga *Noduralia spumigena*, *Remote Sensing of Environment*, **4**, pp. 305–310.

PAPPEL, I. K. (1983). Vertical structure of the sea in coastal zone (in Russian), in *Study of Variability of Optical Properties of the Baltic Sea*, Tallinn, Valgus, pp. 144–168.

PATHAK, P. N. (1982). Comparison of sea surface temperature observations from Tiros-N and ships in the North Indian Ocean during MONEX (May–July 1979), *Rem. Sens. Env.*, **12**, No. 5, pp. 363–369.

PECK, T. M., SWEET, R. J. M., SOUTHGATE, H. N., BOXALL, S., MATTHEWS, A., NASH, R., AIKEN, J., BOTTRELL, H. (1994a). COAST (Coastal Earth Observation Application for Sediment Transport), in Proceedings of EARSeL Workshop on Remote Sensing and GIS for Coastal Zone Management. Rijkswaterstaat Survey Department (24–26 October 1994, Delft, The Netherlands), pp. 260–266.

PECK, T. M., WARD, S., SWEET, R. J. M. (1994b). COAST (Coastal Earth Observation Application for Sediment Transport), in Oceanology International '94. Conference Proceedings (8–11 March 1994, Brighton, UK), vol. 2.

PERERA, L. K., TATEISHI, R. (1995). Do remote sensing and GIS have a practical applicability in developing countries? (including some Sri Lankan experiences), *Int. J. Remote Sensing*, **16**, No. 1, pp. 35–51.

PICHEL, W., WEAKS, M., SAPPER, J., TADEPALLI, K., JANDHAYALA, KETINENI, S. (1991). Satellite mapped imagery for CoastWatch, in *Coastal Zone '91*. Proceedings of the Seventh Symposium on Coastal and Ocean Management (ASCE/Long Beach, CA, 8–12 July 1991), 2531–2545.

POPULUS, J., MOREAU, F., CODQUELET, D., XAVIER, J.-P. (1994). An assessment of environmental sensitivity to marine pollution: solutions with remote sensing and geographic information systems, in Proceedings of the Second Thematic Conference *Remote Sensing for Marine and Coastal Environments* (New Orleans, USA, 31 January–2 February 1994), **1**, p. 461–472.

Processes of Sedimentation in the Gdansk Basin (Baltic Sea) (1987). (In Russian), Atlantic branch of USSR Academy of Sciences Institute of Oceanology, Moscow, 275 pp.

Radar Studies of the Earth from Space (1990). (In Russian), compiled by M. Nazirov, A. Pichugin and Yu. Spiridonov, ed. L. Mitnik and S. Victorov, Leningrad, Gidrometeoizdat, 200 pp.

RADZIEJEWSKA, T., SHIMMIELD, G. (1994). Historical and recent records of River Oder/Pdra discharge in sediments and biota of the Pomeranian Bay and Southeastern Baltic coastal lagoons, in abstracts of papers presented at the Nineteenth Conference of Baltic Oceanographers (Sopot, Poland, 29 August–1 September), p. 177.

REMOTE SENSING FROM RESEARCH TO OPERATION (1992). Proceedings of the eighteenth Annual Conference of the Remote Sensing Society (University of Dundee, 15–17 September 1992), ed. A. P. Cracknell and R. A. Vaughan.

REPORT ON SATELLITE SYSTEMS AND CAPABILITIES (1995). Joint WMO/IOC (CMM-IGOSS-IODE) Subgroup on Ocean Satellites and Remote Sensing (in press).

REPORT OF THE TOGA WORKSHOP ON SEA SURFACE TEMPERATURE AND NET SURFACE RADIATION (1984) (La Jolla, California, USA, 28–30 March 1984). Paper WCP-92, December 1984.

RICHARDSON, P. L. (1980). Gulf Stream ring trajectories, *J. Geophys. Res.*, **10**, pp. 90–114.

RINNE, I., MELVASALO, T., et al. (1981). Studies of nitrogen fixation in the Baltic Sea (in Russian), in Proceedings of the Thirteenth Conference of Baltic Oceanographers, Leningrad, Gidrometeoizdat, pp. 389–396.

ROBAKIEWICZ, M., KARELSE, M. (1994) Hydrodynamics of Gdansk Bay by 3D model, in abstracts of papers presented at the Nineteenth Conference of Baltic Oceanographers (Sopot, Poland, 29 August–1 September), p. 126.

ROCZNIK HYDROGRAPHICZNY MORZA BALTYCKIEGO (1951–1970) (1953–1972). Warszawa.

RODENHUIS, G. S. (1992). The St. Petersburg Barrier, paper presented at the International Seminar *Ecological Problems of Metropolitan Areas: Key Indicators, Modelling and Decision Making*, Centre INENCO (St. Petersburg, Russia, 17–21 February 1992), 35 pp.

ROOZEKRANS, J. N. (1993). The operational use of NOAA-AVHRR image-products in the marine environment, in Proceedings of International Symposium *Operationalization of Remote Sensing* (19–23 April 1993, Enschede, The Netherlands). Volume 7, pp. 141–149.

RUD, O., KAHRU, M. (1994). Long-term series of NOAA AVHRR imagery reveals large interannual variations of surface cyanobacterial accumulations in the Baltic Sea, in Proceedings of the EARSeL workshop *Remote Sensing and GIS for Coastal Zone Management* (Delft, The Netherlands, 24–26 October 1994), pp. 287–293.

RUFENACH, C. L., SCHUMAN, R. A., MALINAS, N. P., JOHANNESSEN, J. A. (1991). Ocean wave spectral distorsion in airborne synthetic aperture radar imagery during the Continental Shelf Experiment of 1988, *J. Geophys. Res.*, **96**, No. C6.

SALGANIK, P. O., EFREMOV, G. A., et al. (1990). Earth survey from the KOSMOS-1870 radar (in Russian), *Earth Research from Space*, **1**, pp. 70–79.

SANDVEN, S., JOHANNESSEN, O. M., KLOSTER, K. (1993). Pre-operational use of Synthetic Aperture Radar (SAR) images from the ERS-1 satellite in sea ice monitoring, in Proceedings of International Symposium *Operationalization of Remote Sensing* (19–23

April 1993, Enschede, The Netherlands). Volume 7, pp. 161-172.
SCHMITZ, J. E., VASTANO, A. C. (1977). Decay of a shoaling Gulf Stream cyclonic ring, *J. Phys. Oceanogr.*, **7**, No. 3, pp. 479-481.
SERAFIN, R., ZALESKY, J. (1988). Baltic Europe, Great Lakes America and ecosystem redevelopment, *AMBIO*, **17**, No. 2, pp. 99-105.
SHERMAN, J. W. III (1993). Ocean satellite programmes in North America, in Proceedings of the Commission for Marine Meteorology Technical Conference on *Ocean Remote Sensing* (26 April 1993, Lisbon, Portugal). Report No. 28. WMO Technical Document No. 604, pp. 29-38.
 (1994). Where is satellite oceanography going? *International Marine Science Newsletter*, No. 69/70, 1st semester 1994, p. 14. UNESCO.
SHUBERT, L. E. (ed.) (1984). *Algae as Ecological Indicators*, Academic Press, 434 pp.
SINGHASANEH, P. (1994). Marine surveillance and information system, in *Remote Sensing and Global Climate Change* (ed. A. Vaughan, A. P. Cracknell). Proceedings of the NATO Advanced Study Institute Summer School on Remote Sensing and Global Climate Change (Dundee, Scotland, July 19-August 8, 1992). NATO ASI Series I, **24**, Springer-Verlag, pp. 439-443.
SLOGGETT, D. R. (1994). An operational, satellite-based infrastructure for the monitoring of oil slicks in European maritime basins (abstract), in Proceedings of EARSeL Workshop on Remote Sensing and GIS for Coastal Zone Management. Rijkswaterstaat Survey Department (24-26 October 1994, Delft, The Netherlands), p. 296.
SMED, J. (1990). Hydrographic investigations in the North sea, the Kattegat and the Baltic before ICES, *Deutsche Hydrographische Zeitschrift, Enganzungsheft Reihe B*, **22**, pp. 357-365.
SORENSEN, B. M. (1979). *Recommendations from the International Workshop on Remote Sensing Sea Truth Data*, International Council for the Exploration of the Sea Council Meeting, Warsaw, Poland, Paper C: 17, 8 pp.
SPACE ARROW (1974). (In Russian), Moscow, 328 pp.
SPENCE, T., LEGECKIS, R. (1981). Satellite and hydrographic observations of low-frequency wave motions associated with a cold core Gulf Stream ring, *J. Geoph. Res.*, **86**, No. C3, pp. 1945-1953.
SPITZER, D., DIRKS, R. W. (1987). Bottom influence on the reflectance of the sea, *Int. J. Remote Sensing*, **8**, No. 3, pp. 279-290.
STERN, M. (1995). Resurs-01. The Russian Monitoring Satellite for the Swedish Space Corporation. Remote Sensing (information from the Swedish Space Corporation), No. 26, February 1995, pp. 18-19.
STEWART, R. H. (1984). Oceanography from space, *Annual Review of the Earth and Planetary Sciences*, **12**, pp. 61-82.
STJERNHOLM, M., WINDOLF, J. (1994). Embedding remote sensing in environmental management in Denmark, in Proceedings of EARSeL Workshop on Remote Sensing and GIS for Coastal Zone Management, Rijkswaterstaat Survey Department (24-26 October 1994, Delft, The Netherlands), pp. 297-302.
STUMPF, R. P., GELFENBAUM, G., PENNOCK, J. R., SCHROEDER, W. W. (1994). Multi-year remote sensing observations of the Alabama-Mississippi Coast (summary), in Proceedings of the Second Thematic Conference *Remote Sensing for Marine and Coastal Environments* (New Orleans, USA, 31 January-2 February 1994), **1**, p. 181.
STURM, B. (1994). Water optical models: their role in the interpretation of ocean colour data and primary productivity analysis, in Abstracts of papers presented at the Nineteenth Conference of Baltic Oceanographers (Sopot, Poland, 29 August-1 September), p. 42.
SUKHACHEVA, L. L. (1987). Using satellite data for determination of optical inhomogeneities in the upper layer of the seas, in *Instruction Manual for Complex Usage of Satellite Data in Studies of the Seas* (in Russian), Leningrad, Gidrometeoizdat, pp. 49-58.
SUKHACHEVA, L. L., TRONIN, A. A. (1994). The Neva Bay investigation from JERS-1, in

Abstracts of papers presented at the Nineteenth Conference of Baltic Oceanographers (Sopot, Poland, 29 August–1 September), p. 104.

SUKHACHEVA, L. L., VICTOROV, S. V. (1990). Analysis of remotely sensed data for the Aral Sea and Caspian Sea (in Russian) (unpublished).

— (1993). Hydrological phenomena in the Caspian Sea as recorded in satellite data of visible and thermal infra-red bands, paper presented at the First Eurasian Symposium on Space Sciences and Technologies (25–28 October 1993, Marmara, Turkey), *Turkish J. Phys.*, **19**, No. 8, pp. 1049–1054.

— (1994). Remote sensing and the 'The Gulf of Finland Year 1996', in Abstracts of papers presented at the Nineteenth Conference of Baltic Oceanographers (Sopot, Poland, 29 August–1 September), p. 93.

SURFACE TEMPERATURE AND SALINITY RECORDS ALONG THE COAST OF FINLAND July 1940–June 1952, 1954, Helsinki.

SUURSAAR, Ü., KULLAS, T. (1994). Viire Kurk Strait, *hibernatus*, in abstracts of papers presented at the Nineteenth Conference of Baltic Oceanographers (Sopot, Poland, 29 August–1 September), p. 136.

SZEKIELDA, K.-H. (1976). Spacecraft oceanography, *Oceanography and Marine Biology Annual Review*, **25**, 4, pp. 403–410.

TAMSALU, R. E. (1979). *Modelling the Dynamics and the Structure of Baltic Sea Waters* (in Russian), Riga, Zvaigzne, 152 pp.

TEMPORARY MANUALS ON ACQUISITION PROCESSING AND USAGE OF SATELLITE IR-INFORMATION ON SURFACE TEMPERATURE OF THE SEAS AND THE OCEANS (1985) Leningrad, Gidrometeoizdat, 128 pp.

TEREKHIN, YU. V., KUNITSA, V. E., ORLOV, A. K., SCHETININ, YU. T. (1983). Some results of processing aerial and satellite-borne images with microcomputers (in Russian), in *Methods of Processing of Space-borne Oceanological Information*, Sevastopol, pp. 94–98.

The Earth's Nature from Space (1984). (In Russian), compiled by A. Tishenko and S. Victorov, ed. N. Kozlov, Leningrad, Gidrometeoizdat, 151 pp.

THOMPSON, T., HAKANSSON, B., ULANDER, L., CARLSTROM, A. (1993). Experiences from the Swedish sea ice programme during BEERS-92, in Proceedings of the First ERS-1 Symposium *Space at the Service of our Environment* (Cannes, France, 4–6 November 1992). Document ESA SP-359 (March 1993), pp. 313–318.

THOMSON, R. E., VACHON, P. W., BORSTAD, G. A. (1992). Airborne Synthetic Aperture Radar imagery of atmospheric gravity waves, *J. Geophys. Res.*, **97**, No. C9, pp. 14249–14257.

TIKHOMIROV, A. I. (1968). Temperature regime and heat storage in the Ladoga Lake (in Russian), in *Thermal Regime of the Ladoga Lake*, Leningrad, LGU Publishers, pp. 144–217.

— (1982). *Thermal Regime of Large Lakes* (in Russian), Leningrad, Nauka, 232 pp.

TITTLEY, B., SOLOMON, S. M., BJERKELUND, C. (1994). The integration of Landsat TM, SPOT PLA, and ERS-1 C-Band SAR for coastal studies in the Mackenzie River Delta, NWT, Canada: a preliminary assessment, in Proceedings of the Second Thematic Conference *Remote Sensing for Marine and Coastal Environments* (New Orleans, USA, 31 January–2 February 1994), vol. 1, pp. 225–236.

TIURYAKOV, B. I., KUZNETSOVA, L. N., CHEVSKIJ, V. N. (1983). Peculiar features of vertical motions in the Baltic Sea under the influence of meteorological conditions (in Russian), in Proceedings of the Leningrad Hydrometeorological Observatory, No. 13.

TROOST, D., BLACKBURN, C., ROBINSON, I., CALLISON, R., DOBSON, M., BLACKBURN, D. (1994). Marine and coastal remote sensing training through an interactive, global network, in Proceedings of EARSeL Workshop on Remote Sensing and GIS for Coastal Zone Management. Rijkswaterstaat Survey Department (24–26 October 1994, Delft, The Netherlands), pp. 303–313.

TSIBAN, A. V. (ed.) (1981). *Study of the Baltic Sea Ecosystem* (in Russian), Leningrad, Gidrometeoizdat, 197 pp.
ULBRICHT, K. A. (1983). Landsat image of blue-green algae in the Baltic Sea, *Int. J. Remote Sensing*, **4**, 4, pp. 801–802.
— (1984). Examples of use of digital processing of remote sensing photography of various phenomena (in Russian), in *Remote Sensing in Meteorology, Oceanography and Hydrology*, Moscow, Mir Publishers, pp. 270–279.
URBANSKI, J. (1994). Using SST AVHRR satellite images for planning the positions of monitoring stations, in abstracts of papers presented at the Nineteenth Conference of Baltic Oceanographers (Sopot, Poland, 29 August–1 September), p. 94.
URVOIS, M., GARREAU, P., JEGOU, A. M., PIRIOU, J. Y., WATREMEZ, P. (1994). Mapping eutrophication sensitive zones: a nearshore information system along the coast of Brittany, France, in Proceedings of EARSeL Workshop on Remote Sensing and GIS for Coastal Zone Management. Rijkswaterstaat Survey Department (24–26 October 1994, Delft, The Netherlands), pp. 155–164.
USANOV, B. P., MIKHAILENKO, R. R., SUKHACHEVA, L. L. VICTOROV, S. V., TSVETKOVA, L. I. (1994a). Some results of experimental investigations aimed at amelioration of ecological state inside and outside the Neva Bay by means of the barrier gateway operation, in abstracts of papers presented at the Nineteenth Conference of Baltic Oceanographers (Sopot, Poland, 29 August–1 September) p. 87.
USANOV, B. P., MIKHAILENKO, R. R. VICTOROV, S. V., SUKHACHEVA, L. L. (1994b). Regional environmental GIS 'The Neva Bay': databases of conventional and remote-sensed data, in abstracts of papers presented at the Nineteenth Conference of Baltic Oceanographers (Sopot, Poland, 29 August–1 September), p. 106.
USE OF SATELLITE DATA FOR ENVIRONMENTAL PURPOSES IN EUROPE (1994). Scot Conseil and Smith Systems Engineering, Final Report.
USSR–GDR (1985). Complex Subsatellite Oceanographic Experiment in the Baltic Sea (ed. S. V. Victorov) (in Russian), Leningrad, Gidrometeoizdat, 103 pp.
VAN DER VAT, M. P. (1994). Modelling of eutrophication of the Bay of Gdansk as a tool for decision makers, in abstracts of papers presented at the Nineteenth Conference of Baltic Oceanographers (Sopot, Poland, 29 August–1 September), p. 127.
VAN HEUVEL, TJ., HILLEN, R. (1994). Coastal management with GIS in the Netherlands, in Proceedings of EARSeL Workshop on Remote Sensing and GIS for Coastal Zone Management. Rijkswaterstaat Survey Department (24–26 October 1994, Delft, The Netherlands), pp. 155–164.
VAUGHAN, R. (1994). Recent advances in imaging coastal zones, in Proceedings of EARSeL Workshop on Remote Sensing and GIS for Coastal Zone Management. Rijkswaterstaat Survey Department (24–26 October 1994, Delft, The Netherlands), pp. 339–348.
VETLOV, I. P., JOHNSON, D. S. (1978). *Role of Satellites in WMO Programmes in the 80s*. Planning Report No. 36. WMP Paper No. 494, Geneva.
VICTOROV, S. V. (1980a). Airborne equipment for oceanographic research in optical band (in Russian), in *Optics of Ocean and Atmosphere (Abstracts of Papers)*, Baku, USSR.
— (1980b). Present stage in the use of remote sensing data for marine pollution estimation, in Proceedings of the United Nations Training Seminar on Remote Sensing Applications (USSR, Baku, 17–29 November 1980). Baku, part 2, pp. 116–132. Second edition: Baku, 1982, pp. 110–128.
— (1982). Satellite oceanography: definition, state-of-the-art and prospects (in Russian), in *Satellite and Aerial Oceanography (Proceedings of the State Oceanographic Institute)*, **166**, Leningrad, Gidrometeoizdat, pp. 4–23.
— (1983). Studies in satellite and aerial oceanography (in Russian), in *Exploration of Oceans and Seas*, Moscow, Gidrometeoizdat, pp. 185–202.
— (1984a). Problems of complex investigation of the Baltic Sea with remote sensing data, in Proceedings of the Fourteenth Conference of the Baltic Oceanographers (Gdynya, Poland, September 1984), **1**, pp. 391–401.

(1984b). Studies in the area of satellite and airborne oceanography (in Russian), in *Studies of Oceans and Seas*, Moscow, Gidrometeiozdat, pp. 185–202.

(1985). Methodology and some results of integrated satellite monitoring (in Russian), in abstracts of papers presented at the Third International Symposium on Integrated Global Monitoring of the State of the Biosphere (USSR, Tashkent, 13–20 October 1985), Moscow, Gidrometeoizdat, pp. 16–17.

(1986a). Complex subsatellite experiments as a stage in development of spaceborne oceanography (in Russian), in *Remote Methods and Techniques for Measurement of Oceanographic Parameters*, Proceedings of the Fifth All-Union Seminar (Moscow, 20–23 September 1983), Moscow, Gidrometeoizdat, pp. 45–48.

(1986b). Complex studies of the Baltic Sea using satellite data (in Russian), *Marine Hydrophysical Journal*, **4**, pp. 59–64.

(1987). Complex studies of the Baltic Sea based on satellite data, in abstracts of papers presented at the Third Congress of Soviet Oceanologists (Leningrad, 14–19 December 1987), Section *Physics and Chemistry of the Ocean. Polar and Regional Oceanology*, Leningrad, Gidrometeoizdat, pp. 49–50.

(1988a). *Regional Satellite Oceanography* (in Russian), Moscow, Gidrometeoizdat, 16 pp.

(1988b), Complex studies of the Baltic Sea based on satellite information, in Proceedings of the Sixteenth Conference of Baltic Oceanographers (Kiel, FRG, September 1988), **2**, pp. 1079–1090.

(1989a). Sixteenth Conference of the Baltic Oceanographers (in Russian), in *Oceanography from Space*, USSR Academy of Sciences, Leningrad, Express-Information, 4-89, pp. 4–13.

(1989b). Satellite-based studies of the seas (regional satellite oceanography) (in Russian), paper presented at the Fifth All-Union Workshop-seminar on Satellite Hydrophysics (Sevastopol, 1988). Text deposited in All-Union Institute of Science and Technical Information, Moscow. Document No. 3765-89, 39 pp.

(1989c). ECOLOGIUM – regional information and analytical public centre 'environmental and population health' for St. Petersburg metropolitan area (conceptual proposals) (unpublished).

(1990a). Environmental monitoring of the Gulf of Finland: an approach based on remote sensing data, *Statistical Journal of the United Nations Economic Commission for Europe*, **7**, 3, pp. 149–162.

(1990b). International project in regional environmental geoinformatics – BALTIC EUROPE; consultation on the development of a health and environment GIS for the European region. Bilthoven, The Netherlands, 10–12 December 1990. WHO Paper ICP/CEH 90/A/28, 25 pp.

(1991a). Remote sensing data in regional environmental monitoring and assessment: a case study of the Baltic region. Paper presented at the Third International Conference on a Systems Analysis Approach to Environment, Energy and Natural Resource Management in the Baltic Region (Copenhagen, Denmark, 7–10 May 1991).

(1991b). Water dynamics and pollution nearby Leningrad Dam, in *The Interaction between Major Engineering Structures and the Marine Environment* (Proceedings of the International Assocation for Bridge and Structural Engineering Colloquium, Nyborg, Denmark, May 1991). IABSE Reports, **63**, pp. 215–223.

(1992a). Remote sensing applications in oceanographic research in the former USSR, in *Remote Sensing from Research to Operation*. Proceedings of the Eighteenth Annual Conference of the Remote Sensing Society, Dundee, Scotland, UK, 15–17 September 1992, pp. 47–62.

(1992b). International project in regional environmental geoinformatics – BALTIC EUROPE. Part 1. Concept, in *Environmental Management in the Baltic Region*, Collection of papers presented at the International Workshop (Leningrad, USSR, 21–24 November 1989). INENCO Proceedings No. 1, St. Petersburg, pp. 64–96.

(1994a). Russian oceanographic satellites: state-of-the-art, in Proceedings of the *Oceanology*

International 94 Conference (8–11 March 1994, Brighton, UK), **2**.

(1994b). Russia's oceanographic satellites face 'hard times', *Microwave Engineering Europe*, August–September, pp. 31–32, 35–36.

(1994c). Caspian Sea Oil Pollution Airborne Data Base, Project Proposal (unpublished).

(1994d). Former Soviet Union satellites (relevant to ocean observation), in Proceedings of the WMO/IOC Technical Conference on *Space-Based Ocean Observations* (September 1993, Bergen, Norway), pp. 9–31.

(1994e). Some notes on the oceanic aspect of the remote sensing and global climate change issue, in *Remote Sensing and Global Climate Change* (ed. A. Vaughan and A. P. Cracknell), Proceedings of the NATO Advanced Study Institute Summer School on Remote Sensing and Global Climate Change (Dundee, Scotland, July 19–August 8, 1992), NATO ASI Series I, vol. 24. Springer-Verlag, pp. 411–427.

VICTOROV, S. V., BYCHKOVA, I. A., DEMINA, M. D., DORONINA, J. A., LOBANOV, V. YU., LOSINSKIJ, V. N., SMOLYANITSKIJ, V. M., DUBRA, J. J., EMELYANOV, I. V. (1986a). Spatial and temporal variability of sea surface temperature of the Baltic Sea in spring (in Russian), in *Problems and Ways of Rational Use of Natural Resources and Nature Conservation* (abstracts of papers presented at the Eleventh Lithuanian Republican Hydrometeorological Conference, Shaulyaj, 20 May 1986). Section 'Antropogeneous Impact on Ecosystems and Problems of Monitoring', Vilnius, p. 66.

VICTOROV, S. V., BYCHKOVA, I. A., DEMINA, M. D., DREMLJUG, I. V., LOSINSKIJ, V. N., SUKHACHEVA, L. L., DUBRA, J. J., RAZUMOVSKAYA, C. J. (1986b). Studies of hydrophysical fields of the Kurshi Bay based on airborne, spaceborne and *in-situ* data (in Russian), in *Problems and Ways of Rational Use of Natural Resources and Nature Conservation* (abstracts of papers presented at the Eleventh Lithuanian Republican Hydrometeorological Conference, Shaulyaj, 20 May 1986). Section 'Antropogeneous Impact on Ecosystems and Problems of Monitoring', Vilnius, p. 67.

VICTOROV, S. V., BYCHKOVA, I. A., LOBANOV, V. YU., SUKHACHEVA, L. L. (1989a). Problems of satellite environmental monitoring of the Gulf of Finland (in Russian), in *Oceanography from Space*, USSR Academy of Sciences, Leningrad, Express-Information, 4-89, pp. 13–26.

VICTOROV, S., BYCHKOVA, I., SUKHACHEVA, L. (1995). Remote sensing and synergism in post-cold-war era marine sciences, Paper submitted to International Conference *Remote Sensing in Action* (11–14 September 1995, Southampton, UK). Abstract F28113.

VICTOROV, S. V., KARPOV, A. V., NIKITIN, P. A. (1993a). Ocean satellites programme in the Russian federation, in Proceedings of the Commission for Marine Meteorology Technical Conference on Ocean Remote Sensing (26 April 1993, Lisbon, Portugal), Report No. 28, WMO Technical Document No. 604, pp. 9–22.

VICTOROV, S. V., KAZMIN, A. S., SKLYAROV, V. E., GASHKO, V. A., SUKHACHEVA, L. L., ANTONENKO, V. G., SAZHIN, S. M. (1984). Studies of water masses with various optical characteristics. Determination of fields of plankton and suspended matter (in Russian), in *The Earth's Nature from Space*, Leningrad, Gidrometeoizdat, pp. 118–124.

VICTOROV, S. V., LOBANOV, V. YU., SMOLYANITSKIJ, V. M., SUKHACHEVA, L. L., LEBEDEVA, N. I., NEKRASOVA, A. N. (1989b). Seston distribution and elements of water dynamics in eastern part of the Gulf of Finland: studies based on satellite imagery and *in situ* data (abstract), in Proceedings of International Workshop on *Environmental Management in the Baltic Region* (Leningrad, USSR, 21–24 November 1989), Centre INENCO Proceedings No. 1, p. 99.

VICTOROV, S. V., LOBANOV, V. YU., SUKHACHEVA, L. L., *et al.* (1990a). Elements of water dynamics in the eastern part of the Gulf of Finland as recorded in high-resolution satellite imagery, in Proceedings of the Seventeenth Conference of Baltic Oceanographers (Norrkoping, Sweden, September 1990).

VICTOROV, S. V., LOKK, J. F. (1986). Some concepts of setting up marine subsatellite test polygons (in Russian), in *Remote Methods and Techniques for Measurement of Oceano-*

graphic Parameters, Proceedings of the Fifth All-Union Seminar (Moscow, 20–23 September 1983), Moscow, Gidrometeoizdat, pp. 30–32.

VICTOROV, S. V., LOSINSKIJ, V. N., SUKHACHEVA, L. L. (1987). Studies of seasonal bloom of the Baltic Sea and elements of eddy dynamics of the upper layer based on satellite data of visible diapazon (in Russian), in abstracts of papers presented at the Third Congress of Soviet Oceanologists (Leningrad, 14–19 December 1987). Section *Physics and Chemistry of Ocean. Climate, Ocean–Atmosphere Interaction, Spaceborne Oceanology*. Leningrad, Gidrometeoizdat, pp. 77–78.

(1988). Studies of seasonal bloom of the Baltic Sea and elements of eddy dynamics of the upper layer based on satellite data of visible diapazon, in Proceedings of the Sixteenth Conference of Baltic Oceanographers (Kiel, FRG, September 1988), **2**, pp. 1063–1078.

VICTOROV, S. V., SAZHIN, S. M. (1986). Interactive computer processing of satellite imagery of the sea (in Russian), in *Remote Methods and Techniques for Measurement of Oceanographic Parameters*, Proceedings of the Fifth All-Union Seminar (Moscow, 20–23 September 1983), Moscow, Gidrometeoizdat, pp. 227–230.

VICTOROV, S. V., SHAPOSHNIKOVA, G. N. (1995). Caspian Sea oil pollution airborne database, summary of paper submitted to the Third Thematic Conference *Remote Sensing for Marine and Coastal Environments* (Seattle, USA, 18–20 September 1995).

VICTOROV, S. V., SMIRNOV, V. G., SMOLYANISSKY, V. M., BYCHKOVA, I. A., SUKHACHEVA, L. L., FOKINA, M. L., KLYACHKIN, S. V., DEGTYAREV, A. G., PASTOKHOV, G. V. (1993b). Multi-purpose operational system of satellite imagery utilization in the regional environmental monitoring and management of natural resources, Paper presented at the International Symposium *Operationalization of Remote Sensing* (19–23 April 1993, Enschede, The Netherlands) (unpublished).

VICTOROV, S. V., SUKHACHEVA, L. L. (1984). Records of optical inhomogeneities of the upper layer of the Baltic Sea in 'Meteor-Priroda' satellite imagery of visual and near-infrared bands (in Russian), in *Optics of the Sea and Atmosphere* (Abstracts of Symposium, Batumi, October 1984), Leningrad, pp. 203–204.

(1992a). Estimation of the maximal distance of turbid waters spread from the region of St. Petersburg into the Gulf of Finland as revealed from satellite data (in Russian), Abstract of paper submitted to the Russian–Finnish symposium on the environmental issues (St. Petersburg, June 1992).

(1992b). What do the satellite images of the Neva Bay show? (in Russian), *Energy*, **6**, pp. 77–79.

(1994). Satellite environmental monitoring of the coastal zone in the Gulf of Finland: state-of-the-art and needs for development (summary), in Proceedings of the Second Thematic Conference *Remote Sensing for Marine and Coastal Environments* (New Orleans, USA, 31 January–2 February 1994), **2**, p. 257.

VICTOROV, S. V., SUKHACHEVA, L. L. LOBANOV, V. YU., LEBEDEVA, N. I., NEKRASOVA, A. N. (1991). Distribution features of suspended substances in the Neva Bay under various hydrometeorological conditions (in Russian), *Meteorology and Hydrology*, **7**, pp. 80–85.

VICTOROV, S. V., SUKHACHEVA, L. L., VITER, V. V., POSTNIKOV, I. YU., SHIROKOV, P. A. (1990b) Visualization of currents in the Neva Bay in high-resolution radar imagery (in Russian), *Doklady of the USSR Academy of Sciences*, **315**, 2, pp. 337–340.

VICTOROV, S. V., TISHENKO, A. P. (1982). Basic principles of design of *in-situ* database meant for using in interactive systems of processing satellite oceanographic data (in Russian), in *Satellite and Aerial Oceanography* (Proceedings of the State Oceanographic Institute), 166, Leningrad, Gidrometeoizdat, pp. 24–29.

VICTOROV, S. V., VINOGRADOV, V. V., TERZIEV, F. S. (1982a). Studies of the Baltic Sea with the help of IR-diapason Meteor-2 information, in Proceedings of the Thirteenth Conference of Baltic Oceanographers (Helsinki, Finland, September 1982), **1**, pp. 732–741.

VICTOROV, S. V., VINOGRADOV, V. V., VISHNEVSKIJ, A. E., POPOV, S. S.,

Tarasov, V. S., Filimonov, V. I. (1982b). Data input and software for shipborne systems of satellite scanners data processing (in Russian), in *Satellite and Aerial Oceanography* (Proceedings of the State Oceanographic Institute), **166**, Leningrad, Gidrometeoizdat, pp. 41–47.

Victorov, S. V., Zueva, A. N., Nepoklonov, V. B. (1993c). Satellite altimetry: Russian data for oceanographic research. Paper presented at the WMP/IOC Technical Conference on *Space-Based Ocean Observations*, Bergen, Norway, 5–10 September 1993 (WMO/TD-No. 649, p. 302).

— (1993d). Satellite altimetry based on Russian data (case study of Norwegian Sea). Paper presented at the International Conference on *Satellite Altimetry and the Oceans* (Toulouse, France, 29 November–3 December 1993).

Vinogradov, B. V., Kondratyev, K. Ya. (1971). *Space Methods in Earth Sciences* (in Russian), Leningrad, Gidrometeoizdat, 190 pp.

Vinogradov, V. V., Likhachev, I. V., Staritsin, D. K. (1980). Experience of computer processing of satellite oceanographic information recorded in direct transmission mode (in Russian), in Proceedings of Leningrad State University, **403**, pp. 118–129.

Viter, V. V., Postnikov, I. Yu., Shirokov, P. A., Victorov, S. V., Lobanov, V. Yu., Smirnov, V. G., Smolyanitskij, V. M. (1989). Use of side-looking radar high-resolution imagery in problems of regional satellite oceanography and environmental monitoring (abstract), in Proceedings of International Workshop *Environmental Management in the Baltic Region* (Leningrad, USSR, 21–24 November 1989). Centre INENCO Proceedings No. 1, p. 100.

Voitsekhovskij, A. B., Gerchikova, G. V., Leibovich, Ya. E. (1986). Two-channel airborne TV equipment for oceanographic research (in Russian), in *Remote Methods and Techniques for Measurement of Oceanographic Parameters*. Proceedings of the Fifth All-Union Seminar (Moscow, 20–23 September 1983), Moscow, Gidrometeoizdat, pp. 216–219.

Vukovich, F. M. (1974). The detection of nearshore eddy motion and wind-driven currents using NOAA-1 sea surface temperature data, *J. Geophys. Res.*, **79**, No. 6, pp. 853–860.

Wagner, T., Czurylo, S. (1994). STEP and marine science education, in Proceedings of the Second Thematic Conference *Remote Sensing for Marine and Coastal Environments* (New Orleans, USA, 31 January–2 February 1994), **2**, pp. 521–524.

Wahl, T., Eldhuset, K., Skoelv, A. (1993). Ship traffic monitoring and oil spill detection using ERS-1, in Proceedings of International Symposium *Operationalization of remote Sensing* (19–23 April 1993, Enschede, The Netherlands). Vol. 7, pp. 151–160.

Wang, Y., Koopmans, B. N. (1994). Monitoring tidal flat changes using ERS-1 SAR images and GIS in the western Wadden Sea area, The Netherlands, in Proceedings of EARSeL Workshop on Remote Sensing and GIS for Coastal Zone Management. Rijkswaterstaat Survey Department (24–26 October 1994, Delft, The Netherlands), pp. 352-361.

Wei Ji, Johnston, J., Mitchell, L., Sclafani, V., McNiff, M. (1994). Integrated spatial decision support: a new methodology for coastal system management (abstract), in Proceedings of the Second Thematic Conference *Remote Sensing for Marine and Coastal Environments* (New Orleans, USA, 31 January–2 February 1994), **1**, pp. 347–348.

Wensink, G. J., Hesselmans, G. H. F. M., Calkoen, C. J. (1993). The commercial use of satellite observations for bathymetric surveys, in Proceedings of International Symposium *Operationalization of Remote Sensing* (19–23 April 1993, Enschede, The Netherlands). Vol. 7, pp. 29–36.

Werle, D. (ed.) (1993). *Radar Remote Sensing Imagery of Coastal Regions on CD-ROM. User Guide*. AERDE Environmental Research, Halifax, Canada, 70 pp.

— (1994). Radar remote sensing of coastal regions – Part 1: Review of previous investigations, in Proceedings of the Second Thematic Conference *Remote Sensing for Marine and Coastal Environments* (New Orleans, USA, 31 January–2 February 1994), **2**, pp. 336–347.

WHITEHOUSE, B. G., LANDERS, P. M. (1994). Transferring Earth observation imagery to the coastal zone via cellular telephony (abstract), in Proceedings of the Second Thematic Conference *Remote Sensing for Marine and Coastal Environments* (New Orleans, USA, 31 January–2 February 1994), **1**, p. 346.
WMO (WORLD METEOROLOGICAL ORGANISATION) (1992). Executive Council Panel of Experts/CBS Working Group on Satellites. Tenth Session (Geneva, 16–20 March 1992). Final Report; Russian Federation Committee for Hydrometeorology and Monitoring of Environment (ROSCOMGIDROMET) Status Report, pp. 4–5.
YAMAMOTO, S. (1993). Satellite remote sensing and ocean research, in Proceedings of the Commission for Marine Meteorology Technical Conference on Ocean Remote Sensing (26 April 1993, Lisbon, Portugal). Report No. 28. WMO Technical Document No. 604, pp. 23–28.
 (1994). Space-based ocean observation in Japan, in Proceedings of the WMO/IOC Technical Conference on *Space-Based Ocean Observations* (September 1993, Bergen, Norway), pp. 36-41.
YAPA, L. S. (1991). Is GIS appropriate technology? *Int. J. Geogr. Inf. Sys.*, **5**, pp. 42–58.
ZAITSEV, YU. I. (1994). Russian space launches 1993, *Space Bulletin*, **1**, 4, pp. 26–29.
ZALESKY, J., WOJEWODKA, C. (1977). *Baltic Europe* (in Polish), Ossolineum, Warsaw.
ZLOBIN, L. I., KISELEV, V. V., MATIYASEVICH, L. M. (1983). Complex subsatellite experiment technique (in Russian), *Earth Research from Space*, **4**, pp. 103–112.
ZUELICKE, CH., SCHLITTENHARDT, P. (1994). *In-situ* measurements of sea surface temperature and surface heat flux, in abstracts of papers presented at the Nineteenth Conference of Baltic Oceanographers (Sopot, Poland, 29 August–1 September), p. 110.
ZUIDAM, VAN, R. A. (1993). Review of remote sensing applications in coastal zone studies, in Proceedings of International Symposium *Operationalization of Remote Sensing* (19–23 April 1993, Enschede, The Netherlands). Vol. 7, pp. 1–16.
ZUJAR, J. O., RODRIGUEZ, E. S., FERNANDEZ-PALACIOS, A., MADUENO, J. M. M. (1994). The use of satellite images in the hydrodynamic al modelling of the coastal water behavior (Tinto-Odiel Estuary, SW Spain), in Proceedings of EARSeL Workshop on Remote Sensing and GIS for Coastal Zone Management. Rijkswaterstaat Survey Department (24–26 October 1994, Delft, The Netherlands), pp. 250–259.

Index

a priori data 14, 59, 69, 74, 102, 153
accuracy requirement 21–5, 27–8
Admiralty Chart data 135, 136
Adriatic Sea 143
Advanced Earth Observation Satellite (ADEOS) 31, 32
Advanced Very High Resolution Radiometers (AVHRR) 142–3, 269
   case study 153–4, 172, 174, 179, 196–7, 211–12, 220–1, 259, 264–5
   information requirement 23, 29–30, 39, 53, 55
   integrated GIS 122, 126–8, 130–1
   methodology 61, 67, 74, 79–80, 89, 91–3, 96–7
   regional systems 109, 115–18
Advanced Visible and Near-Infrared Radiometer (AVNIR) 31
advective eddies 189–90
AEM-2 satellite 92
alarm scenario 167, 168–9
Alexander von Humboldt (research vessel) 67, 71–2, 74, 85, 185, 200, 202, 265
allowed interval of non-synchronicity (AINS) 74, 80–7, 88, 172, 184
ALMAZ satellites 37, 39–40, 236
altimetry 22, 24, 48–9, 50
American Swordfish Association 149
animation sequence technique 106
anticyclonic eddies 191–2, 193, 196
APOLLO algorithms 117
Arabian Sea 82, 128, 134
Aral Sea 143, 144
Arctic regions 109–17 *passim* 132
ARGOS system 115, 272
ASEAN–Australian programme 126–7
Atlantic Ocean 58, 66–7
ATSR (infrared radiometer) 23
Australian–ASEAN programme 126–7
Automated Picture Transmission mode 6, 30, 36, 39, 53, 93, 172
autonomous satellite data acquisition 10, 21, 23, 51–6, 80, 93, 97
Azov Sea 61, 62, 63, 65

BALTEX project 172
Baltic Declaration 171
Baltic Europe 88, 155–72
Baltic Marine Environment Protection Commission 157–8
Baltic Monitoring Programme 160
Baltic Oceanographers 155, 158–9
Baltic Proper 63, 74, 174, 205, 218
Baltic Sea 15–16, 19, 61–4
   biological phenomena 214–25
   case study 14, 153–266
   cloud-free situations 91–5, 116
   COSE series 17, 66–7, 74, 76, 78–80, 82, 84, 87, 147
bathymetry 134–6
Beaufort Sea (Canada) 132
bio-optical patterns 215–21
biological phenomena (Baltic) 214–25
biomass 10, 215, 217
Black Sea 61–3, 66, 143, 149
blue-green algae 118, 184, 190, 197–8, 214–16, 217–18, 219–20
Bornholm Island 189, 190
BOSEX experiment (1977) 69
bottom topography 131–6
British Meteorological Office 117
British National Space Centre 146
Brittany (information system) 127–8

Caribbean Sea 143, 145
Caspian Sea 61–3, 65–6, 141–2, 147, 149
Centre for International Environmental Cooperation 155
CEOS 25, 27
chlorophyll 18, 136, 154, 201, 217, 219
cloud situations 17, 91–9
COAST project 140–1
coast topography 131–6
Coastal Assessment Framework 121
Coastal Earth Observation Application for Sediment Transport 140–1

301

coastal environment of Neva Bay (monitoring) 225–64
coastal living resources programme (ASEAN–Australian) 126–7
Coastal Zone Color Scanner 18, 30, 39, 121–3, 131, 142–3, 153, 207, 215–16, 269
coastal zone environment 131–51
coastal zone management 26, 121–31
Coastwatch 120–1, 124–5, 129
communication (GIS technology) 128, 130
Compact Airborne Spectrographic Imager (CASI) 141
complex oceanographic subsatellite experiments 5, 7, 14, 17, 60, 147
 AINS evaluation 80–7, 88
 case study 172, 184–7, 200, 202, 207, 209, 211
 objectives (Baltic Sea) 66–7
 programme (of first) 67–72
 programme development 72–4
 some findings 74–80
Conference of Baltic Oceanographers 154–5, 158–9, 221
contrast enhancement technique 101
Cooperative Synoptic Investigation of the Baltic 154
coral reef studies 126
CORINE project 156, 163–4
Coriolis parameter 182, 188, 192, 208, 210
cyclonic eddies 195–6, 203

Danish Science Centre 171
Danish Straits 188, 198–9
data
 acquisition systems 3, 10, 21, 23, 51–6, 97
 calibration 9, 80, 134
 collection (sea-truth) 12–13, 57–9
 manipulation/analysis 168–9
 processing 9, 99–106
 regional oceanographic research 29–51
 subsatellite 57–61, 63, 82, 87, 153
 topic-oriented use 131–51
'data vectors' 11
databases 9, 102–5, 121–31, 167–8
deep water 75, 184–8
Defence Research Agency 146
Delft Hydraulics 133, 135, 262, 265, 273
DELTA-2 46
DELWAQ software package 265
density slicing technique 100
DGXII-D-4 25
diatomic algae 214
'dipole eddies' 196
discharge lenses 211–12
dispersion analysis 77, 86–7, 176–7
DISTERM system 54
DMSP SSM/I data 109, 115–16
Dutch operational system 117–18

EARSeL 18, 127, 266
Earth Observation data 140–1
Earth Observation Sciences 146
Earth Observing System 30, 129
EBRD 263
ECOLOGIUM 170–1, 262
eddies 16, 75, 183–96, 216

edge enhancement technique 101
Ekman transport effect 173, 177, 181
EKRAN (Moscow) 115
EKSPERIMENTARIUM 171
electromagnetic radiation 4, 18
energy spectrum 160, 163
environmental database (US West Coast) 125–6
Environmental Monitoring and Assessment Program (EMAP-E) 124
environmental monitoring of Neva Bay 225–64
Environmental Research Institute of Michigan (ERIM) 274
environmental sensitivity index 136–8
ENVISAT 129, 151
ERDAS system 139
ERS-1 32, 109, 118–19, 130–3, 136–7, 146–8, 150–1, 265, 273
Estonian Marine Institute 265
ethics/ethical regulation 20
EURICA software package 106
European Committee on Ocean and Polar Sciences 266
European Environment Agency 146
European Polar Platform 31
European Science Foundation 266
European Space Agency 21, 32, 118, 129, 143, 151
eutrophication sensitive zone 127, 214

Federal Agency for Hydrometeorology and Monitoring of Environment 47, 106
fisheries forecast activities 128
floods, see St Petersburg Flood Barrier
Fragment scanner 62, 67, 71, 73, 141
friction eddies 186, 190–4
frontal eddies 188–9, 190
'frozen field' assumption 61, 62

Gdansk Bay 208, 265
geolocation information 128, 129–30
geodetic satellites 35, 48–50
GEOIK series 48–9
Geographical Information Systems 66, 115, 138, 269, 271, 273
 Baltic Europe 155–72
 case study 221, 239–40, 261, 265
 databases and 14, 102–6
 integrated 105–6, 121–31, 270, 274
 knowledge base 14, 103, 105, 168
Georgiy Ushakov (research vessel) 54
Geostationary Operational Meteorological Satellite 43–6, 47, 53
German Remote Sensing Centre 129
Global Ocean Observing System (GOOS) 12, 21–2, 108
Global Positioning System 115, 129–30
global satellite oceanography 6–11
Glomma River plumes 143
Golden Horn of Istanbul 138
GORIZONT 115
Great Belt Link (Denmark) 262
'ground–satellite–ground' radio link 3
Groupement pour le Développement de la Télédétection Aérospatiale 270
Gulf of Alaska 58

Gulf of Bothnia 93, 118–19, 153, 181
Gulf of Finland 63, 93, 153–4, 212, 259, 269, 272
  eddies 190–1, 195, 216
  sediments/pollutants 221–5
  upwelling 95–9, 176, 179, 260
Gulf of Mexico 121–5, 129–30, 269
Gulf of Riga 76–7, 93, 174, 178, 179, 223, 265
Gulf of St Lawrence 133
Gulf Stream 73, 91, 148–9
Gulf of Thailand 272

HELCOM databases 172
Helsinki Commission 155, 157–8
High-Resolution Picture Transmission 30, 39, 55, 117–18, 196, 220
high-resolution SAR 35, 37–9
high-spectral-resolution scanners 25
High Accuracy Reference Networks (HARN) 129–30
histograms (image processing) 101
horizontal resolution 22–5, 28
Human Resources Mobility Programme 266
hydrodynamics 145, 184
Hydrographical Observations 179
'hydrological spring' 89, 91, 173
hydrometeorological factors 239, 247–8
hydrometeorological stations 82, 85, 90, 98, 173
hydrophysical fields 17, 81, 192, 260
hydrostatic stability 203–4

ice 13, 24, 109–15, 118–19, 184
Ice Centre 119–20
icebreakers (US Coastguard) 115–17
Iceplott system 119
image processing 99–102, 105–6, 238–9
in-flight tests (payload) 60, 76
in situ data 3, 12, 17, 91, 108–9, 117–18, 134–5
  GIS 14, 105–6, 122, 126, 172
  sea-truth 4–5, 9, 59, 61, 63, 66, 80, 87–8, 102, 119, 122
  subsatellite 59–61, 63, 66, 74, 80, 85, 87–8
India (fisheries forecasting) 128
inertial period (upwellings) 174, 178
information
  regional satellite oceanography 21–56
  sources/structure (Baltic) 164–7
  system, nearshore (Brittany) 127–8
  see also data; Geographical Information Systems
infrared bands 2, 36, 47, 81
  calibration of 89–91
  instruments in 29–31
  use of satellite imagery 91–9
infrared radiometers 23, 24
INMARSAT 109, 115
inshore water circulation 197–8
Institute for Automation and Problems of Control (Vladivostock) 54
Institute for Baltic Sea Research 58, 265
Institute for Marine Research 160
Institute of Oceanology of the Polish Academy of Sciences 58, 265
Institute for Remote Sensing Applications of the Joint Centre of the European Communities 58, 143, 265

Institute of Sea Research 66
integrated GIS 105–6, 121–31, 270, 274
Integrated Water Management System (St Petersburg) 240, 263
Intelligent Satellite Data Information System (ISIS) 129
Inter-ministerial Workshop on Satellite Oceanography (Leningrad) 79
Intergovernmental Oceanographic Commission 20, 22, 29, 268
International Association for Bridge and Structural Engineering 226
International Baltic Year 69, 154
International Commission of Experts 262–3
International Council for the Exploration of the Sea 158–9, 162
International Focal Points 274–5
International Project in Regional Environmental Geoinformatics 156
International Science Foundation 18
International Workshop on Environmental Management in the Baltic Region 155
IR/UV sensors 146
isopicnic surface 192, 193

JAA-2N system 54–5
Japanese Earth Resources Satellite (JERS-1) 31, 109, 131
JGOFS Project 10
Joint CMM–IGOOS–IODE Subgroup 22
Joint Complex Oceanographic Satellite Experiment 160

K-par 125
Kaiyo Maru (Japanese ship) 115
Klaipeda Marine Laboratory 63, 67, 72
Klaipeda Strait 74, 208
Kochren criteria 77
Koporskaya Bay 222, 224–5, 248
KOSMOS series 34–7, 39–41, 55, 58, 113, 199, 212, 235–6, 243, 247, 273
Kotlin Island 39, 229, 240, 248, 251–2
Kronstadt 39
Kunda Bay 199
Kurshi Bay 63–4, 154, 174, 200–1, 205, 206–11, 221

Laboratory for Satellite Oceanography and Airborne Methods 14, 18, 19, 53, 64, 91, 93, 188, 212, 264
Ladoga Lake 54, 66, 88–91, 223, 228, 263
lake areas 87–91
Land Remote Sensing Commercialisaton Act 267
LANDSAT data 55, 142, 215, 217, 221, 228, 240–1, 252
  MSS 31, 131, 133
  TM 31, 124, 126, 131–5, 137–9, 143, 196
Leningrad Polytechnic Institute 54
Lindenberg Observatory 67
local area coverage mode 30
local wind condition 197, 198
Lockheed system 267–8
LUNOKHOD lunar vehicles 74

Major Engineering Structures (MES) 261, 262
mangrove studies 126
Margulis equation 209–10
marine ecosystem 172, 214
Marine Hydrophysical Institute 35, 54
marine environment (Neva Bay) 225–64
Marine Observation Satellite 31
marine pollution
  emissions in Baltic 157, 161–2, 231
  monitoring 10, 136–8, 214–25, 230–1, 261
  oil slicks 64–6, 136–7, 145–8
medium-resolution radar satellite subprogramme 36–7
Medium Resolution Imaging Spectrometer (MERIS) 31
METEOR series 34, 53–4, 71, 84–5, 87, 141–2, 150, 191, 199–200, 212, 235
  PRIRODA 39, 67, 69, 188, 198, 215–16
  -2 data 17, 39–40, 42, 47, 67, 76–80, 110
  -3 data, 39–40, 42–3, 44, 47
meteorological satellites 35, 36
microwave
  band 32–3
  radiometers 22–4, 36, 38
Moderate Resolution Imaging Spectro-Radiometer (MODIS) 30–1
MONEX 1979 experiment 82
MSU-E 39, 41, 43, 46–7, 62, 196, 218, 235, 243, 247, 255–7, 269
MSU-L 247
MSU-M 46–7, 62, 142, 216–17, 219, 240
MSU-S 39, 41, 43, 47, 61–2, 142, 144, 150, 217, 219, 240, 242
MSU-SK 39, 41, 46–7, 62, 235, 243, 247, 269
MSU-V 41, 46
multi-channel satellite sensors 3
Multi-Channel Scanner 31, 131, 133
Multi-Media Decision Support System 131
multi-year satellite images 124–5, 240–59
'mushroom-like' dynamical structures 74, 75, 195–206, 209, 239
Musson (research vessel) 54

Nansen Environment and Remote Sensing Centre (NERSC) 109
NASA 30, 32, 42, 129, 273–4
National Aerospace Laboratory 273
National Science Education Standards (USA) 274
National Wetlands Research Center 123
navy/naval research (USA) 109, 122–3
NESDIS 129
The Netherlands 117, 134, 273
Neva Bay 39–40, 154, 221–2
  environmental monitoring 225–64
  multi-year images 240–59
  St Petersburg Flood Barrier 227–35
Nimbus/CZCS 131, 153
Nimbus-7 18, 34, 122, 153, 188–9, 207, 215, 216–17, 221
NOAA 16, 17, 142
  Baltic case study 153–266 *passim*
  Coastwatch 120–1, 124–5, 129
  information requirements 23, 29–30, 39, 53, 54–5

integrated GIS, 121, 124–5, 127–31
methodologies 67, 71, 79–80, 85, 88–93, 96–8
regional systems 109, 113, 115–18
non-synchronicity 80–7, 88
Normalised Difference Vegetation Index (NDVI) 117
North-Western Regional Administration (USSR) 179
North Sea 15, 117–18, 136, 146, 148, 188
Norway 107–9, 151, 265, 272
NSCATT 32
numerical hydrodynamical models 145
nutrient emissions 161, 231

object-anomaly phenomenon 74
ocean colour 13, 31
Ocean European Archive Network project 143, 221
Ocean Colour and Temperature Scanner 31
Ocean Remote Sensing Group Meets (1994) 268
Ocean Science for the Year 2000 20
ocean topography 13
ocean wave spectra 13
oceanographers, satellites and 1–6
oceanographic observations 80–7
oceanographic satellites, civilian 35
oceanography, remote sensing and 1–20
OCEANOR (of Norway) 272
OCEANS '83 Conference 267
ODER Project 265
Odessa Polytechnic Institute 54
oil emissions (Baltic) 162
oil slicks 64–6, 136–7, 145–8
OKEAN 34, 36–8, 46–7, 53, 113, 144
Onega Lake 54, 89
Ontario Science Park 171
Operational Line Scanner 115
operational systems for marine applications 107–9, 117–20
operationalisation of remote sensing 19, 269
Optical Sensor 31
Orbita (research vessel) 98
Orbital Sciences Corporation 30
output subsystems 169–71

Pacific Ocean 149
PARSEK system 54
passive tracers 15
PH-MNZ aircraft 147
photographic satellites 17, 35, 47–9
phytoplankton 10, 127, 214, 217, 219, 260
polar icebreakers 115
pollutants, transport of 221–5
pollution, *see* marine pollution
polygons 4, 58, 59, 66, 127
principal synoptic situations 181–3
Professor Albrecht Penk (research vessel) 211
PROSAT system 118
PV-Wave 123

quasi-synchronous activities 80, 82, 87, 192, 217
QuickLook software 123, 129

radar imagery 36
RADARSAT 24, 32, 151, 267, 273

raw satellite data 7, 17
red–green–blue pictures 118
Redang Island archipelago 126–7, 270
regional integrated GIS 121–31
regional marine environment (topic-oriented use of data) 131–51
regional satellite oceanography
  Baltic case study 153–266
  concepts 11–20
  global and 6–11
  information used 21–56
  methodological aspects 57–106
  worldwide activities 107–51
regional systems (examples) 107–21
remote sensing 34, 126–7
  Baltic case study 153–266
  image processing 99–102, 105–6
  integrated GIS 128–31
  oceanography and (development) 1–20
  topic-oriented use of data 131–51
Report on Satellite Systems and Capabilities (1995) 21–2, 29, 268
RESURS satellites 34, 39–43, 47–9, 53, 55, 100, 269
'ring' (Gulf Stream) 73, 91, 148–9
river
  discharges 197, 199
  plumes 141–5, 206–11
ROSKOMGIDROMET 47
Rossby radius 188, 191, 203, 208, 210
Royal Netherlands Meteorological Institute (KNMI) 117
Rudolf Samoilovich (research vessel) 73, 191–3
Russian Agency for Hydrometeorology and Environmental Monitoring 53
Russian geodetic satellites 48–50
Russian space-based sea ice information system 109–15

St Petersburg Flood Barrier 39, 197, 226, 248–9, 251–2, 256, 259–60
  development 262–4
  lessons learned 261–2
  Neva Bay and 227–35
satellite–aircraft system 85–7
satellite–ship system 82–5
Satellite Altimetry Applications in Oceanography project 49–50
satellite
  data 16
    acquisition systems 3, 10, 21, 23, 51–6, 97
    processing 9, 99–106
    regional oceanographic research 29–51
    topic-oriented use 131–51
  databases 121–31
  imagery
    in cloudy situations 91–9
    dynamical processes 172–214
    'mushroom-like' structures 196–200
    monitoring of biological phenomena (Baltic Sea) 214–25
  oceanography 2–3
    global 6–11
    regional, *see* regional satellite oceanography
  payload 3–5, 11, 34, 46–7, 60

satellites
  oceanographers and 1–6
  sensors, *see* sensors
SCANSAR mode 32
scatterometer 22, 32, 36
sea-truth
  data collection 57–9, 122
  field activities 61–6, 119
  subsatellite data 57–9, 61, 63, 80, 87–8, 153
  test areas 4–5, 9, 59, 61–6, 87–9
sea ice 13, 24
Sea of Rains on the Moon 74
sea surface 89–90
  Baltic case study 174–5, 179, 184–5, 188–9, 202, 205–13, 223, 259, 265
  cloudy situations 91, 93, 95–9
  COSE 67–71, 76–88
  data collection 12–13, 57–9
  dispersion analysis 77, 86–7, 176–7
  GIS 103, 127–8, 172, 221, 265
  information requirements 22–3, 30–1, 54–5
  test areas 59, 61–3
  worldwide activities 117, 119, 127, 142, 148–9
Seasat 1, 3, 34, 37, 58, 149
seasonal alongshore fronts 173
seasonal bloom 214–21
SeaStar satellite 30, 55, 58, 267
SeaWatch/Thailand monitoring 272
SeaWiFS 25, 30, 55, 58, 121–3, 141, 145, 267, 273
sediment transport 138–41, 221–5
sensors 7, 9, 11
  calibration 61, 63, 65, 87–91
  data requirements 21–9
  of former USSR 33–50
  satellite data 29–33
shipborne data 19, 58, 150–1
  COSE 67–8, 69–75
  'mushroom-like' structure 200–6
  see also individual vessels
SHOMAN tool 131
shore topography 131–6
'signals' 2
Silicon Graphics Crimson workstation 123
Slupsk Bank 188–9
Solar Irradiance Field 122
Southern Machinebuilding Enterprise 35
Soviet–German Complex Oceanographic Subsatellite Experiments 212
Space Technology Education Program 274
spatial coherent analysis 17–18, 95–9
spatial resolution 11, 21–5, 28–31, 41–2, 61, 69
special interest groups 266
spectral band 11, 41–2
spectrophotometer 73
SPOT satellite 31, 55, 124, 126, 131–4, 137, 139, 265
State Centre for Natural Resources Research (Moscow) 66, 106
State Oceanographic Institute 54, 66, 82–4, 97, 160, 235
  Laboratory for Satellite Oceanography and Airborne Methods 14, 18–19, 53, 64, 91, 93, 188, 212, 264
'still zone' 212, 225
Strelets (research vessel) 217

subsatellite data 57–63, 82, 87–8, 153
  see also complex oceanographic subsatellite experiments
Sun-synchronous orbit 30–2, 48
surface cyanobacterial accumulations 154, 220–1
surface topography 27
suspended sediment 18, 101, 125, 221–4
  in Neva Bay 240–5, 248–9, 252, 257–9
suspensions (marine pollution) 136–7
Swedish Meteorological and Hydrological Institute 18–19
Swedish operational system for marine applications 118–20
Swedish Space Corporation 268–9
synchronicity 80–7
synergy/synergetic approach 18, 100
synoptic scale 257, 259
Synthetic Aperture Radar 24, 40, 134–5, 149, 267
  ERS-1 32, 109, 118–19, 130–3, 136–7, 146–8, 150–1, 265, 273
  high-resolution 35, 37–9

't-structures' 196
temperature, natural standard of 87–9
temporal resolution 21–8 *passim*
Temporary Manuals on Acquisition 54
TERAGON image processor 118
TeraScan system 115
test areas (sea-truth) 4–5, 9, 59, 61–6, 87–9
test site (subsatellite data) 60
Thematic Mapper 31, 124, 126, 131–5, 137–9, 143, 196
thermal pollution 223–4
thermoactive zone 88, 89
thermobar 88–91, 173, 218
thermocline 192, 260
'thermodynamical temperature' 81
thermohydrodynamical interaction 10, 88–9
thermoinertial zone 88, 89–91
three-dimensional data 186–7, 203, 206, 265
threshold dispersion approach 61–3
Tinto–Odiel estuary 139–40
Tiros Operational Vertical Sounder 115
TOBIS (data buoys) 272
TOGA Project 8, 10, 12
TOPEX/POSEIDON mission 50
topic-oriented use of satellite data 131–51
topogenic eddy 188
Total Ozone Mapping Spectrometer 42
Total Suspended Matter 117
TRISULA software package 265
Tromso Satellite Station 109, 119, 150
Tropical Oceans and Global Atmosphere project 8, 10, 12
turbidity determination 138–41, 221
turn-key data systems 55
two-dimensional data 2–3, 61, 183–4, 205

Ulva (seaweed) 127
UN Economic Commission for Europe 155, 157
UNEP 172
UNESCO 20, 22, 29, 268
upwelling events
  Baltic Sea 16, 173–83, 260
  Gulf of Finland 95–9, 176, 179, 260
  Neva Bay 98, 99
US Coast Guard 115–17
US CoastWatch Program 120–1, 124, 129
US Environmental Protection Agency 124
US Geological Survey 124, 129
US Mission to Planet Earth 30
US West Coast 125–6
Use of Satellite Data for Environmental Purposes in Europe (1994) 25–9, 39
USSR Hydrometeorological Service 179
USSR sensors and satellites 33–50

VAX-II/GPX workstation 117
VAX/VMS computer 118
VEB 'ROBOTRON' 106
visible bands 29–31
Visible and Infrared Scanning Radiometers (VIRSR) 24, 29
Visible and Thermal Infrared Radiometer (VTIR) 17, 31
Visiting Investigator Program 273
Vistula Bay 205, 206, 207

waste water 230–3
water
  circulation, inshore 197–8
  colour 22, 25
  dynamics 131, 148–50
  exchange variability 198–9
  level changes 243, 245–7
  line procedure 136
  motion (visualisation) 183–4
  quality 26, 140–1
WHO/Europe 155, 156
Wijsmuller Engineering B.V. 273
wind
  direction/strength 243, 245, 247
  vector 13, 22–3
WOCE Project 10
Working Group on Oceanic Satellites and Remote Sensing 268
World Meteorological Organisation 22, 29, 43, 47, 268
World View system 267–8

'yellow substance' in Baltic Sea 18

Zeeland project 135–6
Zeepipe Development project 136
'zero-solution approach' 262
Zvezda Baltiki (research vessel) 219